ANIMAL POPULATIONS
IN RELATION TO
THEIR FOOD RESOURCES

THE BRITISH ECOLOGICAL SOCIETY
SYMPOSIUM NUMBER TEN

ANIMAL POPULATIONS
IN RELATION TO
THEIR FOOD RESOURCES

A Symposium of
THE BRITISH ECOLOGICAL SOCIETY
Aberdeen 24-28 March 1969

Edited by
ADAM WATSON
B.Sc. Ph.D. D.Sc.

The Nature Conservancy
Blackhall, Banchory
Scotland

BLACKWELL SCIENTIFIC PUBLICATIONS
OXFORD AND EDINBURGH

SBN 632 06220 7

FIRST PUBLISHED 1970

Printed in Great Britain by
SPOTTISWOODE, BALLANTYNE AND CO LTD
LONDON AND COLCHESTER
and bound by
THE KEMP HALL BINDERY, OXFORD

CONTENTS

v

PART II · THE IMPORTANCE OF BEHAVIOUR MECHANISMS IN RELATING ANIMAL POPULATIONS TO THEIR FOOD RESOURCES

PART III · POPULATION PROCESSES IN RELATION TO THE QUANTITY, QUALITY AND AVAILABILITY OF THE FOOD RESOURCES

PART IV · SUMMING UP AND GENERAL DISCUSSION

PREFACE

A welcome total of 270 people took part in some or all of the sessions of the British Ecological Society's Tenth Symposium on 'Animal Populations in relation to their Food Resources', at the University of Aberdeen. They came from many different countries, including as far away as Alaska and Australia. We had 3 days at the Symposium sessions and finally a Friday for excursions. Most of the participants stayed together at one of the University residences, Dunbar Hall, so it was easy to have informal discussions without having to travel.

Aberdeen in March is not a place that appeals to most people thinking of a 4-day visit; the grey North Sea, biting raw wind and grey granite usually combine with grey skies to make the place bleak, perhaps better expressed in local dialect as 'dreich'. In fact, it turned out very differently. People who had travelled to Aberdeen commented that there was dirty weather all over western Europe, but in Aberdeen the sun shone from virtually cloudless skies throughout the sessions.

The Committee considered early on that there were two possibilities for this Symposium. One was to invite well-known workers from the various taxonomic groups, e.g. one from fish, one from insects, one from ungulates and so on, and hope that a useful general discussion would arise spontaneously to illustrate problems and suggest lines for new research. The second approach was to consider a series of possible questions interlocking in some general framework, and invite papers specifically to fit these questions, even if some of the workers might not be well known and might be young ecologists pioneering some new field. At the same time, a few general reviews were sought, to illuminate in more detail and over a wider field the problems raised by these questions. We thought the first approach would be safer, but the second, though more risky, would be more likely to be rewarding. The result of the second approach, given in this volume, is therefore a mixture of preliminary progress reports on new or current lines of work, general reviews, and detailed reviews of long-term research projects. Papers were also invited from agriculturalists, nutritionists and behaviourists since each can contribute towards a better understanding of the role of food in influencing populations. The Symposium was partly intended to promote a fuller appreciation amongst ecologists of recent advances in these other fields. The Symposium will have been worth while for our science—as distinct from our personal enjoyment of the actual meeting—if any new co-operation and new lines of work do in fact result from it in the future.

The title of the Symposium covers such a vast field that we decided at the beginning to try to get a more fruitful concentrated discussion by greatly limiting the subjects tackled. Such matters as energy flow in populations and ecosystems, the timing of breeding seasons, the migration of birds and other animals, or the physiological mechanisms involved with breeding, migration and so on, could each be a subject for a symposium on its own. These aspects were therefore excluded, and some papers that were offered to the Committee were turned down for this reason—not, may it be emphasized, because they were not good papers in their own right. The subject of the Symposium is one where speculation has been rife, often without much factual basis. Because of this, the main body of papers, in Sections I–III, was limited to research where people had in fact measured at least two and preferably three of the main aspects of populations, nutrition, and behaviour.

Grateful thanks are due to the Principal and the University of Aberdeen for being our hosts, for providing accommodation and meals at Dunbar Hall, for the use of a lecture theatre, and for their evening reception given to the Society in the Elphinstone Hall. The Vice-Principal, Prof. G. M. Burnett, welcomed us at the reception and also attended the Society dinner. Prof. V. C. Wynne-Edwards, who originally suggested that this Symposium should be held in Aberdeen, helped greatly in preliminary arrangements with the University, and also opened the Symposium, welcoming the Society to Aberdeen on behalf of the University. Dr C. E. Lucas, F.R.S., Department of Agriculture and Fisheries for Scotland, kindly allowed us to visit the Marine Laboratory at Torry during one of the excursions on the Friday. Dr W. J. Eggeling, Director, The Nature Conservancy, Scotland, agreed to have administrative work concerning bookings for accommodation and meals completed at the Conservancy's Edinburgh office, and the Society is grateful to Mr W. B. Prior and others who carried out this task. Dr K. L. Blaxter, Dr J. C. Coulson, Dr M. W. Holdgate, Prof. A. Macfadyen, Mr H. N. Southern and Prof. V. C. Wynne-Edwards kindly agreed to act as Chairmen of the various sessions. Thanks are also due to those members of the Society who opened the discussions, and to those institutes and universities who made it possible for some speakers and participants to attend the Symposium. We must also thank the following Discussion Recorders, who must have given up some of the enjoyment of attending this Symposium by having to do this work: Section I—Mr B. W. Staines, II—Dr C. J. Feare, IIIa—Dr I. Newton, IIIb and Summing up—Dr H. J. Milne.

The Symposium Committee comprised: Convener—Dr G. R. Miller, Editor—Dr A. Watson, organizers of Sections I—Dr R. Moss and Dr J. M. Hinton, II—Dr I. J. Patterson, III—Dr D. Jenkins.

Adam Watson

INTRODUCTION

By David Lack

Edward Grey Institute,
Botanic Garden, Oxford

It is an honour to be invited to speak first at this conference. It perhaps means that the views I published in 1954 have completed the slow, and in the early stages hard, climb to respectability, but the latter level is, in science, but briefly occupied before the descent to obsolescence. This is as it should be, and I am happy to open a conference which helps to accelerate such a descent in order that others may climb to the peaks ahead. I expect also that this meeting with Professor Wynne-Edwards will be as happy as our last in these parts, in late July on a neighbouring mountain, where our sandwich lunch was followed by finding a new Scottish site for *Saxifraga rivularis*. Our meeting today should be followed by much more exciting ecological discoveries than that.

In my book of 1954, I advocated that each species of animal reproduces as fast as it is able, and that its numbers are limited by a consequent density-dependent death-rate. In those birds which feed their young, the reproductive limit—the clutch-size—has been evolved through natural selection in relation to the maximum number of young for which the parents can provide enough food without detriment to themselves. This is a special case, but in other animals also, the reproductive rate has likewise been evolved, through natural selection, in relation to, among other factors, the availability of food for the breeding adult. Indeed such views evidently apply, suitably modified, even to plants, in which connection I might cite the title of a recent paper '" Clutch-size" in the buttercup' by Johnson & Cook (1968). In birds, the same quantity of food per day could serve to raise fewer young quickly or more young slowly, and the compromise evolved through natural selection will depend on the length of time for which food for the young remains abundant, and on the rate of predation on the nestlings. Similarly in other animals, and in seed-bearing plants, the same quantity of food reserves could be used to make many eggs or seeds, each with small food reserves, or fewer eggs with larger food reserves, and the compromise evolved by natural selection will depend on the conditions in which the embryo emerges and other factors. The compromise reached however, will be that which, on average, results in the greatest production of surviving offspring.

The situation is not, of course, so simple as this bare outline might suggest. Thus there are both hereditary and phenotypic variations in the number and size of the eggs or seeds, and the capacity to vary phenotypically with the conditions is presumably itself subject to hereditary variations, as this capacity differs greatly in different species. There are also further complications in relation to predation, since hungry nestling birds call more loudly than well-fed ones, so are more liable to predation, and large eggs or a large clutch tend to be more conspicuous than small ones, so the birds may have to evolve special means of concealment or defence. In invertebrates and flowering plants, likewise, the problems of both protection and transport are different for large and small eggs or seeds. A further complication is that, if there is only a brief season in which food for the young is abundant, there will be an advantage in breeding as early as possible, but this may mean that the female has to form her eggs immediately after the period of winter food shortage. This holds, for instance, in the great tit (*Parus major*) in England (Perrins 1965). In tropical forest, with much less marked, though definite, seasonal variations in the abundance of insect foods, breeding may be more difficult for birds, as their adult numbers stay close to the limit set by food, and in at least one species the female uses proteins in its wing muscles to form its eggs (Ward 1969). Hence the breeding seasons of birds, like their clutch-size, have been evolved as a compromise between counteracting advantages. Various different, and partly counteracting, factors likewise influence the laying or spawning seasons of other animals, and the flowering times of plants.

At least in birds, the reproductive rate varies inversely with population density only to a small extent, an extent which present evidence suggests is too small to be important in the regulation of numbers. Almost certainly, therefore, though this is still not proven, the main density-dependent control of numbers in birds is the mortality, and I would suppose that the same holds in most other animals. It has long been agreed that the main density-dependent mortality factors are food shortage, predation and disease. I have often been quoted as saying that all animal populations are limited by food. But I have never said this, and though I think it almost certainly true for at least most bird populations, I specifically excluded from such a statement the herbivorous mammals and the phytophagous insects, and the latter, after all, comprise the great majority of the world's

higher animals. The available evidence suggests that, under natural conditions, both herbivorous mammals and phytophagous insects may be held
down by predators, including insect parasitoids, below the level set by
food, though this does not necessarily mean that food has no influence on
numbers in such circumstances. In addition, where carnivorous mammals
have been removed by man, and in occasional years in forests (at least in
coniferous monocultures), deer or caterpillars respectively may eat out
their food supplies. The possible role of disease in regulating numbers in
nature needs further study, as myxomatosis has held down the numbers of
the rabbit for a long time, and Warner (1968) has recently shown that
disease may have eliminated certain island birds.

Other workers have sometimes stated that I have underestimated the
importance of behaviour in regulating animal numbers. However, the
chapters of my 1954 book dealing with this problem were short, not
because I thought the subject unimportant, but because at that time so
little was known. Bird numbers are greatly modified by behaviour, for
instance through peck-order fighting, and especially through movements,
but their influence is local and secondary, and on a world view they do not,
in themselves, cause any changes in numbers. Bird movements include
not only regular migrations and irregular irruptions (between which there
is no clear line, Svärdson 1957), but the inconspicuous avoidance, especially
for breeding, of areas of high population density. In this last connection,
both the precise role and the ecological value of territorial behaviour have
remained the big unknown, and the factors determining the size and spacing
of breeding colonies in colonial species are equally puzzling. In other animals, also, movements may modify numbers, for various mammals, fish
and insects carry out long-distance migrations and others disperse irregularly. However, I consider that all such behaviour is secondary to an ultimate
limit set by mortality, and that it has been evolved through natural selection
only where it assists the survival, normally in relation to food, of the
individuals concerned and/or their offspring. Many other invertebrates
avoid a season of temporary food shortage by temporarily contracting out,
so to speak, from the living world, through hibernation, aestivation or a
diapause.

In summary, I hold that food supplies determine, through natural
selection, the reproductive rates of all higher animals (and plants), and set,
through density-dependent mortality, an upper limit to the numbers of
animals, this limit being frequently reached in most birds and certain
other groups, but rarely reached in many species that are usually limited by
predation (or perhaps disease); further, that the prevalence of food shortage

has led to the evolution of various important types of behaviour in animals. It would, however, be quite wrong to put this forward as a purely personal view. It owes a great deal to others, both predecessors and later workers, including many of those who have worked at the Edward Grey Institute at Oxford since the war.

All this is by way of introduction, to set, I hope, this conference in context with general theories of population regulation in animals, as I understand them. But we are primarily concerned here with advancing the subject further. What most needs to be done?

First, the basic ideas outlined here, which followed the initial presentation by mathematical ecologists such as A. J. Nicholson, have been worked out in the field primarily in birds. Birds are specialized organisms, so may give a one-sided picture and there is an urgent need to test and develop these ideas for other animals. But as you can see from the programme for this meeting, we are here asked to consider a pleasing diversity of animals, including herbivorous and carnivorous mammals, insects of various orders, terrestrial and marine mollusca, triclads and protozoa.

Secondly, if my initial outline is correct, one may expect food to have different effects on (a) the reproductive rate in the breeding season, (b) the death rate of those animals whose numbers are limited by density-dependent food shortage, and (c) those animals whose numbers are rarely limited by food.

Thirdly, an animal's nutritional requirements may vary greatly at different times of year. For instance, a bird presumably takes not only different quantities, but to some extent different types, of food according to whether its prime need of the moment is to survive a long winter night, to lay its own weight in eggs, to feed a large number of growing nestlings, to moult all its feathers, or to fly continuously at high altitude for 24 or 48 hours. Similar considerations apply in other animals.

Moreover, not only do the needs of the animal vary with the time of year, but so does the quality of particular food resources. For a herbivorous mammal and for a caterpillar, a young leaf is more nutritious than an old one, as we know ourselves from the lettuces in our gardens. But the problem goes much deeper than that, and we need full studies of the seasonal and other variations in the quality of the foods available to particular species. The titles of the talks to be presented here suggest that this subject is now advancing rapidly.

An animal selects particular prey species in relation not only to its own physiological needs and to the varying nutritional value of its foods, but also to the ease with which different species can be obtained. It seems to be advantageous for a hunting bird to specialize temporarily on a particular type of prey, as food is then found more quickly than if it searches simultaneously for a variety of species. Given a choice of prey, it will be advantageous for it to select that species which provides it with most food in relation to hunting time and effort, and this, in turn, will be related to the abundance, the size, the food quality and the ease of discovery and capture of the prey species available to it at the time in question. The subject is complex, as discussed recently in relation to birds by MacArthur & Pianka (1966) and Royama (1966). Moreover in a highly specialized feeder like the oystercatcher (*Haematopus ostralegus*), different individuals may restrict themselves for long periods to particular prey species, which are often very different from each other and require very different means of capture (Norton-Griffiths 1968). Much more research is needed, in both birds and other animals, on the behaviour involved in food selection, and again, the titles in the programme promise some discussion of the problem.

There is the related point, to be discussed by at least one contributor, of food selection in relation to competitive exclusion. Two bird species which are limited in numbers by food shortage cannot depend on the same food resources. As shown theoretically by MacArthur & Levins (1964, 1967), the nature of these resources determines whether two species with similar feeding habits are more likely to be separated by habitat or by feeding methods, and also whether two more specialized species are likely to displace, or to be displaced by, a single more generalized one, but their findings need checking in the field. Moreover, while the survival value to each species of selecting a different habitat or different types of prey is clear, the means by which this is achieved, and the parts played respectively by hereditary factors and learning, have scarcely been studied.

This leads on to the general problem of behaviour in relation to food supplies. Birds, having powerful flight, probably move to a greater extent than most other types of animals in order to escape the threat of starvation, but movements are certainly important in other animals. The most critical problem in birds is probably the significance of territory. It is now becoming clear, in part through work to be reported later in this symposium, that territorial owners really do prevent other individuals from settling on their ground, and thus limit the population density. But the precise functions of this behaviour are still uncertain, perhaps because they differ in different types of birds. In some species, territorial fighting serves primarily to obtain a mate or a nesting site, and appears to have no function with respect to food. In others, it might be linked with the need for solitary hunting, and hence it possibly affects local spacing rather than

2

absolute density. In yet others, it certainly affects population density, but whether the size of territory defended is directly related to the availability of food for the adult or for its young and, if so, how the species achieves this, is still far from clear. This, incidentally, is one of the few problems of avian ecology in which a few simple field experiments have yielded a much more decisive answer than field observations, but much work remains to be done. Perhaps studies of other types of animals may throw valuable light on the significance of territory to birds.

I have here concentrated on problems of autecology, but in parallel with such essential studies, other workers are now looking afresh at the grand view of the ecosystem, and are studying the flow of nutrient resources and of energy through it, with the complex interrelationships which this entails, and we are to hear at least one contribution under this head.

Lastly, it may be asked whether the conclusions of a conference such as this are likely to throw light on problems of practical interest to man. So far as man's own population is concerned, the answer is probably no, for once the basic principle has been grasped that, in a stable population, the birth rate equals the death rate, the similarities are far smaller than the differences between human populations and those of wild animals. But indirectly, so far as the control of pests and the conservation of rare animals are concerned, the types of problem that we are studying here are of basic importance. Even at this date, perhaps particularly in birds, the ecologist may still find himself horrified by the ways in which the practical man tends to ignore both fundamental principles and the ecology of the species in which he is interested.

In regard to bird pests, for instance, the ordinary man still thinks that if he shoots them they will become scarcer. While this may be true in some cases, it is certainly not in others, as shown for the wood pigeon (*Columba palumbus*) by Murton (1965, 1968), and his findings have been followed by an extremely welcome, if belated, change in the official policy of the Ministry of Agriculture, Fisheries and Food who employ him. But the same body has not as yet put out a generally convincing study of the factors limiting either the oystercatcher, or the cockle (*Cardium edule*) on which it often preys, before recommending the destruction of thousands of these birds in the supposed interests of a small group of cockle-fishermen; at least, the case as presented last year aroused strong opposition from many ecologists who heard it (Davidson 1968, and subsequent discussion). Again, the owners of grouse moors have normally ordered all raptorial birds to be destroyed there, although theoretical ecologists long ago pointed out that this might not increase the numbers of red grouse (*Lagopus*

lagopus) shot. The ecologists have now been proved right by the Unit of Grouse and Moorland Ecology in the Nature Conservancy (Jenkins, Watson & Miller 1964), but not before many of Britains' raptors had been destroyed over huge areas.

The conservationist has to learn similar lessons. His initial task, rightly, was to pass laws and to set aside reserves to undo direct destruction by man, but this is often insufficient to preserve a species. For instance, the Royal Society for the Protection of Birds purchased the main breeding site when the once persecuted avocet (*Recurvirostra avosetta*) returned to England, and thus preserved it from collectors, but after a while the avocet ceased to increase and they have now sensibly appointed a research biologist to discover the possible causes. This same body (in *Bird* for Jan. 1969) reported that Dutch conservationists, troubled by the marked decline of the white stork (*Ciconia ciconia*) in Holland, are planning to release there birds bred in captivity. But the current decrease in Holland is part of a general decline in western Europe, so in the long term the released birds will presumably suffer the fate of the wild ones. Surely, ecologists ought first to determine the factors responsible for this widespread decline, knowledge of which might well lead to very different measures of conservation. Many further examples could be quoted, but I will now conclude by saying that, while the primary aim of this gathering is to uncover fundamental principles of ecology, our discoveries are likely to be of great value in practice.

REFERENCES

DAVIDSON P.E. (and Discussion) (1968) The oystercatcher—a pest of shellfish. *The Problems of Birds as Pests* (Ed. by R. K. Murton and E. N. Wright), pp. 141–55 and Discussion pp. 174–80. London & New York.

JENKINS D., WATSON A. & MILLER G.R. (1964) Predation and red grouse populations. *J. appl. Ecol.* 1, 183–95.

JOHNSON M.P. & COOK S.A. (1968) 'Clutch size' in buttercups. *Am. Nat.* 102, 405–11.

LACK D. (1954) *The Natural Regulation of Animal Numbers.* Oxford.

MACARTHUR R.H. & LEVINS R. (1964) Competition, habitat selection and character displacement in a patchy environment. *Proc. natn. Acad. Sci. U.S.A.* 51, 1207–10.

MACARTHUR R.H. & LEVINS R. (1967) The limiting similarity, convergence, and divergence of coexisting species. *Am. Nat.* 101, 377–85.

MACARTHUR R.H. & PIANKA E.R. (1966) On optimal use of a patchy environment. *Am. Nat.* 100, 603–9.

MURTON R.K. (1965) Natural and artificial population control in the woodpigeon. *Ann. appl. Biol.* 55, 177–92.

MURTON R.K. (1968) Some predator-prey relationships in bird damage and population control. *The Problems of Birds as Pests* (Ed. by R. K. Murton and E. N. Wright), pp. 157–69. London and New York.

NORTON-GRIFFITHS M. (1968) The feeding behaviour of the oystercatcher *Haematopus ostralegus*. D.Phil. thesis, Oxford University.

PERRINS C.M. (1965) Population fluctuations and clutch-size in the great tit *Parus major* L. *J. Anim. Ecol.* **34**, 601–47.

ROYAMA, T. (1966) The breeding biology of the great tit *Parus major*, with reference to food. D.Phil. thesis, Oxford University.

SVÄRDSON G. (1957) The 'invasion' type of bird migration. *Br. Birds*, **50**, 314–43.

WARD P. (1969) The annual cycle of the yellow-vented bulbul *Pycnonotus goiavier* in a humid equatorial environment. *J. Zool. Lond.* **157**, 25–45.

WARNER R.E. (1968) The role of introduced diseases in the extinction of the endemic Hawaiian avifauna. *Condor*, **70**, 101–20.

SYNOPSIS OF SYMPOSIUM

The purpose of the Symposium was to consider the influence of the quantity, quality and availability of food on the processes regulating animal populations. In terms of these three aspects, food is often difficult to define because animals feed selectively and several papers emphasize this point. Food is not always a proximate factor affecting populations and the meeting also considered the role of social behaviour as an important intermediary between food and population numbers. There are three sections:

I The relevance of food selection and utilization to population processes.
II The importance of behaviour mechanisms in relating animal populations to their food resources.
III Population processes in relation to the quantity, quality and availability of the food resources.

PART I · THE RELEVANCE OF FOOD
SELECTION AND UTILIZATION TO
POPULATION PROCESSES

Before we can discuss the relationships between animal populations and their food resources in any detail, we must be able to define and measure food. The aim of Section I was to present studies which illustrate some of the problems involved in such measurements and which contribute to our knowledge of factors affecting food supplies.

The first review paper (Eadie) provides a background of detailed knowledge about the relationships between the numbers and performance of domestic sheep and pasture utilization under different systems of management and in areas of different fertility. The main problem in hill-sheep farming is that stocks and performance tend to be limited by poor food in winter, and that the vast production of good food in early summer cannot be utilized before it declines in nutritive value in late summer. Our more scanty knowledge about wild animals may be compared with the sheep.

A second review (Klein) considers North American deer, which illustrate a range of situations: (i) low density, selection of good-quality food, good breeding performance; (ii) medium density, heavy utilization of good food plus use of poorer foods, reduced breeding performance; (iii) high density, disappearance of good food, poor breeding, and over-utilization; followed by (iv) population crash. Concurrent effects on population structure are described, and the importance of regulatory mechanisms such as starvation, natural predation and hunting is discussed.

The next three papers consider food selection in the context of nutrition, palatability and availability. A number of factors interact to result in the food preferences which all animals show. Icelandic ptarmigan (Gardarsson & Moss) usually have available a vast excess of material which is edible. When the bulk of this material is of low nutritive value relative to their requirements they choose to eat only the most nutritious food. The physical structure of a plant food interacts with its palatability to define its acceptability to snails (Grime, Blythe & Thornton). This paper is also a good example illustrating how well-designed yet simple experiments in the

3

laboratory can elucidate our understanding of feeding, behaviour and even animal distribution in the wild. The numerical density of the invertebrate prey of redshank (Goss-Custard) is not an important determinant of the bird's feeding rate. Some other factors such as the size of the prey, the behaviour of the prey at different temperatures, and the physical structure of the mud in which the prey lives, are probably important. This paper also illustrates the complexity of the problem of feeding behaviour and selection, and describes current laboratory experiments to test the preliminary findings from the field.

The following two papers deal with interactions between animals in relation to a common food source. The physical structure of a grass sward is not static, but is affected by grazing. On heavily grazed African grasslands (Bell), different species of large ungulates successively graze on the same area of grassland in a definite order. Each changes the structure of the sward and thus facilitates its utilization by the next species in the succession. Each species feeds in a different way and is adapted to its place in the succession. Although different species of animals may thus facilitate the use of a food supply by each other, they may also compete for it (Reynoldson & Davies). Competitive exclusion of two triclad species occurs if their food habits are similar. Co-existence occurs if their food habits are different enough.

A food supply is not independent of its consumer. A grass sward may change in species composition, and thus in feeding value, in response to different levels of grazing and plant selection by sheep and cattle (Nicholson, Paterson & Currie). Thus animals may change the quality as well as the quantity of a food supply by exploiting it, and this may possibly have effects in turn on the performance of the animals.

Interactions between predators (amoebae) and prey (microflora) are described in the last paper (Heal & Felton). Edible bacteria that support growth and reproduction of amoebae produce exudates which stimulate the activity of amoebae and inhibit encystment. Exudates of inedible bacteria stimulate encystment and in some cases kill amoebae.

R. Moss and A. Watson

CHAIRMAN'S REMARKS

By K. L. Blaxter

The relation between animals and their food supply is extremely complex. Much of this complexity has been uncovered in laboratory experiment. We know, for example, a great deal about the quantitative nutrient requirements of a few animal species—some laboratory and farm mammals and birds, some insects, some protozoans, and one or two representatives of other phyla. In these species we know how nutrient needs vary in relation to demands of growth and reproduction, how absolute shortages of food and specific nutrient deficiencies affect these processes, and how nutrient needs are modified directly or indirectly by environmental factors such as climate and the presence of other animals such as symbionts, parasites and pathogens.

Such studies have defined in a quantitative way what these species require in food to grow and multiply. We can state categorically from information of this sort the relationship between the amount and nutritional adequacy of the diet consumed and reproductive performance, that is the time rate of appearance of new sexually mature animals per 100 or 1000 of a population. These studies indeed extend and define the ecological principle of 'effective food' (H.G. Andrewartha & L.C. Birch, 1954, *The Distribution and Abundance of Animals*, p. 498. Chicago) in more precise terms. Even so, extension of the laboratory findings for this favoured few species to field situations is fraught with many difficulties. These arise not only because of the difficulties of measurement of what is eaten—difficulties which with ingenuity can be overcome—nor because the environment is highly variable in time, but because a new dimension is involved. This dimension consists of a continuum of interactions between the animal concerned and other animals of its own kind or other species, and, with herbivores, interactions between the animals, the plants on which they feed, and the organisms which are components of the so-called detritus chain. Experimentation, in the sense that determinants can be isolated and measured, is extemely difficult in these circumstances even when one is armed with sound laboratory information on nutritional needs.

5

When we attempt to describe the effect of food on reproductive performance of species for which we have little or no background of experimental study to define quantitative need, we have to exercise great care. Obviously we must first know what is eaten and under what circumstances it is eaten. If, however, we are concerned with the effect of food on reproduction, eventually we must know how much is consumed, the time pattern of consumption, and the nutritive adequacy of what is consumed.

The relation between the food available and the animal population supported and, in particular, the equilibrium relationships involved are very subtle. Food controls populations through an all or none mechanism of starvation except but rarely. Rather, the relationship is a continuum: indeed, one could devise an interesting theory and support it by much anecdotal information to show that the fundamental nutrient needs of species are determinants of animal dispersion and their spatial and temporal distributions, but then we possibly have too many theories in ecology. Experimental approaches are far more rewarding.

SHEEP PRODUCTION AND PASTORAL RESOURCES

By J. Eadie

Hill Farming Research Organisation, Edinburgh

INTRODUCTION

Sheep production in the United Kingdom takes place in a wide range of pastoral environments, from the intensively stocked, highly fertilized lowlands, to the extensively grazed pastures of the hills and mountains.

Systems of hill sheep farming vary from region to region. In some areas the sheep remain on the hills the year round, and in other areas flocks are brought down to enclosed sown pasture for varying periods and at different times, such as tupping, during late winter, and at lambing. In some areas the sheep are seldom, if ever, hand-fed. In others feeding ranges from the provision of roughage during periods of snowstorm and deep snow cover to regular supplementing of the diet during the latter part of pregnancy. But, by and large, hill sheep production is characterized by a high dependence on the resources of the hill pasture and the grazed pasture.

In the more intensive production systems for lowland sheep, conserved grass, mainly in the form of hay but quite often as silage, is fed for much of the winter, and supplementary concentrate feed is much more important in the sheep's diet, particularly in late pregnancy. Even so, many quite intensive production systems for lowland sheep can be regarded as pasture-based, with conserved grass as a major part of the system.

PRODUCTION PER AREA AND PRODUCTION PER SHEEP

Between the extremes of complete dependence on grazed pasture all the year round, and a prime dependence on conserved grass during the winter, the whole spectrum of possibilities is to be found in practice.

Sheep production per unit area varies enormously across the range of pastoral environments and husbandry systems. At the extreme, hill farms may sell as little as 2–3 kg lamb liveweight per hectare (2–3 lb per acre) per annum, but the better hill farms may sell up to 25 kg per ha per annum. In very intensive lowland enterprises, production may exceed 600 kg/ha/ annum.

This order of difference in production per hectare arises from differences in overall stocking rate and differences in the individual performance of the sheep .

Stocking rates on the hills may be lower than one ewe per 4 ha or as high as 1 ewe to 0·5 ha. Stocking rates in intensive lowland systems may exceed 12 ewes per hectare on a year-round basis.

Weaning percentages (i.e. per 100 ewes) in hill flocks vary from 60% to over 100% depending on the hill environment. Lowland sheep commonly achieve weaning percentages in excess of 150% and not infrequently 180–190%.

The growth rates of hill lambs (singles) are often around 0·22 kg/head/day up to marking at 5–6 weeks of age; growth rates steadily decline until weaning, to give weaning weights in the range 20 to 27 kg. Ewes with twin lambs are seldom run on the hills during lactation because of the poor growth rates so produced. The growth rates of lowland lambs are often in the region of 0·27 to 0·36 kg/head/day giving weights of 36 to 41 kg at 12–16 weeks of age.

Two sets of data in Table 1 indicate the relative importance of stocking rate and the components of individual sheep production as determinants of

TABLE 1

Stocking rates and animal performance per ewe and per hectare, in two environments

	Hill	Lowland
Stocking rate in ewes/ha	0·8	10·0
No. of lambs weaned/100 ewes	80	170
Mean wt of weaned lamb (kg)	24·0	36·0
Weaned lamb/ewe (kg)	19·2	61·2
Weaned lamb/ha (kg)	15·3	612

production per hectare in the hills and in the lowlands. These data are hypothetical, but are presented as reasonably typical (in so far as any set of figures may be said to be typical) of both hill and lowland systems. In any event neither is in any way extreme.

Hill and lowland systems of sheep husbandry differ in the way in which flock replacements are made. On the hills the better ewe lambs are retained from each year's lamb crop, in numbers which depend on the number of lamb crops taken before the ewes are sold as cast ewes, and on the ewe death rate. These replacement females are run on the hills with the ewes during their first and second summers and are usually mated at around 18 months of age. Replacements to lowland flocks are generally bought as gimmers prior to first mating. Because of these husbandry differences, lamb production and lamb sales are one and the same thing in lowland flocks, but this is not so on the hills. The appropriate data for comparison here are the figures for production and not for sale, and it is these that are shown in Table 1.

In the examples given in the table, stocking rate in the lowlands exceeds that in the hills by a factor of 12, production per ewe by a factor of 3, and production per hectare by a factor of 40.

PASTURE PRODUCTION

In the first place it is necessary to see to what extent these differences in animal production between the hills and the lowlands may be explained on the grounds of differences in pasture production.

Generally speaking, unfertilized lowland pasture will yield some 2000 to 3000 kg dry matter per ha per annum in the United Kingdom. Under conditions of reasonable pasture management, swards which contain some clover and which are given some nitrogen fertilizer (< 100 kg N/ha/annum) will yield some 5000 to 7000 kg dry matter per ha per annum, and responses to further nitrogen fertilization are of the order of 10–15 kg dry matter per ha per kg of N. Yields of 10–15 000 kg dry matter per ha per annum have been recorded where 300 to 500 kg N/ha per annum have been applied (Holmes 1968).

Yields of hill pastures are poorly documented, no doubt largely for the reason that estimates of pasture production made by cutting techniques are likely to be somewhat at variance with actual pasture production in conditions of extensive grazing.

Milton (1940) and Milton & Davies (1947) measured the yields of *Festuca-Agrostis* and *Molinia* pastures in Wales. Their results, together with experience in the Hill Farming Research Organisation would indicate that *Agrostis-Festuca* type pastures yield some 2200 to 2750 kg dry matter/ha/ annum with some types approaching 3200 kg/ha/annum. Yields from pasture dominated by *Molinia* and *Nardus* are poorer, with production levels of 1300–1750 kg dry matter/ha/annum.

Miss S. Grant of H.F.R.O. is measuring the annual dry matter production of what are thought to be the edible portions of heather (*Calluna
vulgaris*); and depending upon the heather cover, the age of the stand and the
type of grazing management she records yields in the range 1000 to 2500
kg/ha/annum. These figures are in broad agreement with those of Miller
(1966).

It is reasonable to suggest that the levels of animal production cited in
Table 1 would be obtained from a lowland pasture yielding in the region
of 9–11 000 kg dry matter/ha/annum, and from a hill pasture yielding some
2000 kg/ha/annum. These figures indicate a difference in pasture production
of some fivefold between the two environments,' which is substantially
less than the differences in stocking rate, and which falls a very long way
short of explaining the differences in animal production per hectare.

To understand why this should be, it is necessary to examine the processes
of animal production from pasture in some detail, but an essential preliminary to this examination is a brief discussion of the factors influencing
nutrient intake, diet selection in grazing sheep, and pasture utilization.

NUTRIENT INTAKE

The quality of the forage or herbage ingested by ruminants may be
expressed, if not best, then at least conveniently and adequately for present
purposes, in terms of its apparent digestibility.

$$\text{Apparent digestibility} = \left(\frac{\text{food intake—faeces output}}{\text{food intake}} \times \frac{100}{1} \right) \%$$

The nutrient intake of the pasture-fed ruminant may be regarded as the
product of dry matter intake and the quality of that intake. Ruminants
on forage or herbage diets tend to eat more of high-quality diets than of
low-quality diets. General quantitative relationships between forage
digestibility and voluntary intake have been described (e.g. Blaxter,
Wainman & Wilson 1961), although it has become clear that there are
marked departures from these general relationships in some pasture species.
Van Soest (1965) has examined relationships between chemical composition,
digestibility and voluntary intake. He has shown that in forages with a low
cell-wall content, voluntary intake and digestibility are not related, whereas
in forages with a high cell-wall content intake is highly correlated with both
digestibility and chemical composition.

Raymond (1963) has discussed what he calls 'extrinsic' factors, that is,
factors of the grazing situation as opposed to those attributes of herbages

which influence intake; availability of herbage (or its lack) is generally regarded as the most important of the former. The evidence would appear to be more convincing in the case of cattle (Johnston-Wallace & Kennedy 1944), than with sheep (Wheeler, Reardon & Lambourne 1963).

Other characteristics of the grazing situation such as the degree of fouling, soiling, etc. might be important in limiting intake in some situations (Raymond 1963).

It is widely recognized that the physiological state of the animal may have a marked effect on food intake. Intake may be inversely correlated with degree of fatness (Ferguson 1956; Hutton 1963). Lactation is associated with quite substantial increases in voluntary intake in sheep (Cook, Mattox & Harris 1961; Davies 1962) and a marked depression in intake in late pregnancy in fat twin-bearing ewes, involving metabolic changes associated with pregnancy and fattening, was observed by Reid & Hinks (1962).

DIET SELECTION

Despite the influence of other factors, however, it is well established that the quality of the herbage ingested by ruminants is a prime determinant of their energy intakes, and this underlines the importance of food-quality selection in grazing sheep.

It is well known that sheep are selective grazers. At the level of a single plant, sheep will eat leaf in preference to stem, and green, or young, material in preference to dry, or old, material. Comparisons of ingested herbage with that on offer show that ingested herbage is higher in nitrogen, phosphorus and gross energy, but lower in fibre (Arnold 1964).

Opportunity for selection does, however, vary from one grazing situation to another, in response to attributes of pasture and to grazing pressure.

The fund of available herbage on a pasture may be characterized in terms of its amount, to which some index of quality is attached. But such a description is often a poor guide to the nutritive worth of that pasture for grazing animals, since no indication is given of the range of quality encountered within the fund of herbage. For example, a large proportion of the fund may consist of a quality not greatly different to that of the mean, as in a regrowth accumulated over a short period of time after a previous close cut or grazing. On the other hand a significant proportion of the available herbage may be of a quality widely different to that of the mean, as in a regrowth accumulated over a lengthy period of time after a previous lax grazing. In the former case selection opportunity is limited. In the latter case because of the marked within-sward gradient of digestibility, considerable opportunities for selective grazing exist.

But the opportunities for diet selection by grazing sheep are also influenced by the amount of herbage available to the sheep. This is best described, over a period of time, in terms of the relationship between herbage production and herbage consumption, to which the term 'grazing pressure' is applied. Thus at a low grazing pressure all animals have a surplus of herbage available to them whatever the stocking rate, and opportunity for selection is considerable. At high grazing pressures, opportunity for selection is much reduced.

The data in Table 2 illustrate the influence of both grazing pressure and the nature of the fund of available herbage. These data refer to a 5-week grazing period on *Agrostis-Festuca* hill pasture beginning in early May. Grazing pressure is calculated as the number of sheep grazing days/ha/day per 1000 kg available green dry matter/ha, and the figures used are means of estimates made on three occasions during the grazing period. The values for digestibility of the ingested pasture are means of five estimates. A comparison of the results from Plots 1, 2 and 3 demonstrates the effect of increasing grazing pressure on diet-quality selection in a pasture where there is a considerable within-sward gradient of quality in the available herbage. Plot 4 shows that an intake of quite high quality may be obtained at quite high grazing pressures where the fund of available herbage contains much less poor-quality dead herbage.

Grazing situations are dynamic, and events which take place at one point in time influence events at future points in time. For example, the low grazing pressure set in the interests of a high level of contemporary nutrition, will, on many pastures, result in a substantial proportion of uneaten herbage whose feeding value declines with time. The accumulated herbage will dilute the quality of the available herbage at some future grazing. It will either depress the quality of the ingested herbage or else influence the grazing pressure which may be attained if a certain level of ingested pasture quality is to be maintained at that grazing.

At any one grazing, therefore, there is often a conflict between the grazing pressure necessary for a high level of contemporary nutrition and that necessary for high levels of utilization. Efficient pasture utilization is not only a desirable end in itself, but is also desirable in that it renders the conflict less acute on a future occasion.

Although there are pastoral environments where the climate and the structure of the sheep population are favourable to a reasonable matching

TABLE 2
Grazing pressure* and ingested herbage quality (organic matter digestibility)

Plot	Dead herbage (kg/ha)	Stocking rate in sheep/ha	Grazing pressure*	Ingested pasture quality (O.M. digestibility)	Range of ingested pasture quality (O.M. digestibility)
1	2240	7·4	5·0	74·9	75·7–73·9
2	2240	12·4	9·0	69·0	70·8–66·9
3	2240	18·5	15·5	68·5	70·2–66·1
4	605	14·8	11·4	76·6	82·3–73·6

*Grazing pressure = $\dfrac{\text{Stocking rate/ha}}{1000 \text{ kg green dry matter available/ha}}$

of pasture supply and demand throughout the year, such as in New Zealand (Hamilton 1956), the extreme seasonality of pasture production in the U.K. exacerbates the whole problem.

The consequences of the above-mentioned conflict on the basis of a whole grazing season are most apparent in grazing systems where fixed stocking rates are used, for example with growing or fattening beef animals. A theoretical discussion of the problem has been set out by Mott (1960) and the problem is summarized in Fig. 1 (after Raymond

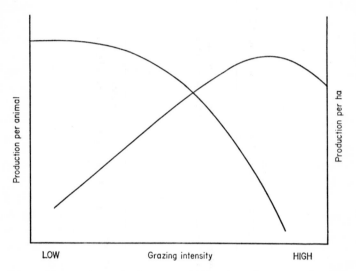

FIG. 1. Changes in production per animal and production per ha with increasing grazing intensity.

1964). Animal production per unit area increases with increased grazing intensity as a consequence of the greater number of animals, but the performance of the individual animal is reduced. Maximum animal production per unit area is almost invariably achieved at grazing intensities which give rise to much less than maximum levels of individual animal performance. Herein lies one of the pastoral agriculturist's greatest problems.

Considering the range of grazing intensities which derive from the range of management systems discussed in this paper, it must be pointed out that the relationship between performance per head and grazing intensity indicated in Fig. 1 is likely to be obtained only in the region of 'optimum' grazing intensity. At the very light grazing intensities characteristic of hill-sheep farming, the performance of the individual animal is poor, and the reasons for this will be outlined presently.

The significance of the problem summarized in Fig. 1 will differ with different kinds of animal production. With growing or fattening cattle, and in certain systems of milk production for example, a reduction in nutrient intake at any point in the season will tend to reduce individual performance. But in breeding ewes, a diet giving rise to no more than maintenance at some point between the end of lactation and the beginning of the mating period may be quite consistent with high levels of performance, provided nutrition is adequate at other times. Again, different classes of animals within the population, e.g. ewes and lambs towards the end of lactation, have different requirements. Both of these characteristics of sheep production may be used to help resolve the difficulty of combining efficient pasture utilization and high levels of performance per head.

PASTURE UTILIZATION

It is now appropriate to consider both hill and lowland systems in more detail.

Hill pasture systems are, in the main, set-stocked, free range, year-long systems. Hill pasture production is highly seasonal, and stocking rates are set at levels which allow of the maintenance of a certain minimum level of winter nutrition. Stocking rates set in this way result in a demand for herbage during the pasture growing season which is low in relation to pasture production and the amount of herbage available. The low grazing pressure thus obtained allows of considerable opportunity for selection and aids the maintenance of a higher-quality diet than would otherwise be the case. But, concurrently, a considerable accumulation of uneaten herbage takes place, whose feeding value declines as the summer proceeds. The consequences are twofold.

Firstly, this fund of accumulated herbage contributes to a marked within-sward gradient of digestibility which effectively limits the proportion of the available food which can be utilized at acceptable levels of nutrition in winter. Some recent studies in the East Cheviots indicate that utilization in excess of some 15% of the available herbage (dry matter) in winter would drive the digestibility of the diet below an acceptable level.

Secondly, despite the processes of plant death and decay, a considerable fund of poor-quality herbage is carried over to the following season, and this effectively dilutes the quality of the ingested herbage despite the low grazing pressure at that time (Eadie & Black 1968).

Hill pastures are heterogeneous in their vegetation and sheep in free-range grazing have access to a range of pasture types. Hunter (1962) has shown that a pasture type tends to support a characteristic seasonal pattern and intensity of grazing. It is a matter of observation that some pasture types, notably *Agrostis-Festuca* pastures, tend to be more heavily grazed and to carry less dead and neglected herbage than others. But sheep range over a variety of pasture types in a day's grazing and are thus given ample opportunity to dilute their diet quality.

It is contended that both set-stocking and free-range grazing are important factors in determining diet quality and in limiting overall pasture utilization in hill pasture systems. It has been calculated that overall utilization is less than 30% of the pasture production (Eadie 1967). This calculation is based on the results of studies in a predominantly grassy hill environment. Evidence for other hill environments is non-existent and it may well be that the nature of the plant material is an important limiting factor to diet quality on blanket bog, for example.

In those hill pasture systems where sheep are removed from the hill for a period between the end of one growing season and the beginning of the next, it is reasonable to suppose that the relationship between overall stocking rate and summer pasture production is closer and that overall efficiency of pasture utilization is greater, but any gains in this respect are obtained at the expense of other resources.

All in all it seems likely that a major factor limiting pasture utilization in most hill situations is the management system, which creates a kind of vicious circle.

By way of contrast, stock numbers in intensive lowland systems are maintained at much higher levels relative to summer pasture production. A part of the explanation for this lies in the various consequences of the cutting of herbage for conservation. Although conservation procedures are far from completely efficient they are substantially more efficient than conservation of herbage *in situ* for the length of a British winter. The quality of the conserved material is initially higher, in that it is cut at a time of year and a stage of growth at which, despite nutrient losses, the resultant product can be fully utilized when fed back. Material conserved *in situ* in a free-range system is, at the beginning of winter, already poor in that the fund contains a substantial proportion of long-standing neglected material and is already depleted by grazing of some of its more highly digestible fraction. The stored material is also much more subject to loss by weathering.

A direct consequence of cutting for conservation is therefore a much more efficient use of that portion of the pasture growth set aside (deliberately or otherwise) for winter requirements. An indirect consequence is the ability to keep over the winter an animal population of a size much more closely related to the supply of summer pasture. A further refinement is the capacity to cope with pasture production which is surplus to immediate require- ments within the pasture's growing season, thus greatly reducing the tendency for undergrazing and its subsequent problems. Periodic cutting is also a means whereby the consequences of any previous under-utilization may be redeemed at little cost to either efficiency of pasture use or the nutrition of the individual animal.

In many systems, of course, requirements in the winter may be met partly by conservation *in situ* and partly by conservation as hay or silage, but in these cases the 'autumn-saved' pasture growth is of comparatively recent origin and comparatively free from neglected herbage. It is also used before weathering has substantially reduced its quality, so that the efficiency of use is greatly in excess of *in situ* conservation in hill pasture systems. Lowland systems also permit the use of the sheep themselves in cleaning up the consequences of previous undergrazing at certain periods, for two reasons (i) because of the readiness with which grazing can be controlled, and (ii) because of the higher nutritional status of lowland sheep. In some cases this may be part of the system, as when lambs (at a time when milk has become a relatively unimportant part of their diet) are given preferential treatment by being allowed to graze ahead of the ewes and so select for themselves a high-quality diet. The ewes follow, inten- sively-stocked, and clean up the pastures. This practice can only be part of a system in which the body condition of the sheep is maintained at a reasonably high level and where good nutrition can be assured when it is required.

The longer growing season of lowland pastures, and the use of nitrogen fertilizer in modifying pasture production and its seasonal distribution, also contribute to an easier matching of pasture supply and demand.

All these manipulations contribute to the fact that a much higher propor- tion of the pasture production is ingested by the sheep in lowland production systems.

SHEEP PERFORMANCE

There are substantial differences in the performance of the individual animal between the hills and the lowlands.

The nutrient requirements of sheep vary from phase to phase of the annual production cycle, and there is an important interdependence between the various phases.

The year-long, free-range, set-stocked pastoral systems characteristic of hill farming, which operate against a background of highly seasonal pasture growth, inevitably result in a cyclical pattern of nutrition. One such cycle has been described (Eadie 1967) in which the peak values which occur in May–June (Fig. 2) are clearly low by lowland standards. From these

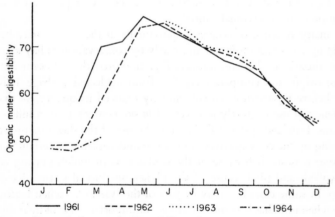

FIG. 2. Seasonal changes in digestibility of pasture consumed by sheep set-stocked on a Cheviot hill.

maximum values, a decline to the end of the pasture-growing season takes place, when the decline accelerates. Sub-maintenance energy intakes are the rule from early November through to March when an improvement, coincident with late pregnancy, and highly variable between years, takes place.

The cycle of energy intake is a reflection of the cycle of ingested pasture quality. Cyclical patterns of nutrition are not, of course, unusual. The important characteristic is the low level at which this cycle operates, giving rise to a low range of body condition (Russell & Eadie 1968).

Current levels of reproductive performance in hill sheep are a function of body condition and nutrition at conception. The quality of the herbage ingested by the ewe declines from the time at which the demands of lactation are becoming less important, so that recuperation of body condition is limited. Mating takes place at a time of declining and sub-maintenance nutrition. Reproductive performance in hill sheep is responsive to both improved premating nutrition (Gunn 1967) and to improved body condition at conception (Gunn, Russell & Doney, in Russell & Eadie 1968).

On present evidence, the fact that hill sheep draw on body reserves over the December to February period is not of great significance to production. The important point is the general inadequacy of these reserves, which is a consequence of inadequate nutrition in summer and autumn.

It is widely recognized that hill ewes are invariably undernourished during late pregnancy, and this is particularly so in twin-bearing ewes. The result is too many poorly developed and weakly lambs at birth which fail to survive.

Growth rates of lambs are well below the potential of hill lambs for growth (Peart 1967). Lamb growth rates up to 6–7 weeks of age are primarily dependent on milk production in the ewe which in turn depends primarily on the nutrition of the ewe during lacation. Improvements in ewe nutrition during lactation have been shown to improve the growth rate of hill lambs at pasture, and particularly the growth rates of twin lambs (Eadie 1967). Lamb growth rates beyond 6–7 weeks of age become increasingly dependent on the quality of the food they get for themselves from their pasture. From this time, growth rates of hill lambs decline steadily to weaning, and this is a function of declining pasture quality.

At each and every stage in the annual cycle of events, nutrition can be shown to be a limiting factor to animal performance on the hills, and hill sheep are performing at well below their genetic potential for the various components of performance.

The nutrient requirements of the lowland ewe are also highly seasonal and do not fit naturally into the pattern of pasture production. It is, however, clear from the previous section of this paper that the manipulations possible in lowland systems go a long way to resolving the conflict between utilization and individual sheep nutrition. Although it would be difficult to sustain the argument that lowland sheep are performing to the limits of their genetic potential there is little doubt that a much more satisfactory matching of nutrient requirements and nutrient provision throughout the annual cycle of events is achieved in reasonably well managed lowland pasture systems.

For example, such good body condition and nutrition at mating can be attained that fertility rates are somewhere near the genetic potential of the sheep. Body reserves throughout pregnancy can be maintained at reasonable levels and the adequate nourishment of the developing foetus can be ensured (the latter with the help of some concentrate feed). Nutrition from pasture is adequate for good rates of growth in twin lambs, and clean good-quality pasture can be provided to maintain quite reasonable post-lactation growth rates.

Whilst it should not be assumed that the needs of lowland sheep are adequately defined, or that the difficulties of providing for these needs are all resolved, the fact remains that lowland sheep under conditions of good pasture management inhabit a nutritional environment in which their genetic potential for performance is much better expressed than do hill sheep. Whilst there is undoubtedly a genetic component in the differences in performance between hill and lowland sheep, the large part of the difference is nutritional in origin.

THE EFFICIENCY OF PASTURE USE

The efficiency with which the pasture that is grown in the hills and lowlands is converted to animal product is markedly different, and some crude calculations will serve to highlight the important points of difference. In truth, the information required to do this with great accuracy is not available, and some assumptions have to be made. It is contended, however, that these assumptions are reasonable, and that the magnitude of the errors involved is unlikely to alter the overall conclusions.

Efficiency of pasture use will be considered in two stages (1) the efficiency with which the pasture grown is ingested by the sheep, and (2) the efficiency with which the ingested material is converted to animal product.

I EFFICIENCY OF UTILIZATION
It has been suggested earlier in this paper that the pasture production basic to the production data given in Table 1 is around 2000 kg/ha/annum in the hills and around 10 000 kg/ha/annum in the lowlands. A reasonable estimate of the annual dry matter consumption of a hill ewe weighing some 50 kg can be based on data from the study of the annual cycle of intake previously referred to and on data for lactating ewes (Eadie, unpublished).

This estimate would be in the region of 500 kg. The weighted mean digestibility of this dry matter would be in the region of 65%. At a stocking rate of 0·8 ewes/ha, the calculated efficiency of utilization of the pasture produced is of the order of 20%.

Lowland ewes are generally heavier than hill ewes. Assuming that a lowland ewe weighs about 68 kg and that the weighted mean of the ingested pasture quality is likely to be at least 5 digestibility units better than that obtained by the hill ewe, the lowland ewe is unlikely to consume less than 720 kg dry matter per annum.

The calculated efficiency of pasture utilization at a stocking rate of 10 ewes/ha in this case is therefore around 70%.

2 THE EFFICIENCY OF FOOD USE

This topic has been discussed by Spedding (1963).

Product output per sheep is influenced by weaning percentage and by the weights of the lambs at disposal.

Food input is dominated by the maintenance requirements of the ewe which are related to body size and which often amount to some 75% of the ewe's total food intake (Coop 1961). The extra requirement of the ewe which bears and rears twin lambs is small relative to that of the ewe which bears and rears a single lamb. Substantial improvements in the efficiency of food use therefore accompany increases in litter size in sheep.

Growth rates of lambs also affect the efficiency of food use, since poor growth rates result in higher intakes of food per unit of gain in the lamb. In the context of the ewe/lamb unit, however, this factor is of minor importance since such increases in intake affect the total food input only slightly. But where poor growth rates result in the disposal of the lambs at comparatively light weights, then, of course, efficiency may be greatly affected.

The efficiency with which the ingested dry matter is converted to product is enhanced in lowland sheep by virtue of the higher weaning percentages and the greater weights at sale of the individual lambs. The lower weights and poorer dry-matter intakes of the hill sheep, however, tend to make them relatively more efficient than a straight comparison of the product output per sheep would indicate.

By expressing efficiency for present purposes in terms of liveweight of lamb produced per 100 kg ingested dry matter, the value for lowland sheep would be 8·5, and for hill sheep 3·8.

These various calculations are summarized in Table 3 and are affected only marginally by the fact that lowland sheep receive food inputs from outside sources to a greater extent than do hill sheep.

TABLE 3
Efficiency of use of pasture dry matter production

I *Ingestion*

	(a) Pasture production kg/ha/annum	(b) Herbage ingested kg/ha/annum	Efficiency $\left(\dfrac{b}{a}\times 100\right)\%$
Hill	2000	400	20·0
Lowland	10 000	7200	72·0

II *Conversion*

	(a) Herbage ingested kg/ha/annum	(b) Lamb production kg/ha/annum	Efficiency $\left(\dfrac{b}{a}\times 100\right)\%$
Hill	400	15·3	3·83
Lowland	7200	612	8·50

III *Overall*

	(a) Pasture production kg/ha/annum	(b) Lamb production kg/ha/annum	Efficiency $\left(\dfrac{b}{a}\times 100\right)\%$
Hill	2000	15·3	0·77
Lowland	10 000	612	6·12

In the examples given, differences in stocking rate exceed the differences in pasture production between the hills and lowlands, because of the widely different efficiencies with which the pasture production is ingested by the animal population. Differences between these environments in the efficiency with which the ingested dry matter is converted to product are somewhat less than the differences in output per sheep.

CONCLUSIONS AND SUMMARY

It is undoubtedly true that an animal population cannot produce beyond the limits of its food supply. For any given level of pasture production there must be some upper limit to animal output. It is, however, clear that the relationship between a sheep population and its pasture resource is complex, and this is especially so in environments in which pasture production is highly seasonal. Where a substantial proportion of the herbage grown is consumed, a situation which, in the U.K. at any rate, is likely to require a high degree of manipulation of animals and pastures, measurements of pasture production and of pasture quality may have some meaning with respect to animal production.

In extensive grazing systems, however, where the proportion of the pasture production which is ingested is low, measurements of pasture production and of the quality of the available herbage are of very limited value. The relation between pasture and animal production can be understood only against a background of knowledge about the processes of animal production. Central to this understanding is a knowledge of the nutritional status of the animal population and of the response of the components of animal performance to changes in nutrition.

REFERENCES

ARNOLD G.W. (1964) Some principles in the investigation of selective grazing. *Proc. Aust. Soc. Anim. Prod.* **5**, 258–10.
BLAXTER K.L., WAINMAN F. & WILSON R.S. (1961) The regulation of food intake by sheep. *Anim. Prod.* **3**, 51–61.
COOK C.W., MATTOX J.E. & HARRIS L.E. (1961) Comparative daily consumption and digestibility of summer range forage by wet and dry ewes. *J. Anim. Sci.* **20**, 866.
COOP I.E. (1961) The energy requirements of sheep. *Proc. N.Z. Soc. Anim. Prod.* **21**, 79–91.
DAVIES H.L. (1962) Intake studies in sheep involving high fluid intake. *Proc. Aust. Soc. Anim. Prod.* **4**, 167.
EADIE J. (1976) The nutrition of grazing hill sheep. *Rep. Hill Fmg Res. Org.* **4**, 38–45.
EADIE J. & BLACK J.S. (1968) Herbage utilisation on hill pastures. *Occ. Symp. Br. Grassld Soc.* **4**, 191–5.
FERGUSON K.A. (1956) The efficiency of wool growth. *Proc. Aust. Soc. Anim. Prod.* **1**, 58.
GUNN R.G. (1967) Life-time performance of the breeding ewe. *Rep. Hill Fmg Res. Org.* **4**, 51–8.
HAMILTON R.A. (1956) Utilisation of grassland. *Outl. Agric.* **1**, 5–11.
HOLMES W. (1968) The use of nitrogen in the management of pasture for cattle. *Herb. Abstr.* **38**, 4.
HUNTER R.F. (1962) Hill sheep and their pasture: a study of sheep-grazing in south-east Scotland. *J. Ecol.* **50**, 651–80.
HUTTON J.B. (1963) The effect of lactation on intake in the dairy cow. *Proc. N.Z. Soc. Anim. Prod.* **23**, 39.

JOHNSTONE-WALLACE D.B. & KENNEDY K. (1944) Grazing management practices and
    their relationship to the behaviour and grazing habits of cattle. *J. Agric. Sci.* **34**,
    190.
MILLER G.R. (1966) Botanical studies. *Prog. Rep. Nature Conservancy Unit of Grouse
    & Moorld Ecol.* **12**, 20–6.
MILTON W.J. (1940) The effect of manuring, grazing and cutting on the yield, botanical
    and chemical composition of natural hill pastures. *J. Ecol.* **28**, 326–56.
MILTON W.J. & DAVIES R.O. (1947) The yield, botanical and chemical composition of
    natural hill herbage under manuring, controlled grazing and hay conditions.
    J. Ecol. **35**, 65.
MOTT G.O. (1960) Grazing pressure and the measurement of pasture production. *Int.
    Grassld Congr.* **8**, 606–12.
PEART J.N. (1967) Lactation and lamb growth. *Rep. Hill Fmg Res. Org.* **4**, 69–76.
RAYMOND F. (1963) Interrelation of quality and intake of herbage. *Symp. Eur. Grassld
    Fed.* **1**, 36–9.
RAYMOND W.F. (1964) The efficient use of grass. *J. Br. Grassld Soc.* **19**, 81–9.
REID R.L. & HINKS N.T. (1962) Studies on the carbohydrate metabolism of sheep.
    XVII: Feed requirements and voluntary feed intake in late pregnancy. *Aust. J.
    agric. Res.* **13**, 1092.
RUSSELL A.J.F. & EADIE J. (1968) Nutrition of the hill ewe. *Occ. Symp. Br. Grassld Soc.*
    4, 184–90.
SPEDDING C.R.W. (1963) The efficiency of animal production. *Wld Conf. Anim. Prod.*
    2, 33.
VAN SOEST P.J. (1965) Voluntary intake in relation to chemical composition and di-
    gestibility. *J. Anim. Sci.* **24**, 834–43.
WHEELER J.L., REARDON T.F. & LAMBOURNE L.J. (1963) The effect of pasture availability
    and shearing stress on herbage intake. *Aust. J. agric. Res.* **14**, 364.

DISCUSSION

A. S. CHEKE: Are mating dates the same in the lowland and in the hill
country?

J. EADIE: They are not the same. The different dates are the best compromise
for lamb growth and survival for the different areas.

FOOD SELECTION BY NORTH AMERICAN DEER AND THEIR RESPONSE TO OVER-UTILIZATION OF PREFERRED PLANT SPECIES

By David R. Klein

Alaska Cooperative Wildlife Research Unit
College, Alaska

INTRODUCTION

Both species of North American deer (*Odocoileus virginianus* and *O. hemionus*), including their several subspecies, characteristically feed on woody browse in the winter or in dry seasons when plants are dormant. During the periods of vegetative growth, forbs, grasses and other green vegetation constitute the major portion of the diet. Numerous authors have reported on regional variations in food habits of deer and Leopold (1933) first proposed classifying deer foods on the basis of preference, availability and apparent quality into five categories: preferred, staple, emergency, stuffing and pastime. It was early recognized by Maynard, Bump, Darrow & Woodward (1935), Hosley & Ziebarth (1935), Clepper (1936) and others that deer have specific food preferences and that these are related to the nutritive quality of plants.

PREFERRED PLANT SPECIES

The now classical example of population eruption, followed by wholesale starvation of mule deer (*O. hemionus hemionus*) on the Kaibab Plateau of Arizona, caused by elimination of natural predators and complete protection from hunting, showed the great effect that a dense deer population can have on the plant ecology of an area (Rasmussen 1941). In this case 4000 deer in 1906 increased to 100 000 in 1924 on 4275 km² (1650 sq miles), virtually eliminated favoured browse species of *Salix* and *Rubus*, and prevented regeneration of aspen (*Populus tremuloides*) forests. Grass, which was little utilized by the deer, increased in density and area. The ability of the area to support deer was reduced to between 5 and 10% of its original capacity.

Although deer show preferences for particular plant species these vary from one region to another. In New York, white cedar (*Chamaecyparis thyoides*), yew (*Taxus canadensis*) and witchhopple (*Viburnum alnifolium*) are preferred winter browse species and are the first to disappear under high densities of white-tailed deer (*O. virginianus*). They are then replaced in the winter diet by hemlock (*Tsuga canadensis*) and other plants of lower nutritive quality (Severinghaus & Brown 1956). In the southern pine type of Louisiana and eastern Texas, species such as ash (*Fraxinus* sp.), greenbrier (*Smilax* sp.) and rattan (*Berchemia scandens*) are highly preferred by deer. When these and other highly palatable plants are eliminated on over-stocked ranges, wax myrtle (*Myrica cerifera*) and red cedar (*Juniperus virginiana*) may appear to be the most preferred plants, although they have a low preference value when the ash, etc., are still present (Lay 1956). Palatability and preference are considered synonymous for the sake of this discussion. In these areas high palatability was frequently, although not always, correlated with high protein and phosphorus content. Controlled burning was found to improve the quality and usually the palatability of browse species, as did removal of the forest overstory through timber harvest.

Dietz, Udall & Yeager (1962) found that big sagebrush (*Artemisia tridentata*) was preferred by mule deer in southwestern Colorado and could withstand heavy browsing pressure. But its palatability depended upon its proximity to other species of high preference, so that a mixed diet could be obtained. Even though the nutritive content of big sagebrush was high, when it was given to captive deer they did poorly unless it was fed in combination with other species. A likely explanation for this phenomenon lies in the work of Nagy, Steinhoff & Ward (1964) which showed a bactericidal effect on the rumen microflora by essential oils from big sagebrush.

Black-tailed deer (*O. hemionus columbianus*) in the northern Coast Range of California select chamise (*Adenostema fasciculatum*) more frequently than all other plant species on both open shrubland and dense chamise chaparral, although various woody plants constitute 50% of the annual diet of deer on shrubland and over 90% on mature chaparral (Taber 1956). Fire plays an important part in maintaining the shrubland type and brings about a more rapid cycling of nutrients in that type than occurs in the chaparral. The result is that plants, including the chamise, are of higher quality in shrubland than in chaparral; the average content of crude protein in the diet of deer in shrubland is 14% and in chaparral 9%. Einarsen (1946), in Oregon, also found wide variation in crude protein content of deer browse species following fire. Protein in samples of vine maple (*Acer circinatum*) averaged 12·8% in September, 3 years after the Tillamook burn, but by 6 years, in the same month and at the same site the average protein content had dropped to 9·3%. On Vancouver Island in British Columbia, the peaks of deer populations are accentuated and cycles are shorter on logged areas that have subsequently been burned than on those not burned (Robinson 1958). Five to six years after fires, the average body size of deer begins to decline but the population continues to increase as factors affecting population size lag behind factors influencing the size of the individual.

SELECTION FOR QUALITY

The foregoing discussion emphasizes the feeding selectivity of deer for certain species of plants of high quality. It is also known that deer are capable of selecting the most nutritious plants among those of the same species, as well as the portions of plants that are of highest quality. Swift (1948) showed that deer feeding in a cultivated field selected wheat and clover which contained 12% more ether extract, 38% more calcium and 34% more phosphorus than was present in wheat and clover in the ungrazed portions of the field. Thomas, Cosper & Bever (1964), using fertilizers on grass pasture, found that deer consumed greater amounts of forage from nitrogen-fertilized plots than from unfertilized ones and further that this preference increased with the level of nitrogen applied. In Alaska, spring and summer feeding behaviour of black-tailed deer (*O. hemionus sitkensis*) and feral reindeer (*Rangifer tarandus*) results in the selection of the highest quality forage available on a given range (Klein, 1962, 1968). This was substantiated through chemical analysis of both forage and rumen contents. Animals abandoned feeding areas as vegetation matured and decreased in quality, and moved to areas where similar plant species were less advanced and of higher quality.

Longhurst, Oh, Jones & Kepner (1968) concluded that deer select forage initially on the basis of smell which is followed up by taste, and they suggested that volatile substances in the plants, acting either to inhibit or attract deer, were largely responsible for determining their palatability.

They thought it unlikely, however, that deer were able to smell differences in nutritive quality of plants.

FACTORS AFFECTING NUTRITIVE QUALITY

Variation in the nutritive value of plants available to wild ruminants has been investigated through physiological studies of plants and a general review of the subject was made by Oelberg (1956). He listed the following factors as of major importance in influencing nutritive quality of plants: (1) the stage of maturity of the vegetation, with highest nutritive quality coinciding with the initiation of growth, (2) edaphic influences such as soil type and quality, (3) climate, through the media of temperature, precipitation and solar radiation, (4) variations within plant species, including differential digestibility, presence of toxic or inhibitory substances, and rate of growth, (5) animal class, which refers to the differential ability of certain species of animals to digest plant components, (6) range condition, here relating to intensity of previous animal use which can alter plant vigour and the successional stage of the vegetation.

QUALITY IN ALPINE AND ARCTIC REGIONS

My own work in Alaska (Klein 1965, 1968) has attempted to identify the relative importance of the factors responsible for high forage quality in alpine and arctic regions, which in turn account for the rapid growth rates of ruminants using the forage. It is known that the short, cool growing season of arctic and alpine regions fosters the evolution of rapidly maturing plants with brief vegetative stages (Bliss 1962), and we know that rapid growth in plants is associated with high nutritive quality. Arctic and alpine vegetation, growing beyond the tree line, is subject to direct solar radiation in contrast to forest floor vegetation. In deer habitat in southeastern Alaska, light intensities may be reduced from in excess of 5000 foot-candles on clear days in alpine areas to less than 100 foot-candles beneath the dense canopy of rain forests. Plants common to the forest floor have a high tolerance for shade but this tolerance is apparently gained by sacrificing growth rate which results in lowered nutritive quality. Vegetation on the forest floor in southeastern Alaska contains less than 15% protein under optimum conditions at onset of growth, in contrast to 20 to 25% for alpine vegetation in the same region. In other parts of North America, physical removal of the forest canopy through logging has markedly improved the nutritive quality of plants growing on the forest floor (Lay 1956; Robinson 1958; Brown 1961).

Any variation in exposure, altitude and slope from a flat plain at sea level will alter the phenological progression. The onset of plant growth in the spring is progressively delayed as altitude increases, other factors being equal. According to Hopkins' (1920) bioclimatic law this delay is 3 to 4 days for each 100 to 130 m increase in altitude. Similarly, exposure governs the seasonal progression of plant growth. In Alaskan deer habitat, south-facing slopes are frequently 2 to 3 weeks ahead of level areas in growth of vegetation and there is an even wider gap with north-facing slopes. On steep south-facing slopes dark coloured rock ledges and outcrops absorb solar energy and re-radiate it to the immediate surrounding areas. This creates a microclimatic effect resulting in plants beginning to grow in rock crevices and adjacent to outcrops even before the snow has left nearby areas. In areas of limited precipitation the effect of slope or aspect may be more directly related to seasonal availability of moisture for plant growth. In this respect Bissell (1953) found higher protein content in plants growing on north-facing slopes than on drought-prone south slopes in the northern Coast Range of California. Of importance in Alaska is the fact that soils on south-facing slopes receive more insolation both daily and seasonally than cool north slopes, This provides more favourable conditions for the growth of soil-building bacteria and fungi. Consequently, processes of decay are speeded up and there is a rapid release of nutrients to the soil.

Since plants in early physiological stages of growth are most nutritious, the greater the topographic variation, in the form of altitude, slope and exposure, the longer will be the period during the growing season when forage of high quality will be available for selective grazing by herbivores that are capable of ranging over the topographic extremes. This appears to be the crucial factor in accounting for differences in growth rates and body size of deer in Alaska (Klein 1965) and our observations suggest that it is important in the ecology of mountain sheep (*Ovis dalli*) and mountain goats (*Oreamnos americanus*) as well.

During the period of summer growth the average daily period of solar radiation in alpine areas exceeds that at sea level, because of the lower angle that the horizon presents to alpine areas. This becomes more important in arctic and subarctic regions because the summer sun approaches the horizon at a relatively low angle. Long day length in both alpine regions and the arctic not only accounts for a longer daily period of plant growth but it also means decreased night-time periods when catabolic energy

4

losses in plants through respiration are not offset by the anabolism of photosynthesis (Kislyakova 1960).

In summer, alpine nights are cool and brief, therefore carbohydrate levels remain high for a longer portion of the day and decreases are not as large as at lower elevations where night-time temperatures are usually warmer. Animals, such as ruminants, that graze at frequent intervals, including the night and early morning, are able to secure summer forage of more constant nitrogen and carbohydrate levels in alpine areas than in non-alpine areas where nights are longer and warmer. Summer vegetation in arctic regions, under 24-hour daylight conditions, undoubtedly benefits from the absence of catabolic night-time metabolism. The high quality of arctic forage, which is reflected in the unparalleled growth rates of caribou and other arctic herbivores during the brief arctic summers (Kitts, Cowan, Bandy & Wood 1956; Krebs & Cowan 1962), quite likely results, at least in part, from this phenomenon.

High nutritive quality of alpine plants has recently been demonstrated by Johnson, Bezeau & Smoliak (1968), in a study of alpine vegetation in the southeastern Canadian Cordillera. They found exceptionally high nutritive quality of alpine plants in contrast to those of the fescue (*Festuca scabrella*) association of the adjacent lowlands. Alpine grasses contained 50% more protein and 100% more phosphorus than grasses of the fescue association, and grass-like plants contained about twice the protein and phosphorus of similar lowland forms.

EFFECT OF FOOD QUALITY ON BODY SIZE AND GROWTH

It is often possible to make comparisons between deer populations which have food resources of differing quality available to them. This can be done between areas, or comparisons can be made on the same area at different times between which a change in the quality of available food has occurred.

Body weight is one of the more universally used criteria for reflecting the growth rate and nutritive status of deer and there have been numerous examples of large body size being correlated with good forage quality. In the central Adirondack Mountains of New York, high deer densities have changed the composition of forage plants and have reduced the quantity as well. Here, adult deer are 9 to 18% less in body weight than deer from central and western New York where hunting pressure is much heavier and range conditions are considered good (Severinghaus 1955). In Missouri, deer were taken from the poor soil areas of the southern Ozark Mountains where populations were in apparent 'balance' with the food supply and were reintroduced to agricultural areas on good soils to the north where they previously had been exterminated. Five years after the introduction, deer killed by hunters from the expanding population on the good soils were 28% heavier than their equivalents in the Ozarks (Steen 1955). In the southern pine type of East Texas adult bucks were as much as 28% heavier on moderately stocked ranges as on 'over-stocked' ranges where preferred food species had been eliminated during long periods of high deer densities (Lay 1956). Einarsen (1946), in comparing deer habitat in western Oregon, found that adult bucks were over 40% heavier from a logged, and recently burned, area than similar bucks from another study area in mature forest with a closed canopy.

In a comparison of two island populations of deer in the coastal rain forest of southeastern Alaska, deer were consistently heavier on the island where the population was controlled by wolf predation and periodic heavy winter snows than on the island where absence of wolves and mild winter weather allowed the deer to overutilize the preferred forage species of high quality (Klein 1964). Average weights of adult females and of all male age groups with the exception of 1-year-old animals differed significantly on the two islands. Males were 17% heavier on the island with wolves and heavy snows at 1 year of age, 24% at 2 years, 31% at 3 years and 37% at 4 years. Comparison of skeletons of deer from the two islands showed that male deer on the island with better-quality food completed their skeletal growth in 3 years while an additional year was required on the other island.

Antlers apparently have a lower priority for growth than most other body tissues and are therefore more directly affected by limitations in forage quality than body size (Cowan & Long, 1962). Park & Day (1942) found a 50% decrease in antler volume and a 5% decrease in body weight in yearling bucks from 1936 to 1939 in Pennsylvania, associated with increasing population pressure on the food supply.

In my own observations on an introduced population of feral reindeer on an island in the Bering Sea, body weights greatly exceeded those of reindeer in domestic herds in 1957, during the early expanding phase of the population (Klein 1968). Six years later, just before a crash die-off occurred, when the density was more than four times as great and the population was greatly in excess of the available winter food supply, average body weights had decreased by 38% for adult females and 43% for adult males. Although winter food supply, through interaction with climatic factors, was the dominant regulating mechanism for this population, qualitative limitations in summer forage, through intraspecific competition, were primarily responsible for the decrease in body size as the population increased.

POPULATION RESPONSES TO FOOD QUALITY

PRODUCTION AND SURVIVAL OF FAWNS

When the quality of deer food deteriorates because of continued high population pressure, plant successional changes, or other reasons, one of the most commonly observed results is decreased production and survival of fawns. Cheatum & Severinghaus (1950) found strong correlations between the fertility of deer and range quality. Average counts of corpora lutea for adult does were 1·97 on the best ranges and 1·11 for the poorest while comparable embryo counts were 1·71 and 1·06. On the basis of these data, fertility appeared to be more greatly influenced by rate of conception than *in utero* mortality after conception. Later studies showed that 25%–45% of the females breed successfully during their first year of life under good range conditions in contrast to less then 5% under poor range conditions (Severinghaus & Tanck 1964). On the Uwharrie deer range in North Carolina, where continued high deer densities produced a 'browse line' on favoured deer food plants and deer turned to pine and other plants of low palatability, counts of corpora lutea were 1·25 for adult does, 1·0 for yearlings and 0·16 for fawns (Barick 1958). On better ranges in North Carolina corpora lutea counts were 2·0 for adult does, 1·5 for yearlings and 0·35 for fawns. Taber (1956) also found better fawn production on the high quality shrubland range of the northern Coast Range of California than on the climax chaparral. The ratio of fawns to does over two years of age at the time of birth was 1·65 on shrubland in contrast to 0·77 on chaparral. Other similar variations in deer fertility have been associated with food quality in Wisconsin (Dahlberg & Guettinger 1956), Massachusetts (Shaw & McLaughlin 1951) and elsewhere. Elk (*Cervus canadensis*) show variation in productivity with range quality and Cowan (1950) found a twinning rate of less than 1% on over-grazed range in contrast to a rate of 25% on the better ranges. Mech (1966), in a study of wolves and moose on Isle Royal in Lake Superior, found twinning rates of 38% under good range conditions with control of the moose by wolves; but in 1929, before the wolves were present, the moose had over-populated the island and Murie (1934) estimated twinning at 2%.

The importance of the nutritional status of female deer for the production and survival of fawns was studied with captive deer by Verme (1962). In five groups of pregnant does, totalling 171 deer, the following results were obtained: (1) in the control group, fed a high-quality diet throughout the winter and spring of pregnancy, pre- and post-natal losses of fawns from nutritional deficiencies were only 7%, (2) a low-quality winter diet followed by a high-quality spring diet resulted in early death of more than a third of the fawns produced, (3) the mortality rate increased to more than 50% and 90%, respectively, when the low-quality winter diet was followed by the feeding of only moderate or low-quality spring rations. The direct cause of fawn mortality was post-natal nutritive failure, resulting in the loss of more than a third of the ninety-two fawns born to the does on restricted diets. These losses occurred during the first 48 hours after birth and were the result of one or more of the following factors: (1) poor condition of the fawn, (2) the fawn too small to reach the teats of the doe, (3) the doe not permitting the fawn to suckle, and (4) delayed lactation or no lactation by the doe. Except for accidents and disease, all fawns that survived the crucial post-natal period also survived throughout the summer and autumn.

A relationship between fawn survival and the use by deer of an important browse species was investigated by Dasmann & Blaisdell (1954) in the Lassen–Washoe area on the boundary between California and Nevada. They found a strong correlation between fawn survival and the percentage utilization of bitterbrush leader growth by deer in winter. They predicted that in the face of competition for bitterbrush by livestock, a moderate decline in fawn survival could be expected when the cropping of bitterbrush by deer exceeded 25%, and a steep decline when it exceeded 34%.

Social dominance can be an important factor contributing to heavy fawn mortality when foods of good quality are in limited supply. Severinghaus & Tanck (1964) observed that under winter conditions that limit food supply and make possible the baiting of deer into live traps, adult deer eat the food along the bait line and drive off the fawns. As a consequence the adults had to be caught and removed from the area before the fawns could be taken. We observed a similar dominance of adults over fawns in feeding activity during extremely deep snow in Alaska in 1956 which resulted in unusually heavy losses of fawns (Klein & Olson 1960).

AGE RATIOS AND DIFFERENTIAL MORTALITY

Decreased fawn survival associated with limitations in quality and quantity of available food is one of several factors contributing to alteration of age ratios in North American deer. When deer populations are stable or slowly increasing, the age structure assumes a pyramidal shape with each age class slightly larger than the succeeding one. Flattening out of the population pyramid occurs when deer occupy new habitat, or are recovering from over-hunting or past catastrophic reduction in the population, where food quality and quantity are high. Recruitment to the population far exceeds mortality and young age classes dominate the population. When food becomes limiting because of high deer densities, ensuing competition, deterioration of the range, unfavourable plant successional changes, or for other reasons, the age of sexual maturity is delayed, fertility is lowered, fawn survival decreases and the population pyramid assumes a bell shape with older animals dominant in the population. Age structure can of course be altered by hunting which, through regulations and hunter preference, tends to be selective for adult animals. However, in areas where hunting has not been important, age ratios of deer populations have been altered by the qualitative and quantitative aspects of the food supply in the manner described above (Gunvalson, Erickson & Burcalow 1952; Klein 1965; Taber & Dasmann 1957).

Poor physiological condition of deer when food is limiting numbers, as well as the usual associated high densities, often result in heavy parasite burdens because of lowered resistance and greater opportunity for transmission of the infective agent. Fawns and yearlings characteristically have lower resistance to parasites and diseases than adults (Longhurst 1956) and therefore losses from such causes also contribute to the distortion of age ratios on over-stocked ranges.

Another mechanism which may contribute to the disproportionate number of old animals present in populations with limited food supply is delayed physiological aging. Delayed completion of growth and reproductive maturation is apparent under conditions of food limitation among captive and wild deer (Wood, Cowan & Nordan 1962; Klein 1964). Studies with laboratory rats have shown that those animals whose physiological age is delayed by reduced food intake tend to have a greater longevity than animals receiving adequate food (McCay, Maynard, Sperling & Barnes 1939). While increased longevity from this cause has not been found in deer, and would be difficult to show in wild populations, it may nevertheless be important where winter conditions are not severe enough to regularly cull the very old animals from the population. Geist (1966a) has shown that among mountain sheep (Ovis canadensis) the smaller horned, slower growing rams live longer than rams with vigorous body and horn growth.

SEX RATIOS AND DIFFERENTIAL MORTALITY

Sex ratios among deer populations are influenced by variations in the quality and quantity of the food supply. When the quality and quantity of available food are less than adequate, sex ratios tend to be altered toward a high predominance of females over males. Gunvalson *et al.* (1952) observed that adult females were greatly in excess of adult males on deer ranges in Minnesota where deer had been protected from hunting and where starvation losses were high due to over-population of the range. I have found similarly distorted sex ratios on over-crowded ranges in Alaska where hunting did not occur (Klein 1965). On ranges where food was not a limiting factor and deer populations were in a vigorous and expanding state, sex ratios tended more nearly toward equality although females generally tend to outnumber males among natural populations of deer species.

Distorted sex ratios among deer under poor range conditions are primarily the result of differential sex mortality. In most cases this mortality appears to be directly linked to nutritional factors. Robinette, Gashwiler, Low & Jones (1957) made an extensive review of the literature in their discussion of factors influencing differential mortality among mule deer in Utah. They concluded that with poor nutrition, mortality on males, particularly during their first year of life, was greatly accentuated. In California, Taber & Dasmann (1954) found differential mortality heavy towards males among fawns of black-tailed deer during their first winter. Longhurst & Douglas (1953), in California found similar differential mortality among fawns during the summer after birth. Longhurst (1956) also reported on winter losses of fawns during the exceptionally severe winter of 1951–1952. During early winter and midwinter, male fawns died at the rate of 400 to every 100 females, but as the winter progressed more females succumbed to bring the over-winter average to 210 males per 100 females. We also observed heavy losses of male fawns in Alaska during a brief period of exceptionally deep snows in March 1956 (Klein & Olson 1960). It appears that male fawns, perhaps due to their higher metabolic rate resulting from a greater rate of growth, activity, curiosity and independence than occurs among females, are more subject than females to mortality when food becomes limiting. Data which appear to contradict this assumption have been collected by Severinghaus (1956) in New York. Among carcasses of starved deer found in winter 'deer yards' the ratio of male to female fawns was 76 to 100. He assumed that the ratio among fawns killed by hunters in the fall, which was 106 males to 100 females, was representative of the population, and he concluded that a differential mortality, heavier on females, affected the fawn group during winter. It is quite likely however, that there was a bias in the autumn sex ratio because of hunter selectivity for male fawns, on the basis of their larger size. The greater vulnerability of male fawns to hunting has been well documented by Anderson (1953) on roe deer (*Capreolus capreolus*). If this hunting selectivity existed, females may have outnumbered males in the fawn group at the beginning of the winter. Also any early or midwinter mortality which might have occurred before the deer were concentrated in the 'yards' would likely have been missed by the type of mortality surveys done. Experience in the western part of the continent suggests that such mortality is heavy for male fawns.

Males continue to suffer heavier mortality throughout their lives than do females, and this differential mortality is accentuated under poor range conditions. The larger loss of males is related to their greater growth requirements when young, their greater activity and therefore propensity for accidents, and perhaps more importantly their tendency to utilize their fat reserves during the breeding season, which results in their entering the winter in poor condition. Among black-tailed deer in Alaska we found that adult male deer were much more subject to winter mortality than adult females. During the prime ages, from $1\frac{1}{2}$ to $5\frac{1}{2}$ years, when natural losses are normally light, we found that the ratio of males to females dying was 167 to 100 (Klein & Olson 1960). In these age groups the sex ratio in the population is already distorted toward the females. Age ratios in the hunter harvest confirm the longer life expectancy of females. In the 1956 harvest, 7% of the males and 22% of the females shot were over 5 years old. Any bias in this sample due to hunter selectivity would tend to favour the males, as antler size increases with age.

A useful example of the differential vulnerability of the sexes to mortality is available from the reindeer of St Matthew Island (Klein 1968). This introduced herd increased to 6000 by 1963, 19 years after their introduction, and suffered a crash due to mortality in the following winter. The adult sex ratio before die-off was 69 males to 100 females. Among the 42 animals known to survive the mortality all were females except one; this appeared to be an abnormal, non-reproductive male, which may have accounted for its survival when no normal males did.

Differential mortality among the sexes undoubtedly is the major cause for distorted sex ratios among deer populations. However, some variation in sex ratios at birth has been observed and this can also be related to food quality and quantity. In Utah, Robinette et al. (1957) found a foetal sex ratio of 122 males to 100 females among does carrying their first young, while among deer that had had previous births the ratio was 106 males to 100 females. McDowell (1960) has shown that a significant correlation exists between foetal sex ratios and the age of the dam in white-tailed deer. Males greatly predominate (154 males to 100 females) in the fawns carried by deer in their first year of life, while the sex ratios of fawns carried by older does are more nearly equal with a possible tendency towards more female fawns in the very old deer. Under optimum range conditions where growth rates are rapid, sexual maturity may be reached in the first year of life and a high proportion of females give birth to young at 1 year of age. Under such circumstances this age group is usually much larger in number than any others and therefore the total contribution of young to the population could be appreciably distorted toward males. Conversely, where growth is retarded due to nutritional factors, females do not give birth to their first young until they are 2 years old and in some cases not until the following year. Under these conditions the sexes of the new-born fawns would be more nearly equal.

It is known among humans and some other mammals that males suffer heavier *in utero* mortality than females (Scheinfeld 1943; Chapman, Casida & Cote 1938). Since *in utero* mortality is increased in deer under conditions of poor nutrition (Cheatum & Severinghaus 1950) it would appear logical to assume that the proportion of males born to females of similar age would be lower on overcrowded ranges of poor quality than on those of good quality.

CONCLUSIONS

Deer, at least in North America, show direct responses in physiology and in population to qualitative aspects of their food supply. Deer are selective in their feeding habits, usually choosing plants, or parts of plants, which are of highest nutritive quality. Where food becomes limited, either through plant ecological changes or increases in population density, the preferred plant species of high forage equality are reduced and often eliminated by the deer.

Deer respond to limitations in forage quality through restrictions in growth rate and decreases in ultimate body size. Such growth restrictions appear to be more closely related to the quality of the spring and summer forage, which corresponds with the period of most active growth of deer rather than with the quality of the winter forage. Christian (1963) and others have suggested that the body size of some mammals is reduced in crowded populations, due to density-dependent physiological stress. While there is evidence to support depression of growth resulting from stress in laboratory populations of rats and mice (Crew & Mirskaia 1931; Calhoun 1950; Strecker & Emlen 1953), growth of North American deer appears to be more directly related to the qualitative and quantitative aspects of the food supply which may not necessarily be related to the density of the population.

Growth repression in individual deer, as a result of qualitative limitations in the food supply, also affects the dynamics of deer populations. Retarded growth delays sexual maturity and increases the likelihood of fawn mortality during their first winter, both of which reduce the productivity of the population. Poor nutrition can directly affect the productivity of female deer by lowering conception rates, by increasing *in utero* mortality and by increasing the mortality of new-born fawns. Lowered productivity, poor survival of fawns and continued heavy mortality of young growing deer, all contribute to the distorted age ratios—heavy towards old deer— which are characteristic of populations suffering from food limitations.

Sex ratios can also be greatly altered by food quality and quantity. Differential mortality, normally slightly heavier on males than females, is greatly accentuated when populations outgrow their food supply and then stagnate or decrease. This heavier mortality on males is often greatest during the first year of life but it continues throughout all age groups.

The summation of these several responses of deer populations to their food supply is: (1) lowered productivity through reduced birth rate and increased death rate when food quality and quantity are limiting, leading to ultimate stabilization, or more frequently, reduction in the population or, (2) high productivity through maximum fertility and low mortality when food is of good quality and quantity, leading to increasing populations if other population controls, such as predation, hunting, or dispersal, are insufficient to limit numbers.

In the recent discussion in the literature of self-regulatory mechanisms in the control of animal populations, generalizations have been extended to include ungulates (Christian & Davis 1964) as well as the rodents that have provided the more classic examples. Christian, Flyger & Davis (1960) attempted to implicate physiological stress as the mechanism of control in the mass mortality of a herd of sika deer (*Cervus nippon*) on an island in Maryland but their data were inconclusive and in conflict with another published report on the die-off (Hayes & Shotts 1959). I have previously suggested that in the case of ungulates there appears to be a relationship between the self-regulatory ability of their populations and the relative stability of the environments within which they have evolved (Klein 1968). North American deer that are adapted to early successional stages of vegetation, which are of a transitory nature, appear not to have well-developed self-regulatory mechanisms and are normally characterized by wide population fluctuations. On the other hand the roe deer in Europe (Anderson 1961; Kurt 1968) and some bovids, such as the Uganda kob (*Adenota kob thomasi*, Buechner 1963) and the North American mountain sheep (*Ovis canadensis*, Geist 1966b), that are found on relatively stable vegetation types, appear to have evolved behavioural mechanisms that tend to contribute to the stability of their populations.

Of importance in comparing roe deer with North American deer is the significance of dispersal to population control. Roe deer habitat in Europe generally consists of 'islands' of favourable habitat with suitable cover, surrounded by less favourable, and often poor habitat. The favourable habitat often coincides with large private holdings where hunting pressure is light while in the surrounding areas the deer are persistently hunted. In unhunted or lightly hunted areas surplus deer are expelled through behavioural interactions (Kurt 1968) which tend to control the population at a given density apparently in relation to the food supply. Because these areas are generally surrounded by areas of low population density, dispersal of the surplus can readily and continually take place. In North America, deer habitat units are usually far more extensive in area than in Europe and are of more uniform character. Even if behavioural mechanisms exist which tend to cause expulsion of surplus animals, dispersal of these animals in North America cannot usually function in population regulation because population pressures are relatively uniform throughout the habitat and there are few surrounding low-density areas to absorb the surplus. The situation is perhaps comparable to fenced roe deer habitat in Denmark described by Anderson (1961), where self-regulation of the population through dispersal was not possible and winter starvation was a common agent of mortality (Anderson, pers. commun.); starvation is rare among roe deer in Denmark when they are not fenced.

It would appear that population regulation, in those North American deer populations which are not controlled by hunting, is most often accomplished through direct interaction with the food supply. While mechanisms of self-regulation of populations may be present among deer in North America they apparently are not usually effective in controlling numbers because of characteristics of the habitat which include large and uniform expanses of suitable vegetation types and often moderately rapid changes in vegetation through seral succession.

SUMMARY

1.   North American deer show food preferences for certain plant species. These preferences vary regionally and are influenced by fire, logging, drought, and by inter- and intra-species competition. Preferred food species are often eliminated under high deer densities and the remaining less palatable plants are often of lower nutritional quality. Deer can select plants and their parts on the basis of nutritional quality.

2.   The maturity of vegetation greatly affects dietary quality. In alpine and arctic regions long periods of solar radiation during the growing season, rapid growth rates in plants, and wide phenological variation, associated with topographic irregularity, account for the high quality of forage that is available to free-ranging herbivores.

3.   Deer show differences in rates of growth and ultimate body size in relation to quality of food. Numerous examples are available from throughout North America of decreases in body size when high densities of deer change the composition of forage plants on the range. Large body size and rapid growth rates are characteristic of deer from areas of good soil quality, recently burned areas, logged areas, and from areas with wide topographic variation. In Alaska, deer from populations controlled by wolves and heavy winter snows are larger than deer from comparable areas without wolves and with mild winters.

4.   Food quality influences deer productivity and survival of fawns. Where food quality is limiting, fertility appears to be more greatly influenced by rate of conception than by *in utero* mortality after conception. Counts of corpora lutea are consistently lower among deer populations where food quality has been reduced through any of the ecological influences mentioned above than from ranges with high forage quality. Decreased fawn survival has been associated with poor nutrition of the female, and heavy losses of fawns during their first winter are common on 'over-stocked' deer ranges.

5.   When food becomes limiting, because of high deer densities, ensuing competition, deterioration of the range, unfavourable changes in plant succession, or for other reasons, the age of sexual maturity is delayed, fertility is lowered, fawn survival decreases and the population pyramid assumes a bell shape with older animals dominating numbers. Conversely, the population pyramid flattens out when deer occupy new habitat, or are recovering from over-hunting or past catastrophic reduction in the population, where food quality and quantity are high.

6. When the quality and quantity of available food are less than adequate, sex ratios among deer tend towards a high predominance of females. Male fawns suffer heavier mortality than females in severe winters when food is limited. Adult males also are more prone to mortality from predation, accidents, and starvation than are adult females. Sex ratios tend to favour males among births from young female deer, which predominate in an expanding population, while among older females young are born in nearly equal sex ratios.

7. North American deer are adapted to early successional stages of vegetation, which are transitory, and they appear not to have well-developed self-regulatory mechanisms. Their populations are normally characterized by wide fluctuations which result from direct interaction with their food supply. This is in contrast to certain other ungulates, such as the roe deer in Europe, the Uganda kob and the North American mountain sheep, that live on relatively stable vegetation types, and appear to have behavioural mechanisms that contribute to population stability.

REFERENCES

ANDERSEN J. (1953) Analysis of a Danish roe-deer population. *Dan. Rev. Game Biol.* **2**, 127-55.

ANDERSEN J. (1961) Biology and management of roe-deer in Denmark. *La Terre et la Vie,* **1**, 41-53.

BARICK F.B. (1958) A study of deer productivity. *Wildlife in N. Carol.* **22**, 6-10.

BUECHNER H.K. (1963) Territoriality as a behavioral adaptation to environment in Uganda kob. *Proc. 16th Int. Congr. Zool.* **3**, 59-63.

BISSELL H.D. (1953) Nutritional studies on California big game. *Proc. A. Conf. West. Ass. St. Game Fish Commn,* **32**, 178-84.

BLISS L.C. (1962) Adaptations of arctic and alpine plants to environmental conditions. *Arctic,* **15**, 117-44.

BROWN E.R. (1961) The black-tailed deer of western Washington. *Biol. Bull. Wash. Dep. Game,* **13**, 124.

CALHOUN J.B. (1950) The study of wild animals under controlled conditions. *Ann. N.Y. Acad. Sci.* **51**, 1113-22.

CHAPMAN A.B., CASIDA L.E. & COTE A. (1938) Sex ratios of fetal calves. *Rec. Proc. Am. Soc. Anim. Prod.* **1938**, 303-4.

CHEATUM E.L. & SEVERINGHAUS C.W. (1950) Variations in fertility of white-tailed deer related to range conditions. *Trans. N. Am. Wildl. Conf.* **15**, 170-89.

CHRISTIAN J.J. (1963) Endocrine adaptive mechanisms and the phsyiologic regulation of population growth. *Physiological Mammalogy,* Vol. 1 (Ed. by W. V. Mayer and R. G. Van Gelder), pp. 189-353. New York.

CHRISTIAN J.J., FLYGER V. & DAVIS D.E. (1960) Factors in the mass mortality of a herd of sika deer, *Cervus nippon. Chesapeake Sci.* **1**, 79-95.

CHRISTIAN J.J. & DAVIS D.E. (1964) Endocrines, behavior, and population. *Science*, N.Y. **146** (3651), 1550–60.

CLEPPER H.E. (1936) Forest carrying capacity and food problems of deer. *Trans. N. Am. Wildl. Conf.* **1**, 410–16.

COWAN I.McT. (1950) Some vital statistics of big game on overstocked mountain range. *Trans. N. Am. Wildl. Conf.* **15**, 581–8.

COWAN R.L. & LONG T.A. (1962) Studies on antler growth and nutrition of white-tailed deer. *Proc. Nat. White-tailed Deer Disease Symp.* **1**, 54–60.

CREW F.A.E. & MIRSKAIA L. (1931) The effects of density on adult mouse populations. *Biol. Generalis*, **7**, 239–50.

DAHLBERG B.L. & GUETTINGER R.C. (1956) The white-tailed deer in Wisconsin. *Tech. Wildl. Bull. Game Mgmt Div. Wisc.* **14**, 282.

DASMANN W.P. & BLAISDELL J.A. (1954) Deer and forage relationship on the Lassen-Washoe interstate winter deer range. *Calif. Fish Game*, **40**, 215–34.

DIETZ D.R., UDALL R.H. & YEAGER L.E. (1962) Chemical composition and digestibility by mule deer of selected forage species, Cache La Poudre Range, Colorado. *Tech. Publs Colo. Game Fish Dep.* **14**, 89.

EINARSEN A.S. (1964) Crude protein determination of deer food as an applied management technique. *Trans. N. Am. Wildl. Conf.* **11**, 309–12.

GEIST V. (1966a) Does horn size determine life-span? *Animals*, August 1966, 200–3.

GEIST V. (1966b) On the behaviour and evolution of American mountain sheep. Ph.D. thesis p. 251. Univ. Br. Columb.

GUNVALSON V.E., ERICKSON A.G. & BURCALOW D.W. (1952) Hunting season statistics as an index to range conditions and deer population fluctuations in Minnesota. *J. Wildl. Mgmt*, **16**, 121–31.

HAYES F.A. & SHOTTS E.B. (1959) Pine oil poisoning in sika deer. *SEast. Veterinarian Mag.* **10**, 34–9.

HOPKINS A.D. (1920) The bioclimatic law. *J. Wash. Acad. Sci.* **10**, 34–40.

HOSLEY N.W. & ZIEBARTH R.K. (1935) Some winter relations of the white-tailed deer to the forests in north central Massachusetts. *Ecology*, **16**, 535–53.

JOHNSON A., BEZEAU L.M. & SMOLIAK S. (1968) Chemical composition and in vitro digestibility of alpine tundra plants. *J. Wildl. Mgmt*, **32**, 773–7.

KISLYAKOVA T.E. (1960) On the twenty-four hour photosynthesis of plants in the far north. *Fiziologiya Rast.* **7**, 62–6.

KITTS W.D., COWAN I.McT., BANDY J. & WOOD A.J. (1956) The immediate postnatal growth in the Columbian black-tailed deer in relation to the composition of the milk of the doe. *J. Wildl. Mgmt*, **20**, 212–14.

KLEIN D.R. (1962) Rumen contents analysis as an index to range quality. *Trans. N. Am. Wildl. Conf.* **27**, 150–64.

KLEIN D.R. (1964) Range-related differences in growth of deer reflected in skeletal ratios. *J. Mammal.* **45**, 226–35.

KLEIN D.R. (1965) Ecology of deer range in Alaska. *Ecol. Moncgr.* **35**, 259–84.

KLEIN D.R. (1968) The introduction, increase, and crash of reindeer on St. Matthew Island. *J. Wildl. Mgmt*, **32**, 350–67.

KLEIN D.R. & OLSON S.T. (1960) Natural mortality patterns of deer in southeastern Alaska. *J. Wildl. Mgmt*, **24**, 80–8.

KREBS C.J. & COWAN I.McT. (1962) Growth studies of reindeer fawns. *Can. J. Zool.* **40**, 863–9.

KURT F. (1968) Das Sozialverhalten des Rehes (Capreolus*capreolus*L.). *Mammalia Depicta*, 102.

LAY D.W. (1956) Some nutritional problems of deer in the southern pine type. *Proc. A. Conf. Sth. Ass. Game Fish Comm*, 10, 53–8.

LEOPOLD A. (1933) *Game Management*, p. 481. New York.

LONGHURST W.M. (1956) Population dynamics of deer, *Calif. Agric.* July, 9, 10, 12, 14.

LONGHURST W.M. & DOUGLAS J.R. (1953) Parasite interrelationships of domestic sheep and Columbian black-tailed deer. *Trans. N. Am. Wildl. Conf.* 18, 168–88.

LONGHURST W.M., OH H.K., JONES M.B. & KEPNER R.E. (1968) A basis for the palatability of deer forage plants. *Trans. N. Am. Wildl. Nat. Resour. Conf.* 33, 181–92.

McCAY C.M., MAYNARD L.A., SPERLING G. & BARNES L.L. (1939) Retarded growth, life-span, ultimate body size and age changes in the albino rat after feeding diets restricted in calories. *J. Nutr.* 18, 1–14.

McDOWELL R.D. (1960) Relationship of maternal age to prenatal sex ratios in white-tailed deer II. *Proc. NEast Sect. Wildl. Soc., Providence, R.I.* 4 pp. Mimeo.

MAYNARD L.A., BUMP A.G., DARROW R. & WOODWARD J.C. (1935) Food preferences and requirements of the white-tailed deer of New York State. *Jt Bull. N.Y. St. Conserv. Dept. & N.Y. St. College Agric.* 1.

MECH L.D. (1966) The wolves of Isle Royale. *Fauna natn. Pks U.S.* 7, 1–210.

MURIE A. (1934) The moose of Isle Royale. *Misc. Publs Mus. Zool. Univ. Mich.* 25, 44.

NAGY J.G., STEINHOFF H.W. & WARD G.M. (1964) Effects of essential oils of sagebrush on deer rumen microbial function. *J. Wildl. Mgmt*, 28, 785–90.

OELBERG K. (1956) Factors affecting the nutritive value of range forage. *J. Range Mgmt*, 9, 220–5.

PARK B.C. & DAY B.B. (1942) A simplified method for determining the condition of white-tailed deer herds in relation to available forage. *Tech. Bull. U.S. Dep. Agric.* 840, 60.

RASMUSSEN D.I. (1941) Biotic communities of Kaibab Plateau Arizona. *Ecol. Monogr.* 11, 229–75.

ROBINETTE W.L., GASHWILER J.S., LOW J.B. & JONES D.A. (1957) Differential mortality by sex and age among mule deer. *J. Wildl. Mgmt*, 21, 1–16.

ROBINSON D.J. (1958) Forestry and wildlife relationship on Vancouver Island. *For. Chron.* 34, 31–6.

SCHEINFELD A. (1943) Factors influencing the sex ratio. *Hum. Fert.* 8, 33–42.

SEVERINGHAUS C.W. (1955) Deer weights as an index of range conditions on two wilderness areas in the Adirondack region. *N.Y. Fish Game J.* 2, 154–60.

SEVERINGHAUS C.W. (1956) Winterkill of deer 1955–56. *N.Y. St. Conserv.* 11, 11.

SEVERINGHAUS C.W. & BROWN C.P. (1956) History of the white-tailed deer in New York. *N.Y. Fish Game J.* 3, 129–67.

SEVERINGHAUS C.W. & TANCK J.E. (1964) Productivity and growth of white-tailed deer from the Adirondack region of New York. *N.Y. Fish Game J.* 11, 13–27.

SHAW S.P. & McLAUGHLIN C.L. (1951) The management of white-tailed deer in Massachusetts. *Res. Bull. Mass. Div. Fish Game*, 13, 59.

STEEN M.O. (1955) Not how much but how good. *Missouri Conserv.* 16, 1–3.

STRECKER R.L. & EMLEN J.T., JR. (1953) Regulatory mechanisms in house-mouse populations: the effect of limited food supply on a confined population. *Ecology*, **34**, 375–85.

SWIFT R.W. (1948) Deer select most nutritious forages. *J. Wildl. Mgmt*, **12**, 109–10.

TABER R.D. (1956) Deer nutrition and population dynamics in the north Coast Range of California. *Trans. N. Am. Wildl. Conf.* **21**, 159–72.

TABER R.D. & DASMANN R.F. (1954) A sex difference in mortality in young Columbia black-tailed deer. *J. Wildl. Mgmt*, **18**, 309–15.

TABER R.D. & DASMANN R.F. (1957) The dynamics of three natural populations of the deer *Odocoileus hemionus columbianus*. *Ecology*, **38**, 233–46.

THOMAS J.R., COSPER H.R. & BEVER W. (1964) Effects of fertilizers on the growth of grass and its use by deer in the Black Hills of South Dakota. *Agron. J.* **56**, 223–6.

VERME L.J. (1962) Mortality of white-tailed deer fawns in relation to nutrition. *Proc. Nat. White-tailed Deer Disease Symp.* **1**, 15–38.

WOOD A.J., COWAN I. McT. & NORDAN H.C. (1962) Periodicity of growth in ungulates as shown by deer of the genus *Odocoileus*. *Can. J. Zool.* **40**, 593–603.

DISCUSSION

R. C. BIGALKE: Is the usual sex ratio amongst adult North American cervids more or less unity, and does the great preponderance of females appear in response to poor nutrition? In many African bovids such as impala and springbok, the normal adult sex ratio is of the order of two females to one male but there is no evidence that this is the result of food shortage.

D. R. KLEIN: The normal sex ratio at birth is close to unity but slightly favouring males. The effect of hunting pressure, which is usually selective for males, makes it difficult to assess the sex ratio among adults in most populations. However, where hunting does not occur and populations are vigorous (i.e. expanding), adult females only slightly outnumber adult males. Where populations are limited by nutritional factors, sex ratios become greatly distorted towards females. In addition to nutrition, other factors such as predation and accidental mortality (i.e. drowning, falls etc.) also appear to be heavier on males.

R. MOSS: Is it likely that wolves might, before they were eliminated, have controlled deer numbers in the past, thus removing any necessity for the deer to evolve a self-regulatory mechanism?

D. R. KLEIN: In southeast Alaska wolves do tend to 'level out' deer population fluctuations. This is apparent if you compare population fluctuations of deer on islands where both deer and wolves are present with islands supporting deer but no wolves. I think, however, the evolution of self-regulatory mechanisms in North American deer or the lack thereof, has been more directly related to characteristics of the habitat than to predators. Predators, nevertheless could lessen selection pressure for evolution of self-regulatory mechanisms.

H. N. SOUTHERN: You have told us that some of the islands off the west coast of Alaska have populations of deer which fluctuate considerably and are without wolves, and other islands have populations of deer which are more stable owing to the presence of wolves. Is this not an ideal situation to introduce wolves on islands where they are not now present, in order to test whether a violent reaction results before predator and prey settle down to a stable relationship?

D. R. KLEIN: This has already been done. The introduced wolves increased from two pairs to about fifteen. The deer declined almost to extinction, and the wolves have themselves consequently declined and only one now remains. There was evidence of cannibalism among the wolves after the deer population had been reduced. An important point is that the island is only about 80 km² in area. The result on a larger island might have been different, with perhaps not such drastic reductions of the deer population and later of the wolf population.

J. RICHARDS: Is it valid to draw conclusions from the introduction of wolves to an island population of deer where there were previously no predators? The deer may have been particularly vulnerable because of a lack of behavioural defence mechanisms against predators.

D. R. KLEIN: Deer learn very quickly to become wary of wolves. On the islands in southeast Alaska where wolves were introduced, deer were previously not wary of man but soon after the introduction of wolves they became very difficult to see and apparently had adapted their behaviour to the presence of wolves. I don't think that any lack of behavioural response of the deer to the wolves was significant in this study.

K. PAVIOUR-SMITH: Since the deer had depleted the shrubby vegetation on your experimental island much more than on islands where there had always been deer and wolves, what effects did the shortage of escape cover for deer have when the wolves were introduced?

D. R. KLEIN: The reduction of cover by the past browsing pressure of deer on the vegetation was undoubtedly a factor of some significance in making the deer vulnerable to wolf predation. However, there were extensive portions of the island where shrub species, which were not utilized as food by deer, offered sufficient cover. Also, reduction of vegetation by long-term heavy deer browsing was only relative; in the

5

rain forest of this part of Alaska there is generally no shortage of cover for deer.

T. B. REYNOLDSON: A great deal is spoken about bird behaviour in relation to shortages of various kinds but little about corresponding behaviour in grazing mammals. A second point is that I take it you are not using 'self-regulation' synonymously with regulation by intraspecific competition?

D. R. KLEIN: To your first point, the best observations we have on grazing behaviour in relation to food quality are from the feral reindeer population on St. Matthew Island in Alaska. In this case, when the population density was still low the reindeer were free to graze selectively.

We observed that in early summer they consistently fed on those sites where vegetation was beginning to grow and consequently was of high quality. As the vegetation matured the reindeer shifted their feeding sites to north slopes or where snow banks were receding and vegetation was just starting growth. When the population had increased to a very high density the opportunity for selective feeding was lost and the reindeer could be seen feeding on all sites available to them. In other words there was not enough high-quality forage to go around.

In answering your second question I am thinking of 'self-regulation' as intrinsic control of populations without serious debilitation of the welfare of the individuals or of the population, through such mechanisms as behavioural interaction or physiological response. This could include physiological stress as postulated by J. J. Christian and reduced productivity through lowered fertility, without necessarily involving malnutrition.

SELECTION OF FOOD BY ICELANDIC PTARMIGAN IN RELATION TO ITS AVAILABILITY AND NUTRITIVE VALUE

By Arnthor Gardarsson

Museum of Natural History, Reykjavik, Iceland

and Robert Moss

Nature Conservancy, Blackhall, Banchory, Scotland.

INTRODUCTION

Herbivorous mammals are well known to be very selective feeders. This is less well established in the case of birds and the aim of this paper is to present some examples of food selection from our study of Icelandic ptarmigan (*Lagopus mutus*). A variety of foods is available to this species throughout the year, but certain items are preferred. In an attempt to explain these preferences, we present some representative examples of food selection and describe associated variations in the availability and nutritive value of the foods. Other factors are undoubtedly important, but will not be discussed here. The work is still in progress and will be written up in greater detail in future publications.

METHODS

COLLECTION AND ANALYSIS OF CROPS

Most food studies using crops of gallinaceous birds have been based on extensive collections from hunters, and some workers (e.g. Davison 1940; Boag 1963) have stressed the desirability of a large sample of crops because of the great variability they encountered in the contents. However, our experience (Gardarsson, in progress) suggests that the variability is small in series of crops from birds all shot in the same habitat at the same time, and this is the case with every series in this paper.

47

Collected crops were preserved in deep freeze, the contents were separated by species and the components measured volumetrically to the nearest 0·1 ml. The results for each food item are presented as a percentage of the total volume of the whole series. Statistical probabilities in the text were calculated using the U-test of Mann & Whitney (1947). These values refer only to differences between food items in the crops and take no account of differences in the availability of these items in the vegetation.

ANALYSIS OF AVAILABLE VEGETATION

In fairly uniform habitats, the vegetation was described by distributing about twenty 0·5 × 1·0 m quadrats evenly over the area in which the birds had been killed. The cover of each species within a quadrat was assessed by

TABLE I
Modified Braun-Blanquet scale

Index	Cover (%)
1	≤5
2	6–25
3	26–50
4	51–75
5	76–100

eye according to a modified Braun-Blanquet scale (Table 1). Depending on the size of the area, the distance between quadrats varied from 20 to 100 m. In less uniform habitats, or if the series came from a larger area, the main communities were sampled by quadrats and a description was made including an assessment of the relative abundance of each species. Grasses, sedges, mosses and lichens, which were usually abundant, are with one exception not mentioned here, because they formed only a minute proportion of the birds' food. Cover was estimated in summer, because important food species, especially *Salix herbacea*, were likely to be underestimated or missed in winter.

CHEMICAL COMPOSITION OF FOODS

The chemical composition of the various food items was measured in terms of nitrogen ($N \times 6·25$ = crude protein), minerals (phosphorus is probably the most important, Moss 1968) and proximate organic constituents—water-soluble carbohydrates, crude fibre and crude fat (diethyl ether extract). Full details of the methods of analysis are available from the Nature Conservancy Chemical Service, Merlewood, Grange-over-Sands, Lancashire. The *in vitro* digestibility using sheep rumen liquor (Tilley & Terry 1963) was measured for most of the more important foods by Gunnar Olafsson cand. agric. and calculated for others from their contents of nitrogen, soluble carbohydrates and crude fibre (Appendix, p. 68). The digestibility measurements were not always carried out on the same samples as the chemical analyses, but were done on samples which were matched as closely as possible in chemical composition (nitrogen, soluble carbo-hydrates, fibre).

It was not feasible to examine the large numbers of samples which would be necessary to take full account of seasonal, annual and geographical variations in the composition of each food. Some indication of these effects was obtained by taking a small number of samples at different places, at times when the ptarmigan were eating the food in question. To make sure that each sample was representative of the site where it was taken we invariably combined a large number of small subsamples (taken from an area of several hectares) to form a composite sample (Calder & Voss 1957) usually of 100 g. In each Table, only one representative analysis of each food is given although we have analyses for several samples of most of the major foods.

EXAMPLES

EXAMPLE I. INCUBATING HENS

Icelandic ptarmigan breed mostly at low altitudes in dwarf-shrub heaths which are typically dominated by *Empetrum nigrum, sensu lato*. A representa-tive example is the *Empetrum–Betula nana* community at Hella (Fig. 1).

The crop contents of three incubating females killed at this place in early July were closely similar in composition and consisted largely (69%) of *Vaccinium uliginosum* flowers (Fig. 1). The leaves of this species were much more abundant than its flowers but formed a smaller proportion (9%; $P = 0.05$) of the crop contents. Clearly, the flowers were preferred to the leaves.

There was some indication that leaves of *Salix herbacea* were also preferred to leaves of *V. uliginosum* because they were more common in the crop contents (19% vs 9%; $P = 0.1$) although much less common in the vegeta-tion (Fig. 1). The most common potential foods—leaves of *Betula nana* and leaves and shoots of *Empetrum*—were ignored, as was the less common *Calluna vulgaris*.

The birds were clearly being very selective, yet all the plants mentioned, including those which were not eaten, would appear to be perfectly adequate foods, judging by their chemical composition (Table 2). Moreover, ptarmigan eat both *Empetrum* and *Calluna* shoots in summer in Scotland (Watson 1964) and *Empetrum* shoots in Japan (Chiba 1965).

The foods mentioned in Table 2 are not all directly comparable because they were not all taken at the same date. In particular, deciduous leaves

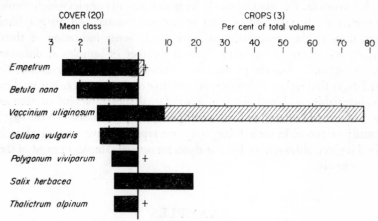

FIG. 1. Vegetation cover and food items from crops of incubating hens at Hella (65° 57′N, 18° 24′W) 3 July 1967. Sample size in parentheses. Hatching indicates reproductive parts.

change rapidly in chemical composition, being most nutritious at the start of growth and declining in nutritive value throughout the summer as they mature. However, leaves and flowers of *V. uliginosum* were sampled at the same time as each other (though not at the same time as the other foods). The preferred flowers were more digestible (49% vs 39%) and a better source of both protein and phosphorus than the leaves. Selection in this instance was clearly associated with a difference in nutritive value.

Of the three species of leaves available, the sample of *Salix herbacea* was hardly distinguishable in digestibility (calculated) and nutrient content (Table 2) from the sample of the slightly less favoured *V. uliginosum* leaves taken at the same time. *B. nana* leaves, although sampled 9 days later than the other species, were less digestible than the other leaf samples and were

TABLE 2

Chemical composition of foods available to incubating females in 1967

	Date	Place	N	P	Ca	Mg	K	Na	Sol CHO	Crude fat	Crude fibre	Ash	*In vitro* digestibility	Conversion g dry wt/100 ml
Vaccinium uliginosum flowers	19 July	Hrisey	2·50	0·30	0·19	0·16	0·94	0·21	18	4·8	—	3·17	(49)	20
Vaccinium uliginosum leaves	19 July	Hrisey	1·94	0·20	0·47	0·25	1·00	0·01	16	4·4	—	3·59	(39)	25
Vaccinium uliginosum leaves, new shoots	19 June	Hrisey	3·19	0·39	0·26	0·22	1·19	0·01	15	4·4	9	2·89	(59)	25
Salix herbacea leaves, new shoots	19 June	Hrisey	3·44	0·32	0·45	0·27	1·69	0·02	19	2·6	14	5·23	(60)	25
Betula nana leaves	28 June	Hella	2·75	0·35	0·47	0·37	0·94	0·02	14	5·5	9	3·57	(54)	25
Empetrum nigrum s.l. leaves and shoots	†		1·0	0·11	0·44	0·19	0·40	0·06	22	17·4	13	2·79	39	35
Calluna vulgaris leaves and shoots	†		1·1	0·10	0·32	0·19	0·30	0·10	22	7·4	16	3·85	42	40

The first two foods are not directly comparable with the next three because they were sampled at a different time. However, *V. uliginosum* leaves may be used as a standard of comparison.

†These are composite data made up as follows: nitrogen and minerals for both species are from the June data of Thorsteinsson & Olafsson (1965); organic constituents and ash from Moss (1968). Because these evergreen shrub species vary relatively little in chemical composition from place to place and change only slowly with season, this composition is probably valid in these cases, though it would not be with deciduous plants.

Calculated digestibility data are in parentheses.

a poorer source of protein. The differences were quite small and we cannot be confident about single samples of such variable material. However, the results suggest that the abundant B. nana was ignored because it was less nutritious than the chosen items, although the preference for S. herbacea over V. uliginosum leaves cannot be explained. Similarly, the fact that the dominant Empetrum and also Calluna were not eaten at all is probably explained by their being the poorest of all the foods available; they were relatively indigestible (39% and 42%) and were poor sources of protein and phosphorus.

Judging by Table 3, leaves of Polygonum viviparum were a good food which, although eaten earlier in spring (Example 7), were now almost ignored. These plants were well dispersed at a low level in the vegetation and appeared to be more difficult to feed upon than the foods actually taken. The incubating females presumably spent as little time away from their eggs as possible and may have discriminated against Polygonum viviparum to save time. We conclude that the preferences shown by the birds can be explained in part by the fact that they were selecting the most nutritious of the readily available foods.

EXAMPLE 2. CHICKS AND ADULTS IN JULY

One of the most striking examples of food selection is the preference shown by both chicks and adults in summer for the spikes and especially bulbils of Polygonum viviparum. Generally, spikes are taken together with the bulbils in the early stages of bulbil growth, but as soon as they are mature the bulbils are stripped from the plants and the spikes left bare. On the island of Hrisey (60° 54' N, 23° 10' W) in Eyjafjördur, in 1966 and 1967, when ptarmigan numbers were high, it was almost impossible to find a complete Polygonum spike in late July, except in exposed places where the birds seldom went.

The series in Fig. 2 was taken at a site where Polygonum was well distribu-buted but only fifth in cover ranking. Nonetheless spikes and bulbils formed 82% of the adults' diet and 63% of the food of the chicks, which also ate arthropods.

Bulbils are an excellent food, containing high levels of protein and phosphorus (Table 3). Although in chemical composition there is apparently little to choose between bulbils, spikes with bulbils, Polygonum leaves, and leaves of Salix phylicifolia, the bulbils are more digestible than these other foods and also more digestible than would be expected from their chemical composition (Appendix, p. 68). This is presumably because the grain-like

TABLE 3
Chemical composition of foods available in July

	Date	Place	N	P	Ca	Mg	K	Na	Sol CHO	Crude fat	Crude fibre	Ash	In vitro digestibility	Conversion g dry wt/100 ml
Polygonum viviparum bulbils spikes with bulbils	July 1968	Hrisey	3·22	0·41	0·30	0·27	0·81	0·01	18	4·8	5	3·09	72	35
leaves and stems	21 July 1968	Hrisey	3·06	0·49	0·15	0·28	1·25	0·01	17	2·1	—	3·59	62	25
leaves and stems	12 July 1968	Hrisey	3·00	0·34	0·73	0·66	1·63	0·01	21	1·5	—	5·76	64	15
Salix phylicifolia leaves	3 July 1967	Myvatn	3·06	0·38	0·64	0·23	1·50	0·01	11	6·3	9	5·10	(55)	—
Thalictrum alpinum leaves and stems	12 July 1968	Hrisey	1·94	0·20	1·13	0·36	2·38	0·02	28	2·7	—	8·00	(52)	25
Vaccinium uliginosum leaves	19 July 1968	Hrisey	1·94	0·20	0·47	0·25	1·00	0·01	16	4·4	—	3·59	39	25
Empetrum shoots	†	†	1·0	0·10	0·44	0·20	0·40	0·08	22	17·4	13	2·79	39	35

†Data from Table 2.
Calculated digestibility data are in parentheses.

bulbils are of a different and more easily digested structure than the other, leafy materials.

Of the other plants available, *Thalictrum alpinum*, *V. uliginosum* and *Empetrum* were relatively poor sources of protein and phosphorus and relatively indigestible. Although we have not analysed *B. nana* from late July, Example 1 shows that it was probably a poorer food than *V. uliginosum* earlier in July and this was presumably still the case later in the month.

Small quantities of green leaves were eaten, especially *Thalictrum*. Perhaps green food was necessary to supply some nutrients that were not

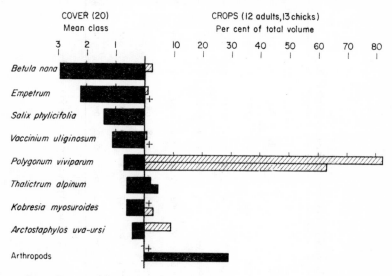

FIG. 2. Cover and food items at Laxardalsheidi (65° 40′N, 17° 15′W) 20 and 21 July 1965. Upper column—adult, lower—young, for other explanations see Fig. 1.

measured here and were possibly in short supply in the brown bulbils, e.g. carotenes, which are precursors of vitamin A. Nonetheless, bulbils were clearly selected and we suggest that this was because they were the most digestible and nutritious of the available foods.

EXAMPLE 3. EARLY WINTER, MOUNTAINS

Most ptarmigan leave the low ground and move to the mountains in September, where they feed in habitats dominated by *Salix herbacea*. The birds stay at high altitudes as long as snow conditions permit, and here the composition of their diet reflects approximately the composition of the vegetation. This is illustrated by two series collected on the mountain Tröllakirkja. In the first series (Fig. 3), many rosettes of evergreen alpine herbs were available, and the birds ate these as well as the *Salix* in roughly the proportions in which they occurred. In the second and more typical

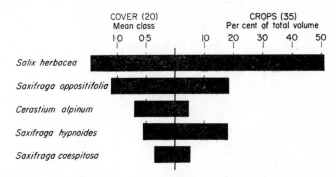

FIG. 3. Cover and food items at Tröllakirkja (65° 00′N, 21° 12′ W—alt. 800 m) 4 October 1965. Sample size in parentheses.

series (Table 4), very little apart from *Salix* was available and this formed the major part of the diet.

Two of the four rosettes available to the birds of the first series (Fig. 3) were analysed (Table 5) as well as a further sample of *Cerastium cerastoides*

TABLE 4

Vegetation cover and main food items in crops collected on 17 October 1966 at Tröllakirkja (65° 00′N, 21° 12′ W) in W. Iceland, at about 600 m altitude

	Cover (n = 30)		Crop contents (n = 36)	
	Mean class	Frequency (%)	Per cent of total volume	Frequency (%)
Salix herbacea	2·1	97	97·4	100
Sibbaldia procumbens	0·4	27	0·7	67
Cerastium cerastoides	0·4	33	trace	33
Saxifraga stellaris	0·3	27	trace	11
Gnaphalium supinum	0·3	27	0.0	0
Veronica alpina	0·2	23	trace	3
Saxifraga rivularis	0·2	17	trace	3
Cardamine pratensis	0·2	17	trace	22
Cerastium alpinum s.l.	0·1	13	0·7	25
Saxifraga caespitosa	0·1	7	trace	3
Equisetum variegatum	0·1	7	trace	6

TABLE 5
Chemical composition of foods available in early winter 1968 in upland habitats

	Date	Place	N	P	Ca	Mg	K	Na	Sol CHO	Crude fat	Crude fibre	Ash	In vitro digestibility	Conversion g dry wt/100 ml
Salix herbacea shoots (current growth)	5 Oct.	Hadegisfjall	3·19	0·45	0·49	0·18	1·13	0·01	35	3·3	10	4·94	54 (74)†	45
Saxifraga hypnoides‡ leaf rosettes	4 Oct.	Austurardalur	1·44	0·35	1·39	0·21	1·25	0·03	35	2·4	6	16·12	(61)	20
Cerastium alpinum‡ leaf rosettes	22 Nov.	Skalafell	2·50	0·35	0·45	0·60	1·63	0·04	33	2·9	—	10·93	(65)	—
Cerastium cerastoides‡ leaf rosettes	22 Nov.	Skalafell	2·44	0·28	0·19	0·26	1·15	0·04	46	2·7	—	7·20	(78)	—

†A calculated figure is given here as well as a measured figure because this sample could not be matched in chemical composition against the poorer samples used for the in vitro determinations.
‡High ash content in these samples was partly due to unavoidable contamination by soil.
Calculated digestibility data are in parentheses.

which was not a major food in this instance. Both *Salix* and the rosettes were well digested and were good sources of nutrients (Table 5) except that *Saxifraga hypnoides* was relatively poor in nitrogen. However, because *Salix* formed the major part of the diet and because the birds' requirements for protein in winter must be relatively low, we may assume that the diet contained sufficient protein. The ptarmigan were presumably selecting highly digestible energy-providing foods and in this respect *Salix* and the rosettes were quite similar.

In this example no clear preference was shown, apparently because all the foods available were approximately equivalent in nutritive value.

EXAMPLE 4. EARLY WINTER, LAVA FIELDS

Not all ptarmigan go to the mountain tops in September and a further series from early winter was taken on a lava field with a complex heath vegetation, where *Empetrum nigrum s.l.* and *Betula nana* were dominant (Fig. 4). *Empetrum* berries were the major food item by volume (36%), but came only third in terms of dry weight because of their high moisture content (Table 6). Vegetative parts of *Empetrum* were largely ignored though they were eaten later in the winter. The other major items in the diet were overwintering leaves of *Dryas octopetala* (29%) and shoots of three species of *Salix* (20%). Although second in cover ranking (Fig. 4) *B. nana* formed only 6% of the diet and was apparently less attractive to the birds than *Dryas* ($P < 0.02$) and the *Salix* ($P < 0.05$) species, which were less important in cover than *B. nana* but were eaten more. *Calluna vulgaris*, *Loiseleuria procumbens* and *Vaccinium uliginosum* were all present in roughly the same proportions as *Dryas* in terms of cover (Fig. 4), but were almost completely ignored except that some berries of *V. uliginosum* were taken.

Ptarmigan in this habitat had a markedly poorer diet than birds which moved to the *Salix herbacea* heath zone in the mountains (contrast Tables 5 and 6).

Shoots of the *Salix* species, *Dryas* leaves and *Empetrum* berries were all more digestible than *B. nana*, to which they were preferred. *B. nana*, however, was a better source of protein and phosphorus than *Dryas* leaves and more especially *Empetrum* berries. However, for the reasons outlined in the last example, it seems probable that the birds were selecting for energy-providing foods, and that enough nutrients to satisfy winter requirements were provided by the *Salix* species and *Dryas*.

Of the completely ignored foods, the leaves and shoots of *Loiseleuria procumbens* and *Empetrum* were poor sources of protein and phosphorus, although they were quite well digested and could probably have formed reasonable sources of energy had other foods not been available. *Calluna vulgaris* was poorly digested and also low in protein and phosphorus. The shoots of *V. uliginosum* contained slightly higher levels of protein and phosphorus than the other uneaten species but were very indigestible.

A difficulty is the fact that the difference in digestibility in Table 6, between *Empetrum* berries and the leaves and shoots of *Empetrum* and *Loiseleuria procumbens*, is negligible and insufficient to explain the obvious

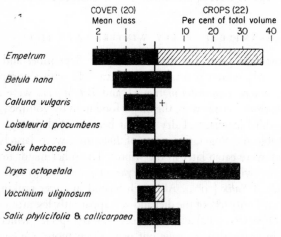

FIG. 4. Cover and food items near Gæsafjöll (64°46′N, 17°00′W) 9 to 12 October 1965. See Fig. 1. for explanations.

difference in preference. Perhaps berries were attractive because they were a rich source of soluble carbohydrates irrespective of their digestibility.

Nonetheless, we may conclude that most of the bird's preferences in the lava field occurred because they were choosing the best foods available.

EXAMPLE 5. LATE WINTER

Later in winter, when most of the mountain-top *Salix herbacea* is inaccessible under hard-packed snow, the birds move down to lower ground again. The next two series illustrate the effects of snow on the birds' food habits on an *Empetrum–Betula nana* heath interspersed with patchy *Betula pubescens* scrub (Table 7).

TABLE 6

Chemical composition of foods available in a lava field habitat in early winter

	Date	Place	N	P	Ca	Mg	K	Na	Sol CHO	Crude fat	Crude fibre	Ash	In vitro digesti-bility	Conversion g dry wt/100 ml
Salix herbacea (current growth)	early May 1965	Hrisey	2·30	0·33	0·89	0·24	0·80	0·06	17	4·4	16	5·42	46	45
Salix phylicifolia shoot tips and side buds	17 Nov. 1965	Armanns-fell	1·83	0·22	0·64	0·21	0·45	0·08	16	3·3	21	3·58	42	45
Empetrum nigrum s.l. berries	early May 1964	Hrisey	0·46	0·09	0·11	0·05	0·70	0·07	35	6·3	20	1·90	40	15
Empetrum shoots (current growth)	early May 1965	Hrisey	0·88	0·10	0·47	0·12	0·32	0·04	22	17·4	13	2·79	39	35
Betula nana catkins	3 Mar. 1967	Bratta-brekka	2·26	0·28	0·30	0·20	0·55	0·14	14	5·0	23	2·91	36	mixture 40
buds	3 Mar. 1967	Bratta-brekka	1·91	0·23	0·45	0·18	0·41	0·13	13	—	17	2·31	—	
wood (current growth)	3 Mar. 1967	Bratta-brekka	1·63	0·20	0·43	0·16	0·39	0·15	16	8·8	17	2·51	29	
Dryas octopetala (overwintering leaves)	early May 1967	Hrisey	1·56	0·16	0·67	0·35	0·45	0·10	18	8·2	13	5·41	(42)	35
Loiseleuria procumbens (current growth)	early May 1966	Hrisey	0·79	0·07	0·45	0·21	0·29	0·04	15	11·9	7	5·75	(38)	40
Calluna vulgaris (current growth)	early May 1965	Hrisey	0·95	0·11	0·34	0·16	0·38	0·15	22	7·4	16	3·85	30	40
Vaccinium uliginosum (current growth)	early May 1965	Hrisey	1·21	0·15	0·34	0·18	0·35	0·06	11	3·9	33	2·80	(13)	50

Calculated digestibility data are in parentheses.

The first series was taken when there was little snow and ptarmigan were feeding on the ground vegetation mainly outside the scrub. *Salix herbacea* was the main food taken, though only fourth and equal with *Vaccinium uliginosum* in abundance rating, closely followed by *Empetrum* shoots which were the most abundant food in the area. *Empetrum* berries still formed 11% of the diet by volume even though they were one of the scarcest foods. Thick snow then covered the ground and there was a considerable change in the birds' diet. They spent much more time in the scrub and *Betula pubescens* catkins now formed half their diet. The main

TABLE 7

Relative availability of main food species, and main food items in crops collected in March 1966 at Hvammur (64° 51′N, 21° 20′W), W. Iceland. The relative availability (RA) of foods is indicated on an arbitrary scale 1 to 5 (most abundant), based on observations during the days when birds were collected as well as on cover estimates.

	RA	Volume (%)	Frequency (%)	RA	Volume (%)	Frequency (%)
		18 and 19 March Little snow 18 crops			20 and 21 March Much snow (> 30 cm) 23 crops	
Empetrum shoots	5	28·3	72	2	7·7	70
Empetrum berries	1	10·8	56	—	trace	4
Betula pubescens	4	1·9	28	4	51·9	87
Betula nana	3	6·7	83	2	36·2	65
Vaccinium uliginosum	2	2·6	28	1	0·5	22
Salix herbacea	2	36·3	78	—	2·9	39
Thymus drucei	1	1·3	56	1	0·1	26
Vaccinium myrtillus	1	1·2	28	—	trace	9
Dryas octopetala	1	5·3	67	1	0·9	26

food eaten on the heath was now *Betula nana* (40%), which projected out of the snow whilst *S. herbacea* and most of the *Empetrum* were covered. *S. herbacea* dropped to 3% and *Empetrum* shoots to 8% of the diet.

S. herbacea was clearly preferred to the two *Betula* species until it was covered by the snow ($P < 0.05$). As in the last example, before the snow made them unavailable, *Dryas* leaves and *Empetrum* berries were also taken in greater amounts than would be anticipated from their abundance, providing additional evidence for their rating as preferred foods.

In contrast to Example 4, however, more *Empetrum* shoots than *B. nana* were eaten before the snow, but *Empetrum* was much more abundant.

After the snow, when the two species were available to about the same extent, more *B. nana* than *Empetrum* was eaten. Neither of these differences was statistically significant but, taken together with a similar tendency in Example 4 (where $P < 0.005$), they suggest that *B. nana* is marginally preferred to *Empetrum*. We do not have enough chemical analyses to draw any firm conclusions about the relative nutritive values of these two species except to point out that *B. nana* is a better source of nutrients than *Empetrum*. As mentioned earlier, however, nutrients are likely to be less important in winter than energy and therefore digestibility.

The main conclusion is the continued preference for *S. herbacea* as long as it was available. This was the most nutritious of the available shrub species (Tables 5, 6 and 9).

EXAMPLE 6. SELECTION WITHIN *Betula pubescens*

Once in the scrub and feeding on *Betula pubescens*, ptarmigan prefer the catkins to the woody twigs. Two series illustrate this (Table 8). The first was taken in December 1966 in scrub at Hvammur, when catkins were still available although twigs were already being eaten. The second series was taken in the same place in late January 1967, when ptarmigan had eaten most of the catkins (data below Table 8), and twigs formed the major part of the diet. Catkins have a higher content of nitrogen, phosphorus and soluble carbohydrate, and a lower fibre content (Table 9); they are clearly a better food than the twigs.

TABLE 8

Two series of crops collected in birch scrub at Hvammur during a period in which ptarmigan were concentrated in the scrub

	7 crops 2 December 1966	14 crops 24 January 1967
	Proportion of total volume (%)	
Betula pubescens catkins	37·6	5·9
Betula pubescens buds, twigs	26·5	82·5
Betula nana	26·5	trace
Empetrum shoots	9·4	8·5
Vaccinium myrtillus	0·0	2·0
Vaccinium uliginosum	0·0	1·0

Mean number of *Betula pubescens* catkins per branch (30 marked branches)
20 November 1966   7 March 1967
22·1   5·9

A. GARDARSSON AND R. MOSS

TABLE 9

Chemical composition of *Betula pubescens*

	N	P	Ca	Mg	K	Na	Sol CHO	Crude fat	Crude fibre	Ash	In vitro digestibility
Catkins	2·51	0·36	0·39	0·18	1·06	0·39	20	10·3	19	4·29	40
Twigs (current growth)	1·23	0·16	0·36	0·17	0·44	0·16	13	5·4	21	2·94	29

These data are from catkins and twigs separated from the same sample taken at Armannsfell on 3 March 1967.

Although catkins in certain preferred areas of *Betula pubescens* scrub are often heavily grazed, it is sometimes possible to find relatively untouched scrub within a few kilometres. Presumably certain features in the habitat make the grazed areas attractive enough to counteract the relatively poor quality of the remaining food.

EXAMPLE 7. SPRING

In spring the ptarmigan return to the lowland heaths to breed. The nutrient requirements of hens in May and early June must be higher than at any other time of year because they at first moult heavily and then lay their eggs. The males do not begin to moult heavily until June.

Table 10 contrasts the diet of a series of birds shot at this season with the vegetation available. *Dryas octopetala* leaves were an important food item (about 18% by volume) for both sexes and were strongly preferred— *Dryas* formed only a small part of the available vegetation (mean cover class 0·1). *Betula nana* was approximately equivalent in volume to *Dryas* in the diet but it was a dominant species in the vegetation and therefore appeared to be less favoured. The various *Salix* species were about half as important as *Betula* (difference not significant) in the diet but were roughly five times less abundant in the vegetation.

Berries (mostly *Empetrum*) formed a major part of the diet of the cock ptarmigan (46%) in terms of volume, but because of their high moisture content (Table 11) were less important in terms of dry weight (about 15%). There was a suggestion that hens ate fewer berries (20% by volume) than cocks ($P = 0\cdot09$). Another difference between the sexes was that hens ate more ($P = 0\cdot004$) freshly growing deciduous herbs (22%, largely *Equisetum arvense*) than males (4%). The cover values for these plants in Table 10 are misleadingly high because they were determined in summer (Methods). In fact these herbs were not noticed on the ground when the birds were shot and were found only when the crops were opened. The birds must have been searching them out with some avidity (cf. Siivonen 1957).

Empetrum was a co-dominant in the vegetation, but its shoots were now completely ignored even though they were eaten in late winter (Example 5) and are a major food of Scottish and Japanese ptarmigan in spring (Watson 1964; Chiba 1965). *Vaccinium uliginosum* was also common but not eaten.

The vegetation is usually growing by the end of May and therefore all foods have improved in nutritive value over their overwintering state.

However, the date when the vegetation begins to grow varies considerably, for example the date when *Betula pubescens* puts out its first leaves may vary by about 4 weeks in different years (Hallgrimsson 1968). In contrast, the ptarmigan always begin to lay within a few days of 20 May (Gudmundsson, pers. commun.). It follows that the quality of the food available to laying hens in the period late May–early June must vary considerably in chemical composition and nutritive value. For example the *Dryas* leaves in Tables 6 and 11 were both taken within a few days of the same date in early May

TABLE 10

Spring foods in the Myvatn area, 23–25 May 1966

	Vegetation cover (5 stands)		Crops			
			♂♂ (*n* = 43)		♀♀ (*n* = 21)	
	Mean class	Presence	Volume (%)	Frequency (%)	Volume (%)	Frequency (%)
Shrubs						
Betula nana	2·5	5	20·2	88	21·5	90
Empetrum nigrum s.l.	2·3	5	0·0	0	0·0	0
Vaccinium uliginosum	1·1	5	0·0	0	0·0	0
Salix phylicifolia	1·0	5	7·2	58	8·6	57
Salix spp.	0·3, 0·2, 0·2	3, 3, 2	3·9	42	3·3	38
Arctostaphylos uva-ursi	0·4	4	0·0	0	0·0	0
Thymus drucei	0·2	4	0·7	35	2·4	38
Dryas octopetala	0·1	1	17·6	60	19·4	38
Evergreen herbs						
Equisetum variegatum	0·4	5	0·3	28	0·7	29
Galium normanii	0·3	4	0·1	19	2·1	43
Selaginella selaginoides	0·2	3				
Armeria maritima	0·1	3	0·1	19	0·3	33
Others						
Deciduous herbs						
Polygonum viviparum	0·7	5	0·8	42	3·2	43
Equisetum arvense	0·1	3	1·4	5	16·4	19
Taraxacum	0·1	3	0·5	12	1·4	19
Thalictrum alpinum	0·7	4				
Bartsia alpina	0·3	5	0·9	26	1·0	33
Others						
Berries						
Empetrum	—	—	38·7	58	16·0	43
Arctostaphylos uva-ursi	—	—	7·4	28	3·6	9
Vaccinium uliginosum	—	—	0·1	2	0·0	0

TABLE II

Chemical composition of some spring foods

	Date	Place	N	P	Ca	Mg	K	Na	Sol CHO	Crude fat	Crude fibre	Ash	In vitro digestibility	Conversion g dry wt/100 ml
Empetrum nigrum s.l. berries	early May 1964	Hrisey	0·46	0·09	0·11	0·05	0·70	0·07	35	6·3	20	1·90	40	15
leaves and shoots	May 1961	†	1·0	0·10	0·58	0·20	0·30	0·06	—	—	—	—	39	34
Dryas octopetala growing leaves	early May 1964‡	Hrisey	2·27	0·17	0·85	0·38	0·40	0·08	14	8·2	14	4·60	(46)	33
Polygonum viviparum leaves	12 June 1966	Hrisey	4·65	0·52	0·38	0·36	2·60	0·02	9	4·3	8	9·06	65	33
Taraxacum leaves	7 June 1966	Hrisey	4·80	0·49	0·59	0·38	3·45	0·09	20	2·4	8	9·55	(71)	20
Equisetum arvense sporophyll heads	27 May 1968	Hrisey	3·81	0·37	0·75	0·54	4·19	0·12	16	2·2	12	12.09	74	30
Betula nana crop contents, mostly catkins	27 May 1966	Myvatn	3·42	—	—	—	—	—	8	—	—	—	38	40

†From Thorsteinsson & Olafsson (1965), May data.

‡This sample was taken in early May but was growing because 1964 was an early spring.

in different years at the same place. However the sample from 1964, which was an early spring, contained 45% more protein than the sample from the late spring of 1967 (Table 11).

The few samples in Table 11 are therefore hardly sufficient to explain all the food preferences shown in spring, because they were taken at different times in different years. However, some obvious points emerge.

Shoots and leaves of the evergreen shrub *Empetrum* change very slowly in chemical composition in spring (Thorsteinsson & Olafsson 1965). By contrast, nutritive value changes more rapidly in the growing leaves and catkins of deciduous shrubs such as *B. nana*, as seen by comparing *B. nana* in Table 6 (winter catkins) and Table 11 (growing catkins). This difference in growth must make the difference in nutritive value between *Empetrum* and *B. nana* much greater than in winter, and may explain why *B. nana* is obviously preferred to *Empetrum* in spring although this preference is slight, if it exists at all, in winter (Examples 4, 5). Also, the fact that *B. nana* is a superior source of protein and phosphorus may assume more importance in spring when the birds' nutrient requirements become greater than in winter.

The deciduous herbs which the ptarmigan eat are excellent sources of nutrients and are highly digestible (Table 11), facts which may explain why the females sought them out with such avidity. On the other hand, *Empetrum* berries are a very poor source of nutrients (Table 11) even though they are rich in soluble carbohydrates. The reason why the hens ate more deciduous herbs and fewer berries than the cocks may have been that their nutrient requirements were greater at this time.

GENERAL DISCUSSION

At all times of year, availability obviously affects the diet of Icelandic ptarmigan. Nonetheless, relatively scarce items often form a large part of the diet and are obviously preferred by the birds to commoner and more readily available foods.

We have not yet examined enough food samples to account for all possible variations in their chemical composition, and the chemical results must be regarded as a preliminary survey. However, the results consistently suggest that the ptarmigan usually select the best-quality foods available.

In winter, ptarmigan strongly prefer *Salix herbacea*, a species which grows in abundance at high altitudes, especially in places where snow accumulates in winter. The chemical composition of high-level *S. herbacea* shows it to be a better food than all other common shrub species in Iceland, and so it is not surprising that the ptarmigan appear to feed on it for as long as possible in winter. Only when *S. herbacea* is inaccessible under thick or crusty snow do the birds turn to other less nutritious but more available shrubs for their main food.

Grazing by Icelandic ptarmigan sometimes leads to the exhaustion of preferred food items such as *Polygonum viviparum* bulbils in summer and *Betula pubescens* catkins in certain areas in late winter, before the birds turn to other less nutritious, but much more available foods. Such selective feeding on relatively scarce high-quality items is presumably advantageous. Abundant species such as *Empetrum* and *Calluna vulgaris* are little used by Icelandic ptarmigan in relation to their availability. These species form a major part of the diet of Scottish ptarmigan (Watson 1964), both would appear to be adequate potential foods (Moss 1968), and the red grouse (*Lagopus lagopus scoticus*) lives almost entirely on *Calluna* (Jenkins, Watson & Miller 1963). It seems reasonable to suggest that selection for the best quality of food available puts individuals at a selective advantage. For example, birds eating better food may be able to lay more eggs (Moss 1968), and be better able to cope with disease (cf. Lack 1954) and other stresses. If true, the present result of this selective process has been to produce a situation where the amount of potential food that is both adequate and available is in vast excess of the needs of the population.

ACKNOWLEDGEMENTS

It is a pleasure to thank S. E. Allen and his colleagues of the Nature Conservancy Chemical Service for performing the chemical analyses and Gunnar Olafsson of the Agricultural Research Institute, Reykjavik, for the *in vitro* digestibility measurements. Drs Finnur Gudmundsson, F. A. Pitelka and Adam Watson made helpful criticisms of the manuscript.

The work was supported in part by U.S. National Science Foundation grant No. GB3326 to the Museum of Natural History, Reykjavik.

CONCLUSIONS AND SUMMARY

1. Icelandic ptarmigan are very selective feeders. Examples show how food selection operates at different times of the year. In summer, bulbils of *Polygonum viviparum* are the most important food and are strongly

preferred. In winter, most birds feed on *Salix herbacea* as long as possible and turn to other shrub species only when *S. herbacea* becomes inaccessible under snow. The food of hens during the laying season differs from that of the cocks in that more freshly growing herbs and fewer berries are taken by hens.

2. Most of these preferences can be explained by the fact that the preferred foods are of greater nutritive value than the less favoured items, even though the latter are quite adequate foods.

3. The adaptive value of food selection is briefly discussed.

APPENDIX

The *in vitro* digestibility (D) of some foods was calculated from their content of nitrogen (N), soluble carbohydrates (S) and crude fibre (F). This equation was based on nineteen samples of a wide variety of foods.

$$D_{NSF} = 9 \cdot 57N + 0 \cdot 784S - 1 \cdot 01F + 26 \cdot 02 \quad (R^2 = 79\%)$$

Equations were also calculated for nitrogen and soluble carbohydrates only and for nitrogen and crude fibre only.

$$D_{NS} = 14 \cdot 71N + 1 \cdot 05S - 6 \cdot 27 \quad (R^2 = 69\%)$$
$$D_{NF} = 3 \cdot 34N - 1 \cdot 48F + 61 \cdot 93 \quad (R^2 = 65\%)$$

R is the coefficient of multiple correlation; R^2 is a measure of the proportion of the variability in D which is accounted for by variations in N, S and F.

REFERENCES

BOAG D.A. (1963) Significance of location, year, sex and age to the autumn diet of blue grouse. *J. Wildl. Mgmt*, **27**, 555–62.

CALDER A.B. & VOSS R.C. (1957) The sampling of hill soils and herbage with particular reference to the determination of trace elements. *Bull. Consultative Comm. for the Development of Spectrographic Work*, **1**.

CHIBA S. (1965) Food analysis of the Japanese ptarmigan. *Misc. Rep. Yamashina Inst. Orn.* **4**, Nos. 23/24, 184–97.

DAVISON V.E. (1940) A field method of analyzing game bird foods. *J. Wildl. Mgmt*, **4**, 105–16.

HALLGRIMSSON H. (1968) On the seasonal alternations of birch forests in Iceland. *Arsr. Skógraektarf. Isl.* 8–22.

JENKINS D., WATSON A. & MILLER G.R. (1963) Population studies on red grouse, *Lagopus lagopus scoticus* (Lath.) in North-east Scotland. *J. Anim. Ecol.* **32**, 317–76.

LACK D. (1964) *The Natural Regulation of Animal Numbers.* Oxford

MANN H.B. & WHITNEY D.R. (1947) On a test of whether one of two random variables is stochastically larger than the other. *Ann. Math. Statist.* **18**, 50–60.

Moss R. (1968) Food selection and nutrition in ptarmigan (*Lagopus mutus*). *Symp. zool. Soc. Lond.* **21**, 207–16.

Thorsteinsson I. & Olafsson G. (1965) The chemical composition and digestibility of some Icelandic range plants. *Rep. Univ. Res, Inst. Dep. Agric.* Ser. A, No. **17**. Reykjavik.

Shvonen L. (1957) The problem of the short-term fluctuations in numbers of tetraonids in Europe. *Riist. Julk.* **19**, 1–44.

Tilley J.M.A. & Terry R.A. (1963) A two-stage technique for the *in vitro* digestion of forage crops. *J. Br. Grassld. Soc.* **18**, 104–11.

Watson A. (1964) The food of ptarmigan (*Lagopus mutus*) in Scotland. *Scott. Nat.* **71**, 60–6.

DISCUSSION

A. Williams: Does one crop contain the whole range of plant species given in one of your tables or are some crops full of uniform material from one species of plant? In other words, what is the variation within crops and between crops?

A. Gardarsson: There is generally a range of species in each crop. Most crops from any one series are similar, although variability amongst crops increases with the number of species eaten.

I. Newton: Is there any indication that selective grazing by ptarmigan will in time reduce the availability of their preferred food plants in the environment? In other words do ptarmigan ever change their habitat to their own disadvantage?

A. Gardarsson: Only short-term exhaustion of some favoured foods occurs. No long-term effect is known to be caused by this.

R. P. Bray: Arthropods were found in the diet of ptarmigan chicks. How important are arthropods in the nutrition of tetraonid chicks?

R. Moss: It probably depends on the alternative foods available. *Polygonum* bulbils are nutritious and are taken by ptarmigan chicks in Iceland, and here arthropods are probably less important than in the red grouse, where the alternative *Calluna* is a poor food.

M. B. Usher: One factor that seems to be of importance and capable of measurement is the intensity of feeding. For example, how selective are cattle when they are actually taking a mouthful of food—can they select plants of only one species? How selective are ptarmigan—can they actually dig through snow to find *Salix herbacea*?

A. Gardarsson: Yes. They will dig down to about 30 cm.;

G. Fryer: Are there any other animals grazing in the same habitat as the ptarmigan? I ask this because it is all very well to show that ptarmigan select the most nutritious foods, but if all herbivorous animals did this they would all tend to seek the same food, which would not be a satisfactory state of affairs. Might your results be related to the relatively simple communities which prevail in Iceland?

A. GARDARSSON: The other main vertebrate herbivore in Iceland is the domestic sheep. In summer the sheep feeds mainly on grasses and sedges, in contrast to ptarmigan. In winter there is some food overlap with ptarmigan but the sheep tend in winter to be confined to relatively small areas near farms where there are usually few ptarmigan.

D. LACK: In connection with the last question, in most of the holarctic there are two species of *Lagopus*, but in Iceland only one. Does *L. mutus* have a wider food range and habitat range in Iceland than elsewhere?

A. GARDARSSON: In general this is true. The Iceland ptarmigan has a broader range of habitats and foods than in areas where the willow grouse or red grouse is present.

J. A. FORSTER: Do male ptarmigan in spring select for foods of high carbohydrate content or do they take food in proportion to its occurrence in the habitat?

A. GARDARSSON: Males tend to select *Empetrum* berries, which are a major part of the crop contents but which form only a small part of the available vegetation. These are very high in soluble carbohydrates but are also very conspicuous. We cannot say whether they are taken because they are obvious or because of their very high sugar content.

C. P. MATTHEWS: I have found in freshwater communities that invertebrates prefer nitrogen-rich litter and seem indifferent to energy content. You place great emphasis on the nitrogen content of food in determining food preferences. Do you think this is more important than energy?

R. MOSS: We assume that the most important nutrient is that which is in the shortest supply relative to requirements. Thus nitrogen is probably more important than energy at some times of the year and vice versa at other times.

A. MACFADYEN: How in fact are digestibility measurements, which I take to be assimilation/consumption ratios, made? Are they simply ratios of dry weight, calculated to have been assimilated, to the dry weight of food consumed? This would imply that the single figure given is a synthesis of figures from a range of food substances and is dependent on the enzymes used.

R. MOSS: Yes. It is a ratio of dry weights. A standard method based on sheep rumen liquor is used.

K.L. BLAXTER: What is the agreement between the *in vitro* digestibility coefficient based on sheep and *in vivo* digestibility in the ptarmigan?

R. MOSS: Preliminary data suggest a broad agreement, but we are only just starting work on this problem.

D. R. KLEIN: Ruminants select plants of high quality on the basis of taste and smell, but ptarmigan are birds, which apparently do not have these senses well developed. Do you have any basis for saying what mechanism is used by ptarmigan in selecting food of high quality?

A. GARDARSSON: We have no evidence on this point for ptarmigan.

FOOD SELECTION BY THE SNAIL
CEPAEA NEMORALIS L.

By J. P. Grime, G. M. Blythe and J. D. Thornton

Department of Botany, University of Sheffield

INTRODUCTION

A majority of the papers in this section are concerned with mammals and birds, and food selection has been examined either by direct observation in the field or by dissection of the gut contents or faeces. With large animals and perhaps also with many small invertebrates, laboratory experiments would appear to be of doubtful value in the study of food selection since the diet in the field may be peculiar to circumstances which are not easily (if at all) simulated in the laboratory.

However, where it can be shown that feeding under controlled conditions is comparable with that occurring in the field, there are advantages to be gained from working in the laboratory. In particular, more accurate measurements of food consumption are possible and it is usually easier to distinguish the separate interactions between plant and animal which control dietary selection.

Because of their convenient size, limited mobility and comparatively unsophisticated feeding behaviour, the larger terrestrial snails are particularly amenable to experimental study. Preliminary investigations strongly suggest that information relevant to ecological situations in the field can be obtained from laboratory experiments.

In the first part of this paper there are described a number of laboratory experiments designed to identify some of the major factors involved in food selection by the snail *Cepaea nemoralis*. In the remainder of the paper an attempt is made to relate the food preferences shown by *C. nemoralis* in the laboratory to food consumption in a natural habitat.

A. LABORATORY STUDIES WITH
CEPAEA NEMORALIS

METHODS FOR ESTIMATING PALATABILITY

In an earlier publication (Grime, MacPherson-Stewart & Dearman 1968) a critical account has been given of the methods used to estimate the palatability of plant material to *Cepaea nemoralis*. Except where there has been innovation, the description of technique is here cut to a minimum; however, the following points would seem to be essential.

COLLECTION OF SNAILS

Mature individuals of *Cepaea nemoralis* (shell diameter approximately 20 mm) were obtained from an area of sheep pasture at the Winnats Pass in Derbyshire, England. The floristic composition of the vegetation at this habitat has been described in an earlier paper (Grime & Blythe 1969). The population sampled is polymorphic with respect to the colour and banding of the shell. Several morphs were used in the single-choice experiments described in this paper although no attempt was made to compare their feeding preferences. In the multiple-choice experiments only the yellow unbanded morph was used.

CARE OF SNAILS

The snails were placed in large plastic containers with glass lids, inside a controlled environment cabinet at 15°C. Over a daylength of 8 h, illumination of 700–800 ft-candles was provided by warm-white fluorescent tubes. Throughout the period of captivity the snails were provided with an excess quantity of palatable foliage for a 24 h period, every 7 days. This was usually followed by a starvation period of at least 24 h before using the snails in experiments.

SOURCE OF PLANT MATERIALS

The leaves and stems of the flowering plants used in the experiments were collected from plants naturally established in the field, and in the short period before use they were stored in polythene bags at 5°C. The plant nomenclature used in this paper follows Clapham, Tutin & Warburg (1962).

EXPERIMENTAL CONDITIONS

The experiments were carried out in the cabinet to which the snails were accustomed. Except where stated, the experiments took place in darkness at 15°C and at a relative humidity exceeding 90%.

ESTIMATION OF LEAF PALATABILITY

Although the experiments involving the consumption of leaves or leaf extracts varied in design, the same basic approach was used to estimate the palatability of the materials presented to the snails. The palatability of the material under test was calculated from the quantity of it consumed, relative to the consumption of a palatable reference material (usually moist filter paper) which was provided initially in an equal quantity and to which the snail had comparable access in the experimental arena. The results were expressed in terms of a palatability index:

$$P.I. = mm^2 \text{ test material consumed}/1\cdot00 \text{ mm}^2 \text{ reference material}$$
consumed.

The palatability index was calculated from the results of both single-choice and multiple-choice experiments.

EXPERIMENTAL DESIGN

SINGLE-CHOICE EXPERIMENTS

In these experiments, individual snails were provided with one sample of test material and an equal area of reference material. Consumption of each was determined by weighing the material remaining at the end of a 16 h period. A major problem in the single-choice experiment arose from variability in the feeding potential of the snails both in time and between individuals. Attempts were made both to minimize and to take account of such variability as follows:

1. During the experiments the snails were maintained in a hungry condition. The quantity of food provided in an experiment was approximately two-thirds the average quantity consumed by an adult snail under the experimental conditions.

2. A continuous record was made of the quantity of food consumed by each snail, so that snails feeding irregularly could be recognized and replaced.

3. After use in three successive experiments, the snails were subjected to a control experiment in which each individual was given a quantity of

palatable material equivalent to the total material supplied in an experiment. These ancillary measurements confirmed that feeding activity was constant and that appetite was sufficiently high over an extended period in captivity. Therefore it is fair to assume that failure to consume a test material indicates low palatability. However, in the absence of measurements of the sequence and rate of consumption of the test and reference materials, the experiments were particularly insensitive to differences in palatability between highly palatable materials. This was not serious in the present investigation because the majority of the leaves examined were found to be relatively unpalatable to the snails.

MULTIPLE-CHOICE EXPERIMENTS

Multiple-choice experiments were designed to avoid some of the limitations of the single-choice experiments and, in particular, to obtain a more direct comparison of palatable materials. For this, ten or more materials of unknown palatability, together with a reference material, were presented simultaneously to a large number of snails in a common container. Consumption of each material was estimated at frequent intervals during the experiment, and from these measurements it was then possible to alter the 'feeding pressure', when required, by adjusting the number of snails in the arena. The multiple-choice experiment thus provides a record of the course of consumption of each material (Fig. 1), and one's judgement is required in deciding at what stage of an experiment the palatability index should be calculated. Consumption rates in the first hour of an experiment are often erratic. Steady rates occur later; these are presumed to arise from the snails' more complete exploration of the arena and their greater discrimination when they have been feeding for some time. It is also necessary to avoid calculations based on the later stages of an experiment if consumption of the more palatable materials has already fallen due to their scarcity in the arena. In the present experiments, all calculations of the palatability index precede 80% consumption of the most palatable material.

MEASUREMENT OF SNAIL MOVEMENT IN
RELATION TO PLANT MATERIALS

In addition to the investigation of palatability, experiments were carried out to examine other snail–plant interactions likely to influence food selection. Multiple-choice experiments were used to analyse the movements of C. nemoralis in relation to the stems of flowering plants. In addition the movements of snails in response to leaf odours were examined in single-choice experiments.

1. EXPERIMENTS WITH LEAVES

SINGLE-CHOICE EXPERIMENTS WITH LEAF DISCS AND LEAF EXTRACTS

This work was carried out as a continuous series of experiments in which the palatability of leaf discs and leaf extracts from fifty-two grassland, wasteland or woody species was measured.

PROCEDURE

At the beginning of an experiment with leaf discs, each of fifteen replicate snails was placed on the bottom of an individual container, with the snail's head between two discs of diameter 1·0 cm, one a reference material cut from the lamina of *Hieracium pilosella*, the other a disc of the material under test. The container was closed by means of a loose-fitting lid lined with polythene. Before inserting the food material and the snail, the inside of the container was moistened with distilled water to maintain a high relative humidity during the experiment.

The experiment was ended after 16 h and the remains of the two food materials were separately dried at 100°C and weighed. Intact portions of leaf matching those originally provided in the experiment were similarly dried and weighed and were used to calculate the quantities consumed in the experiment.

The procedure with leaf extracts was basically the same as for fresh leaves. The reference material was a square of Whatman No. 1 filter paper of area 150 mm² and dry weight 13·8 mg, moistened with distilled water. The test material consisted of an identical square of filter paper treated with aqueous leaf extract.

The extract was prepared by grinding 1 g of fresh leaf material in 10 ml of distilled water and filtering through a Whatman No. 1 filter paper. Squares of filter paper were placed on a clean glass sheet and 0·06 ml of filtered extract was added to each. The filter paper was then dried at a temperature of approximately 40°C. Two further additions of extract were made, bringing the total volume of liquid added to each square to 0·18 ml.

When placed in position in the container, the two pieces of filter paper were moistened with distilled water. Care was necessary in order to avoid transfer of solutes from treated to untreated filter paper. This risk was greatly reduced in the present investigation by using honey jars as containers. The convex floor of the jar caused excess moisture to drain away from the centre of the jar.

RESULTS

(a) Consumption of leaf discs

Of the fifty-two species which were tested, 60% appeared to be highly unpalatable in that the palatability index approximated to zero. Of the remaining leaf samples, half were clearly less palatable than discs of

TABLE I

Comparison of the palatability index (P.I.) within nine classifications of the species investigated

	Fresh leaf			Water extract		
	Mean	No. of species		Mean	No. of species	
Classes	P.I.	P.I.<0·1	P.I.≥0·1	P.I.	P.I.<0·9	P.I.≥0·9
Grasses[1]	0·01	12	1 ***	0·89	4	9 n.s.
Dicotyledons	0·39	15	25	0·81	16	24
Frequent in disturbed habitats	0·48	4	13 **	0·86	5	12 n.s.
Infrequent in disturbed hatitats	0·20	23	13	0·82	15	21
Frequent in disturbed habitats[2]	0·55	2	13 *	0·89	4	11 n.s.
Infrequent in disturbed habitats[2]	0·28	13	12	0·77	12	13
Leaf surface hard	0·04	15	3 **	0·85	6	12 n.s.
Leaf surface soft	0·42	12	23	0·83	14	21
Hairs on both leaf surfaces	0·47	6	12 n.s.	0·85	7	11 n.s.
Leaf glabrous	0·19	18	12	0·80	12	18
Hairs on both leaf surfaces[3]	0·58	4	10 n.s.	0·90	4	10 n.s.
Leaf glabrous[3]	0·35	4	9	0·88	4	9
Cell sap known to be distasteful	0·18	4	4 n.s.	0·50	8	0 ***
Remainder	0·31	23	22	0·89	12	33

[1] Including *Carex flacca*.   ***$P<0.001$.
[2] Excluding grasses.   **$P<0.01$.
[3] Excluding grasses and species known   *$P<0.05$.
    to have distasteful sap.   n.s.—not significant.
                                                 (Differences assessed by Fisher's exact test.)

Hieracium pilosella, and half were highly palatable. The mean palatability index of the grasses was extremely low (0·01) compared with the mean index of 0·39 for the dicotyledons. The proportion of totally unpalatable species among the grasses was significantly higher than among the dicotyledons (Table 1). By the same criterion, leaves with a hard surface were less palatable than soft-textured species. Plants from disturbed situations were more palatable on average than plants restricted to undisturbed habitats and this difference obtained even when grasses, which were more frequent in the latter group, were discounted.

In these experiments, leaf hairiness did not appear to inhibit feeding. In fact, samples with hairs on both surfaces had, on average, a P.I. twice that of the entirely glabrous materials. No significant difference was apparent between hairy and glabrous materials when grasses and distasteful plants were withdrawn from the comparison.

(b) Consumption of leaf extracts on filter paper

In the eight species known to be poisonous or distasteful, both the mean P.I. (0·50) and the proportion of palatable species were considerably lower than in the other materials investigated (Table 1).

Water extracts from fifteen of the fifty-two species induced a significant reduction in the palatability of filter paper. In three species, *Hedera helix*, *Caltha palustris* and *Arrhenatherum elatius*, the filter paper became almost totally unacceptable. Repulsive extracts were also obtained from *Plantago lanceolata*, *Ranunculus acris*, *Scabiosa columbaria*, *Teucrium scorodonia*, *Lotus corniculatus* and *Lathyrus pratensis*. In a number of species there was evidence of a slight reduction in the palatability of treated filter paper. This group included mildly poisonous species, e.g. *Senecio jacobaea* and *Rumex acetosella*, and certain aromatic plants, e.g. *Thymus drucei* and *Anthoxanthum ororatum*.

MULTIPLE-CHOICE EXPERIMENTS WITH INTACT LEAVES, LEAF DISCS AND LEAF EXTRACTS

The experiments described so far do not allow an adequate examination of the relevance to leaf palatability of the interaction between the external surface of the leaf and the snail's mouth parts and associated sensory organs. In particular, the preparation of a leaf disc may involve a breach in the protection afforded by the leaf exterior. Accordingly, experiments were designed to compare the consumption of intact leaves, leaf discs and leaf extracts from the same plant species.

Three experiments were carried out using leaves of ten species of flowering plants. The plants were selected (Table 2) to provide a range of leaf surfaces and included species known to contain unpalatable constituents.

PROCEDURE

Each experiment was carried out in a large rectangular arena. The test materials were pinned to the floor in twenty blocks, each containing one sample from each of the ten plant species together with a reference material of moist filter paper. Within each block, the materials were allocated random positions on a grid. In the experiment with intact leaves, lobes equal in area to the discs were introduced into the arena through slits in the

TABLE 2

The species of flowering plants used in the multiple-choice experiments with stems

| Species | Hairs on upper surface of leaf | | | Existing information on palatability |
	Type	Mean density (No./mm²)	Mean length (mm)	
Cabbage		None		—
Geranium pratense	Simple	10	0·3	Gain (1891)
Geranium robertianum	Glandular	2	0·3	—
	Simple	2	0·3	—
Hedera helix		None		Gain (1891), Long (1924), Kingsbury (1964)
Hieracium pilosella	Simple	4	2·2	Milton (1933)
Leontodon hispidus	Simple, bifid	10	0·6	Milton (1933)
Mercurialis perennis	Simple	7	0·2	Long (1924)
Rumex acetosa		None		Gain (1891), Long (1924), Kingsbury (1964)
Scabiosa columbaria	Simple	30	0·8	Gain (1891), Long (1924)
Urtica dioica	Stinging	2	1·2	Gain (1891), Tansley & Adamson (1925), Milton (1933), Gilham (1955), Myers & Poole (1963)
	Simple	2	0·6	

floor. Because of the time required in setting up the latter experiment the number of replicates was reduced from twenty to ten; in other respects the procedure was the same as that used for discs and leaf extracts.

At the beginning of each experiment, the internal surfaces of the arena were sprayed with distilled water and an individual of *C. nemoralis* was placed in the centre of each block. Although the snails were placed in standard starting positions, they were subsequently free to wander over the inside of the arena. At intervals during the experiment the snails were removed and the area of each replicate which had been consumed was estimated by eye to the nearest eighth. The snails were then returned to their original starting positions and the feeding was allowed to continue until the next estimation.

RESULTS

The curves in Fig. 1 describe the course of consumption of several materials in the experiment involving intact leaves. Errors in visual estimation are apparent but these were small compared with the differences in consump-

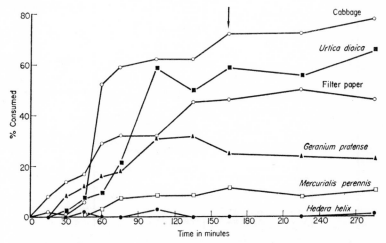

FIG. 1. The consumption of filter paper and of intact leaves from six species of flowering plants in a multiple-choice experiment.
The arrow indicates the stage in the experiment when the palatability index was calculated and when tests of significance (see Fig. 2.) were made.

tion between palatable and unpalatable leaves. Values for the palatability index in this experiment were based on estimations carried out after 165 minutes. Curves were similarly obtained and palatability indices calculated for the experiments with leaf discs and leaf extracts. The results from the three experiments are summarized in Fig. 2. Relative to filter paper, only two of the species examined, cabbage and *Urtica dioica*, were palatable as intact leaves, and with the exception of *Leontodon hispidus* and *Geranium robertianum*, all the remaining species were of rather low palatability. The results for leaf extracts show that three unpalatable species, *Hedera helix*, *Scabiosa columbaria* and *Geranium robertianum*, contained solutes repellent to the snails. There was also evidence that epidermal hairs deterred consumption; three species with a fairly dense covering of leaf hairs, *Leontodon*

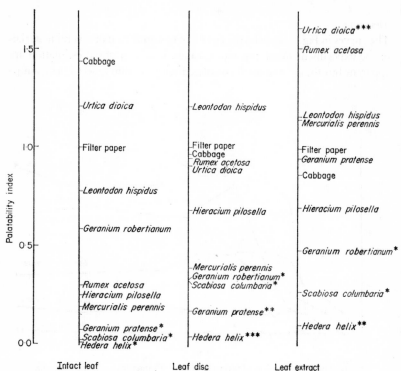

FIG. 2. Comparison of the palatability index of the intact leaf, a leaf disc and a leaf extract, in ten flowering plants.

Asterisks indicate the significance of the difference between a species and filter paper, in the following ratio: number of replicates consumed by more than 50% to number of replicates untouched. $*P<0.05$, $**P<0.01$, $***P<0.001$.

hispidus, *Hieracium pilosella* and *Scabiosa columbaria*, had leaf discs which were more acceptable than the intact leaf. In two species, *Mercurialis perennis* and *Geranium pratense*, which both have stiff hairs, the disc remained relatively unpalatable. Since neither of these species shows evidence of an unpleasant sap in the leaf extract it seems likely that the hairs constitute a major barrier to the snail. In view of these results, it is perhaps surprising to find that *Urtica dioica* had palatable intact leaves, despite the possession of long stinging hairs on both the upper and lower epidermis. However, the results show that the leaf extract from this species is strongly attractive to *Cepaea nemoralis*, and the palatability of the intact leaf appears to depend upon the combination of a repulsive exterior with an extraordinarily attractive interior. No satisfactory explanation has been found for the rather low palatability recorded for the intact leaf of *Rumex acetosa*. The species appears to be succulent and (insofar as the leaf extract provides a reliable indication) the cell sap is palatable.

2. EXPERIMENTS WITH STEMS

From the earlier experiments it is clear that interactions between the snail's head and the external surface of the leaf may have a major effect on palatability. In gastropods, we must consider the additional possibility that stimuli and responses involving the foot of the snail are important in food selection. In the experiments so far described, it was possible for feeding to occur without contact between the foot and the leaf surface. In an attempt to investigate phenomena involving the foot, two series of multiple-choice experiments were carried out. The experiments were designed to

TABLE 3

The species of flowering plants used in the multiple-choice experiments with stems

| | | Stem hairs | |
| | Stem surface texture | Type | Mean length (mm) |
Species			
Bromus ramosus	Hairy	Stiff, descending	1·2
Festuca arundinacea	Glabrous (rough)	—	—
Filipendula ulmaria	Glabrous (smooth)	—	—
Geranium pratense	Hairy	Stiff, descending	0·3
Heracleum sphondylium	Hispid	Stiff, toothlike, many descending	0·4
Mercurialis perennis	Hispid	Toothlike	0·2
Ranunculus acris	Sparsely hairy	Rather flexuous	1·6
Rumex acetosa	Glabrous (smooth)	—	—
Urtica dioica	Hairy, urticaceous	Stinging	0·6
		Simple (descending)	0·3

measure the rate and direction of the snail's movement rather than to estimate food consumption. To minimize the complications resulting from feeding, stems rather than leaves were used. Stems were selected (Table 3) to obtain a range of surfaces comparable to that provided by the leaves in Table 2. Apart from this consideration, the use of stems was of specific interest since there is evidence in the field that the ascent of stems is often a necessary preliminary to feeding (Grime & Blythe 1969).

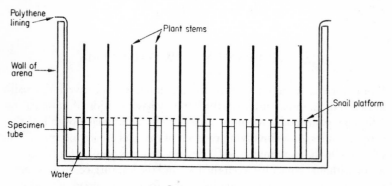

FIG. 3. Diagrammatic transverse section through the arena used for the experiments with stems.

FIRST SERIES OF MULTIPLE-CHOICE
EXPERIMENTS WITH STEMS

PROCEDURE

Experiments were carried out in an illuminated arena in which a hundred stems were supported vertically to a height of 6 cm in a 10×10 Latin square arrangement (Fig. 3). The stems comprised nine plant species together with 'control stems' circular in cross-section and constructed from polythene. The ten replicate lengths of stem representing each species were selected for uniformity and all fell within a range of diameter (including hairs) of 4–7 mm.

At the beginning of each experiment, ten emergent snails were placed individually at random positions on the floor of the arena. After 5 minutes, a record was made of the plant stems ascended, 'ascent' being recognized as loss of contact between the snail and the floor of the arena. The height to which each snail had climbed was also noted. The ten snails were then placed in new random positions on the floor and the procedure repeated. After use in two successive experiments the snails were replaced by a new batch of snails. In all, fifty experiments were carried out.

RESULTS

Compared with the polythene 'stems', the number of ascents (Table 4) was high on *Heracleum sphondylium* (despite the hispid surface on the stems of this species) and low on *Festuca arundinacea*, *Geranium pratense* and *Bromus ramosus*. The last two species both have numerous downwardly-projecting hairs on the stem.

TABLE 4

Comparison of stems from nine species of flowering plants with respect to the height of ascent by *C. nemoralis* in the first series of experiments with stems

Stem	No. ascents	Mean height ascended (cm)
Polythene	39	5·08
Mercurialis perennis	30	2·95**
Rumex acetosa	38	2·50***
Urtica dioica	39	2·53***
Filipendula ulmaria	33	5·01 n.s.
Heracleum sphondylium	50	2·97***
Festuca arundinacea	20	3·80 n.s.
Ranunculus acris	39	3·08**
Geranium pratense	18	2·72***
Bromus ramosus	17	3·06*

Asterisks indicate the significance of the difference between a particular stem and the polythene control 'stem' in mean height of ascent (see notes under Table 1 for levels of significance).

In seven of the plant species tested, the average height of ascent was significantly reduced in comparison with the control 'stems' (Table 4). With the exception of *Rumex acetosa*, all these species had hairs of some description on the stem surface. In two glabrous species, *Filipendula ulmaria* and *Festuca arundinacea*, the heights reached were comparable to those on the plastic 'stems'.

In the absence of information on the behaviour of each snail over the experimental period, it was not possible to recognize specific phenomena affecting either the frequency or height of ascent. The results suggested that stem hairs inhibit climbing but it was not possible to distinguish the role (if any) of the snail's foot in this interaction. A second series of experiments was carried out to investigate the responses in more detail.

SECOND SERIES OF MULTIPLE-CHOICE
EXPERIMENTS WITH STEMS

PROCEDURE

The arena, the arrangement of stems and the plant species selected were the same as in the first experiments. However, on this occasion only two snails were present in the arena at any time and each was continuously observed from the time of release until it successfully climbed a stem. Snails were replaced after use in two successive experiments and the experiments ended when two hundred ascents had been described. For each replicate stem, a record was made of (a) the number of occasions on which the head or antennae passed within 5 mm of the surface without subsequent contact ('passes'), (b) the number of contacts with head or antennae and (c) the number of successful ascents. Again, success in climbing was recognized as complete loss of contact between the snail and the floor of the arena.

RESULTS

In Table 5, the total number of 'contacts' and 'passes' are given for each species. For the plastic stems and for two densely hairy species, *Geranium pratense* and *Bromus ramosus*, the ratio of 'contacts' to 'passes' is approximately 1:1. However, in the remaining species, the relative frequency of 'contacts' was considerably higher. In *Heracleum sphondylium* and *Urtica*

TABLE 5

Comparison of stems from nine species of flowering plants with respect to the ratio of the number of 'passes' to the number of 'contacts' by *C. nemoralis* in the second series of experiments with stems

Stem	Passes	Contacts
Polythene	27	26
Mercurialis perennis	12	26n.s.
Rumex acetosa	12	29*
Urtica dioica	8	55***
Filipendula ulmaria	11	28*
Heracleum sphondylium	6	45***
Festuca arundinacea	10	33**
Ranunculus acris	18	26n.s.
Geranium pratense	21	19n.s.
Bromus ramosus	15	16n.s.

Asterisks indicate the significance of the difference in this ratio between a particular stem and the polythene control 'stem'.

TABLE 6

Comparison of stems from nine species of flowering plants with respect to the ratio of ascents to abortive contacts by *C. nemoralis* in the second series of experiments with stems

Stem	Ascents	Abortive contacts
Polythene	18	8
Mercurialis perennis	12	14n.s.
Rumex acetosa	10	19**
Urtica dioica	27	28n.s.
Filipendula ulmaria	17	11n.s.
Heracleum sphondylium	24	21n.s.
Festuca arundinacea	21	12n.s.
Ranunculus acris	11	15n.s.
Geranium pratense	5	14**
Bromus ramosus	2	14***

Asterisks indicate the significance of the difference in this ratio between a particular stem and the polythene control 'stem'.

dioica the ratio was approximately 8:1 and the sum of 'contacts' and 'passes' was exceptionally high in both species. The results indicate that stems of these two species were attractive to the snails and that the snails showed a preference before actual physical contact.

Table 6 shows the total number of ascents and abortive contacts for each plant species. On the plastic stems and on two glabrous species, *Filipendula ulmaria* and *Festuca arundinacea*, the ratio of ascents to abortive contacts was approximately 2:1. On the stems of the remaining species there were relatively fewer climbs, and in three species, *Rumex acetosa* (1:2), *Geranium pratense* (1:3), and *Bromus ramosus* (1:7), the ratio was significantly different from that obtained for the control stems. The latter two species have a covering of long stiff descending hairs on the stem. During the course of the experiment snails were seen to experience considerable difficulty in climbing the stems of both species and, on frequent occasions, snails became detached when the burden of the shell was shifted from the floor of the arena to the foot–stem interface.

3. EXPERIMENTS WITH PLANT ODOURS

In the experiments with stems, there was strong evidence that *C. nemoralis* was attracted towards stems of certain species. As this occurred in light and darkness, the possibility is raised that plant odours may be important both in finding and selecting food. A series of single-choice experiments was designed to examine snail movements in relation to the source of plant odours.

PROCEDURE

The experimental arena consisted of a closed rectangular plastic box with a false floor of perforated plastic mesh. At the beginning of each experiment, crushed foliage of the test species was placed beneath the floor at one end of the box and ten hungry emergent snails were placed within a narrow (3·0 cm) central zone across the width of the arena. The interior of the box was moistened thoroughly and placed in darkness at 15°C. After 30 minutes, separate counts were made of the number of snails which had moved out of the starting zone towards the leaf material and towards the opposite end of the arena. Five leaf materials were tested in this way and a control experiment was carried out in which no plant materials were inserted into the box. The experiments were arranged in replicate series in each of which the leaf materials were presented in a different random sequence.

RESULTS

Table 7 contains the totals obtained from seven successive series of experiments using the same set of ten snails. With the control and with leaves of three species, the ratio of snails moving towards leaves to snails moving away is approximately 1:1. There is evidence (not however significant statistically) of attraction towards *Urtica dioica* and of repulsion by *Hedera helix*.

TABLE 7

Comparison of five species of flowering plants with respect to the capacity of leaf odour to influence the direction of movement of *C. nemoralis* in a closed arena (see text)

Species	Number of migrating snails	
	Away from leaves	Towards leaves
Control	28	27
Urtica dioica	21	31
Stellaria media	28	25
Rumex acetosa	26	26
Scabiosa columbaria	27	31
Hedera helix	30	20

None of the departures from 1:1, in the ratio of the number of snails moving towards the source of odour : number of snails moving away, are statistically significant.

TABLE 8

Movement of C. *nemoralis* in relation to the source of leaf odours

Minutes	Number of snails	
	Over *U. dioica*	Over *H. helix*
0	15	15
5	15	15
10	14	16
15	14	16
20	16	14
25	18	12
30	24	6
35	20	10
40	23	7
45	20	10
50	24	6
55	24	6
60	26	4

Note. Crushed leaves of two species were placed at opposite ends under a false floor and counts of snail numbers were made in each half of the rectangular arena at intervals of 5 minutes.

Confirmation of this response to leaf odours was obtained by examining the movement of a larger number of snails in an arena containing leaves of both *Urtica dioica* and *Hedera helix*. The results of this additional experiment (Table 8) showed that over the period of an hour there was a progressive and highly significant movement of snails away from *Hedera helix* and towards *Urtica dioica*.

B. FOOD SELECTION BY *CEPAEA NEMORALIS* IN THE LABORATORY COMPARED WITH FOOD SELECTION IN A NATURAL HABITAT

In Britain, *Cepaea nemoralis* occurs in a variety of habitats including woodlands but is perhaps most familiar as an inhabitant of grassland on limestone or chalk. In the North of England, it prefers slopes of southern aspect; this is particularly marked at the Winnats Pass, a limestone gorge in North Derbyshire which has been the subject of detailed study. Quadrat data collected from an area in which the pass extends from east to west confirmed that *Cepaea nemoralis* is strictly confined to the south-facing side (Table 9). In passing, it is interesting to note that the north-facing slope is occupied by another large snail, *Arianta arbustorum*, a species absent from the south-facing slope.

TABLE 9

Estimates of the frequency of occurrence and mean density of two species of snails on opposite sides of the Winnats Pass, 25 August 1965 (only specimens with a shell diameter exceeding 10 mm included)

	Cepaea nemoralis		*Arianta arbustorum*	
	North aspect	South aspect	North aspect	South aspect
No. square metre quadrats containing species†	0	19	9	0
Total no. snails observed	0	50	87	0
Mean density of snails in quadrats containing species	0	2·8	9·7	0

†Maximum possible is twenty.

On the south-facing slope at the Winnats Pass, the vegetation is xeromorphic and sparse compared with the opposite slope where there is a closed turf including taller grasses and herbs and several plants of common occurrence in woods or in streamside vegetation (Grime & Blythe 1969). At the Winnats Pass therefore, it is clear that a marked discontinuity in the distribution of *Cepaea nemoralis* corresponds with a change in topography and vegetation, and it is of interest to determine how far the distribution of the snail arises from food preferences. This problem has been investigated by comparing the potential vegetable food sources in the area with respect to (a) their palatability to *C. nemoralis* in a multiple-choice experiment in the laboratory, and (b) their frequency of occurrence in the faeces of snails in the field.

(a) FOOD SELECTION IN THE LABORATORY

PROCEDURE

The materials investigated (Table 10) were the main vegetation components at the Winnats Pass. Green and senescent leaves were tested separately and the materials chosen were those present in sufficient quantity in the field to be considered as a possible major food source for *Cepaea nemoralis*. The snails were freshly collected from the field and were starved for 5 days before use in the experiment. The plant materials were arranged in twenty 10 × 6·5 cm blocks on the floor of the arena. Each block was subdivided into twenty 3·25 × 1·00 cm compartments and the materials were allocated at random to the compartments. Individual fragments of fresh or senescent leaves were obtained from different leaves, templates being used to standardize the area of each replicate. Where possible, leaf discs were used but in the case of grasses it was necessary to use standard lengths of leaf blade. The area of the sample was adjusted to compensate for differences in leaf thickness but all samples fell within the range 50–70 mm².

TABLE 10

Consumption of the main vegetation components from the Winnats Pass by C. nemoralis in a multiple-choice experiment

		Hours		
Source	Species	0·5	7·5	16·5
South	Living foliage			
facing	Festuca ovina	0·00	0·00	0·00
slope	Festuca rubra	0·00	0·00	0·00
	Arrhenatherum elatius	0·00	0·00	0·00
	Dactylis glomerata	0·00	0·00	0·00
	Poa pratensis	0·00	0·00	0·00
	Senescent foliage			
	Festuca ovina	0·00	0·00	0·00
	Festuca rubra	0·00	0·00	0·00
	Arrhenatherum elatius	6·25	12·70	13·30
	Dactylis glomerata	0·00	0·05	0·65
North	Living foliage			
facing	Festuca rubra	0·00	0·00	0·00
slope	Holcus lanatus	0·00	0·30	0·45
	Poa trivialis	0·00	0·05	0·05
	Mercurialis perennis	0·35	0·43	0·53
	Urtica dioica	8·35	17·30	17·80
	Rumex acetosa	0·40	0·55	0·75
	Polemonium caeruleum	0·00	0·00	0·00
	Senescent foliage			
	Festuca rubra	1·65	1·90	2·20
	Holcus lanatus	0·00	0·10	0·30
	Mercurialis perennis	44·40	88·90	88·90
	Urtica dioica	2·75	20·00	20·60

The tabulated values refer to the mean percentage leaf area consumed.

The leaf materials were held in position by pins or staples. Securing the 400 fragments was a lengthy operation; to prevent deterioration, they were covered by filter paper moistened periodically with chilled water until starting the experiment.

At intervals during the experiment, consumption of the plant materials was estimated by eye using a hand-lens where necessary. Each interval extended over 30 minutes to give enough time for accurate estimations. Over the initial 150 minutes of feeding, estimations were made every 30 minutes and the number of snails in the arena was twenty. For the remainder of the experiment the number of snails was increased to forty and the intervals between estimations were lengthened first to 1 hour and later to 3 hours.

RESULTS

Table 10 illustrates the course of consumption of each material over the period of the experiment. Total consumption was most rapid during the first 100 minutes. After this, despite a doubling of the number of snails in the arena, there was a fairly abrupt decline in feeding which eventually stabilized at the negligible rate of 0·02% of the total leaf area/h.

When consumption of the individual materials is considered it is clear that, with the exception of *Urtica dioica*, non-senescent foliage from a majority of the commoner species at the Winnats Pass was unpalatable to *Cepaea nemoralis*. The material selected in the laboratory was mainly senescent foliage from *Mercurialis perennis*, *Urtica dioica* and *Arrhenatherum elatius*. Except for senescent leaves of *Arrhenatherum elatius*, grass components which form the main part of the aerial biomass at the Winnats Pass, were avoided in the experiment. Of the materials consumed in quantity, only the senescent leaves of *Arrhenatherum elatius* are frequent on the south-facing side of the gorge. In view of the strong preference for *Mercurialis perennis* and *Urtica dioica* it is perhaps surprising to find *Cepaea nemoralis* absent from the north-facing slope where both these plant species are locally abundant.

(b) FOOD SELECTION IN THE FIELD

Food selection in the field was inferred from the frequency of various components in the faeces of individuals of *Cepaea nemoralis* removed from specific habitats at the Winnats Pass. Faecal analysis was also carried out on *Arianta arbustorum* taken from the same locality.

TABLE II

Dissection of faeces from six collections of snails at the Winnats Pass (all snails were mature individuals and were collected in daylight on 13 October 1966)

Collection	Species	Habitat	Aspect	No. of snails examined	Proportion containing chlorophyll (%)	Proportion containing dead plant remains (%)	Distribution of chlorophyll		Packed with large hairs or stings	Containing seeds of *Urtica dioica*
							Diffuse	In bryophytes		
1	*Cepaea nemoralis*	*Festuca* grassland	S	26	46	92	11	3	0	0
2	*C. nemoralis*	m² *Urtica* patch	S	16	69	69	11	2	8	0
3	*Arianta arbustorum*†	Tall herbs	N	25	88	48	22	0	22	3
4	*A. arbustorum*‡	Tall herbs	N	25	76	64	19	3	18	2

†Snails removed from vegetation.
‡Snails removed from ground surface.
Except where indicated, the values refer to the number of snails with the attribute concerned.

8

PROCEDURE

In each collection, the snails taken were the first mature individuals encountered at the site. Each snail was immediately placed in a separate container and was provided with a large piece of moistened filter paper. In the laboratory, faeces were collected from each snail until the appearance of filter paper in the faeces indicated evacuation of the material most recently ingested before collection. The faecal material was then dissected under a binocular microscope.

The collections for which results are given in Table 11, were as follows: (1) *Cepaea nemoralis* from grassland on the south-facing slope, (2) *Cepaea nemoralis* from a small isolated patch of stunted nettles in a hollow on the south-facing slope near the site of the first collection, (3) and (4) *Arianta arbustorum* from a stand of *Urtica dioica* and *Mercurialis perennis* on the north-facing slope. The snails were removed from foliage in collection 3 and from the ground surface in collection 4.

RESULTS

The faeces of *Cepaea nemoralis* taken from grassland (collection 1) were made up of undigested fragments of plant material, the bulk of which was clearly recognizable as the senescent leaves of grasses. Chlorophyll occurred as residual traces in old leaves or in isolated fragments of moss shoots. By comparison, chlorophyll was abundant in the faeces from *Arianta arbustorum* and also in some of the *Cepaea nemoralis* removed from the nettle patch. In marked contrast to the faecal matter of *Cepaea nemoralis*, that from both collections of *Arianta arbustorum* was mainly composed of a bright green amorphous mass packed with the stinging hairs of *Urtica dioica* and with simple hairs similar to those on the leaves of *Urtica dioica* and *Mercurialis perennis*. Both ripe and immature seeds of *Urtica dioica* were recovered from the faeces of *Arianta arbustorum*. Remains of *Urtica dioica* were also identified in the faeces of *Cepaea nemoralis* from the nettle patch on the south-facing slope (collection 2).

CONCLUSIONS

The results of the laboratory experiments with *Cepaea nemoralis* are of interest in relation to three problems, (a) recognition of the phenomena which influence food selection in snails, (b) explanation of the food selection of *Cepaea nemoralis* in the field, and (c) the importance of diet in regulating snail density.

(a) MECHANISMS OF FOOD SELECTION IN SNAILS

With few exceptions, e.g. *Urtica dioica*, it appears that native species of flowering plants are unpalatable to *Cepaea nemoralis*. However, palatability varies with the age of the leaf, and in species such as *Mercurialis perennis* and *Arrhenatherum elatius* senescence is associated with a spectacular increase in palatability (Table 10).

The experiments show that low palatability in a non-senescent leaf may be due to a variety of causes. In some plant species, the snail is deterred by the leaf exterior which may be hard (e.g. many grasses) or hispid (e.g. *Geranium pratense*) or may possess both of these characteristics to some extent (e.g. *Mercurialis perennis*). Alternatively in species such as *Hedera helix* and *Caltha palustris* the cell contents may render the leaf unacceptable. The experiments also show that there are species such as *Arrhenatherum elatius* and *Scabiosa columbaria* in which low palatability of the intact leaf is due to the combined effects of internal and external features. Perhaps the most interesting combination of attributes is that which determines palatability of the leaf in *Urtica dioica*. In this species the leaf surface is protected by stinging hairs but feeding is induced by a chemical stimulus. Recent experiments, in which consumption of filter paper has been found to be enhanced by the introduction of odours from the leaf of *Urtica dioica*, show that this stimulus is, at least in part, olfactory.

Palatability is perhaps most usefully regarded not as a plant characteristic, but as a function of plant and animal attributes under particular environmental conditions (Tribe 1950). Hence, in these studies, palatability has been influenced by snail characteristics of which we are largely ignorant. For example, there is need to take account of the changing structure and functioning of the mouth parts and sensory organs of snails as they reach different stages of maturity.

Food selection by snails involves considerations apart from palatability. The experiments show that plant odours may attract or repel *Cepaea nemoralis* and it seems quite possible that snails may be attracted towards and even induced to climb stems of aromatic species such as *Heracleum sphondylium*.

Cepaea nemoralis appears to be able to ascend most types of plant stem although it has considerable difficulty in maintaining adherence and traction on hispid surfaces especially where downwardly-projecting hairs are characteristic of the species (e.g. *Bromus ramosus*). In addition to stem structure, a number of plant characteristics may be expected to influence food selection through their effect on leaf accessibility. Flowering plants differ in phenology and, even within the same locality, they may occupy different microhabitats. These are subtleties outside the scope of the present laboratory experiments.

(b) FOOD SELECTION BY *Cepaea nemoralis* AT THE WINNATS PASS

At the Winnats Pass, *Cepaea nemoralis* is restricted to the south-facing slope. The results of laboratory experiments (Grime, MacPherson-Stewart & Dearman 1968; Table 10 in the present paper) suggest that almost all the living foliage on this slope is unpalatable to the snail and it is not surprising therefore that the diet is composed mainly of senescent foliage. In the multiple-choice feeding experiment in Table 10, only *Arrhenatherum elatius* among the species abundant on the south-facing slope had palatable senescent foliage. *A. elatius* would seem to be a productive source of senescent material since in contrast to the other grasses on the slope, it is a species of potentially fast growth-rate and the foliage dies back completely in autumn.

The senescent leaves of *Mercurialis perennis* and to a lesser extent, the green and senescent foliage from *Urtica dioica* are palatable to *Cepaea nemoralis* (Table 10). Both of these species, and *Arrhenatherum elatius* also, are present in abundance on the north-facing slope which in terms of food supply appears to be a most suitable habitat for *Cepaea nemoralis*. To explain the diet of this animal it appears to be necessary, first, to explain its confinement to one side of the gorge.

Cepaea nemoralis may be excluded from the *Urtica dioica–Mercurialis perennis* communities of the north-facing slope by competition from the snail *Arianta arbustorum*, whose behaviour appears more closely adapted to exploitation of tall herbs (Grime & Blythe 1969). This would not explain the absence of *Cepaea nemoralis* from the grassland areas of the same slope; here conceivably there may be competition from slugs. However, the most likely explanation for the failure of *Cepaea nemoralis* to exploit the north-facing slope may be found in the direct effects of climate upon the animal. *Cepaea nemoralis* is a species of southern distribution and it is quite probable that its life history depends on a warm summer.

(c) DIET AND POPULATION DENSITY AT THE WINNATS PASS

Compared with the populations of *Arianta arbustorum* on the north-facing slope, the densities of *Cepaea nemoralis* are low (Table 9). There is little evidence of predation and it is probable that snail numbers are related to diet. Confinement of *Cepaea nemoralis* to the south-facing slope has resulted in occupation of a habitat where the most acceptable materials are senescent leaves. The resulting diet would appear to be of low nutritive value especially in comparison with that of *Arianta arbustorum* feeding on the opposite slope.

Shallow rendzinas such as those on the south-facing side of the Winnats Pass are often extremely deficient in phosphorus (Park, Rawes & Allen 1962; Grime 1963a, 1963b, 1965; Davison 1964). It may be suggested that phosphorus, either through a deficiency in the diet or by reducing the contribution of palatable species to the vegetation, is a limiting factor on snail numbers. In accord with this suggestion, is the occurrence of high densities of the snail *Arianta arbustorum* in stands of *Urtica dioica*. The studies of Pigott & Taylor (1964) show that, in addition to its palatability for snails, *Urtica dioica* is associated with habitats that have high levels of phosphorus.

SUMMARY

1. By means of single-choice and multiple-choice laboratory experiments, the attempt has been made to identify plant characteristics affecting the palatability of leaves to the terrestrial snail, *Cepaea nemoralis*. In order to estimate the relative importance of physical and chemical features, comparison has been made of the rates of consumption of intact leaves, leaf discs and leaf extracts (on filter paper).

2. In further experiments the frequency and rate of ascent of stems has been found to vary according to plant species and there is evidence which suggests that stem hairs play an important part in these responses. Experiments have shown that, in the direction of its movements, *C. nemoralis* may respond to leaf odours.

3. In the second part of the paper, food selection in the laboratory is compared with feeding (inferred from faecal analysis) in a natural environment. The results indicate that, in the area examined, diet influences snail density rather than snail distribution.

REFERENCES

CLAPHAM A.R., TUTIN T.G. & WARBURG E.F. (1962) *Flora of the British Isles*, 2nd edit. Cambridge.

DAVISON A.W. (1964) Some factors affecting seedling establishment on calcareous soils. Ph.D. thesis, Univ. of Sheffield.

GAIN W.A. (1891) Notes on the food of some of the British molluscs. *J. Conch., Lond.* **6**, 349–60.

GILHAM M.E. (1955) Ecology of the Pembrokeshire Islands. III. The effect of grazing on the vegetation. *J. Ecol.* **43**, 172–206.

GRIME J.P. (1963a) Factors determining the occurrence of calcifuge species on shallow soils over calcareous substrata. *J. Ecol.* **51**, 375–90.

GRIME J.P. (1963b) An ecological investigation at a junction between two plant communities in Coombsdale. *J. Ecol.* **51**, 391–402.

GRIME J.P. (1965) The ecological significance of lime-chlorosis: an experiment with two species of *Lathyrus*. *New Phytol.* **64**, 477–87.

GRIME J.P. & BLYTHE G.M. (1969) Relationships between snails and vegetation at the Winnats Pass. *J. Ecol.* **57**, 45–66.

GRIME J.P., MacPHERSON-STEWART S.F. & DEARMAN R.S. (1968) An investigation of leaf palatability using the snail *Cepaea nemoralis*. *J. Ecol.* **56**, 405–20.

KINGSBURY J.M. (1964) *Poisonous Plants of the United Kingdom and Canada*. New York.

LONG H.C. (1924) *Plants Poisonous to Livestock*. 2nd edit. Cambridge.

MILTON W.E. (1933) The palatability of self-establishing species in grassland. *Emp. J. exp. Agric.* **1**, 347–60.

MYERS K. & POOLE W.E. (1963) A study of the biology of the wild rabbit, *Oryctolagus cuniculus* (L)., in confined populations. IV. The effects of rabbit grazing in sown pastures. *J. Ecol.* **51**, 435–51.

PARK K.J.F., RAWES M. & ALLEN S.E. (1962) Grassland studies at the Moor House National Nature Reserve. *J. Ecol.* **50**, 53–62.

PIGOTT C.D. & TAYLOR K. (1964) The distribution of some woodland herbs in relation to the supply of nitrogen and phosphorus in the soil. *J. Ecol.* **52**, (Suppl.), 175–85.

TANSLEY A.G. & ADAMSON R.S. (1925) Studies of the vegetation of the English Chalk. III. The Chalk grasslands of the Hampshire–Sussex Border. *J. Ecol.* **13**, 177–223.

TRIBE D.E. (1950) The behaviour of the grazing animal—a critical review of present knowledge. *J. Br. Grassld Soc.* **5**, 200–14.

DISCUSSION

J. M. CHERRETT: In the choice chamber trials for individual snails was there evidence of individual variability in preferences? Such variability might confer dietary flexibility on a population important in a changing habitat.

J. P. GRIME: In general the replicates gave similar results, and although some of the individual animals died during the experiments there was no evidence that their preferences had been significantly different from those that lived.

I. N. HEALEY: Do you have any evidence of a role for micro-organisms and detritus on the surface of leaves in determining their palatability to snails?

J. P. GRIME: This must be a possibility, but is probably small. We are not in a position to answer this at present.

L. BELLAMY: There has been some work done in the USA on freshwater triclads in which the previous presence of the animal's mucus trails markedly affected the outcome of choice experiments. Was any comparable effect found with the snails in the stem selection experiment?

J. P. GRIME: We recognized that the passage of snails might facilitate climbing and could therefore exaggerate the early results. However, I do not believe that this occurred. Each snail was placed in a different random starting-position in the arena, and, more to the point, very similar results were obtained in the two series of experiments described.

J. GREENWOOD: Strictly speaking the statistical significance of this would be decreased, in that purely random original 'choices' by the snails would be magnified if later snails tended to climb steps with mucus on.

J. P. GRIME: I agree.

FACTORS AFFECTING THE DIET
AND FEEDING RATE OF THE
REDSHANK (*TRINGA TOTANUS*)

By J. D. Goss-Custard

Department of Psychology, University of Bristol

INTRODUCTION

Recent laboratory and field work suggests that as the density of a prey increases, the number taken by a predator rises to a plateau beyond which further increases in prey density have no effect on the number of prey consumed (Holling 1965; Murton 1968). This 'functional response' (Solomon 1949) is important in studying food in relation to animal populations (Solomon 1964; Holling 1965; Hassel 1966). In the field it is often difficult to measure the density of prey encountered by predators because not all the prey may be vulnerable and environmental factors may cause the proportion that is vulnerable to vary. Furthermore, the predation risk of different classes of a prey population may also vary. The following study was of the redshank (*Tringa totanus*), a small wading bird inhabiting muddy parts of estuaries and feeding on the rich intertidal macrofauna there. The aim was to study food selection and rate of food intake in relation to prey density, size and availability. This paper summarizes field results reported in full elsewhere (Goss-Custard 1969, 1970) and discusses current experiments on how redshank exploit a food supply consisting of numerous prey items varying greatly in size.

STUDY AREA AND METHODS

The field work was done on the Ythan estuary in Aberdeenshire during two winters (1964–1966). Two areas (A1 and B1) were studied in detail and four others less intensively (A2, A3, A4 and B4). The areas measured 100 m by 60–100 m. Observations were made in each area throughout the first season (October to April) and in areas A1–A4 during 2 months in the second season. The 1964/65 season was divided into four periods each about 7 weeks long. The size-composition and density of the prey populations were measured in each area at the mid point of each period. All measurements of diet and feeding rate during a period were combined and compared with measurements of the prey populations at the centre of the period.

The peck rate of redshank was easily measured by stop-watch. To determine feeding rate (number of prey taken per minute) it was also necessary to measure feeding success as not all pecks were successful. To do this I observed birds at close quarters and recorded the proportion of pecks followed by swallowing. The diet was determined partly by field observation and partly from the gizzard contents of birds shot while feeding. The numbers of each prey taken per minute could then be calculated reasonably accurately (Goss-Custard 1969). Prey size was obtained from the gizzard contents by measuring either intact animals or fragments which bore a known size relationship to total body length.

RESULTS

FEEDING METHODS AND THE MAIN PREY SPECIES

Feeding redshank moved at about 12 m a minute and appeared to detect their prey by sight. They fed by making rapid pecks and usually only the tip of the bill was inserted into the mud. They were recorded taking eight species of prey but four predominated. The most important was *Corophium volutator* (Pallas), an amphipod crustacean reaching 9 mm long and living in burrows down to about 8 cm. The other three were a small gastropod mollusc (*Hydrobia ulvae* (Penn.)), a burrowing bivalve mollusc (*Macoma balthica* (L.)) and the common ragworm (*Nereis diversicolor* F. O. Müller).

Mud temperature affected the diet and feeding rate, probably because of its effect on the behaviour and thence availability of the prey (Goss-Custard 1969). Between 6°C and 15°C, diet and feeding rate in the two main areas were independent of temperature and *Corophium* was the main prey in terms of numbers (87%+) and biomass (82%+). In 1964/65, the feeding rate decreased as the temperature decreased, and the diet changed so that at −1°C to +1°C *Macoma* contributed 76% of the biomass ingested in area A1 and *Nereis* contributed 54% in B1. The difference between areas in the low-temperature diet reflected the different relative densities of the prey in the mud. Low temperature had a much less marked effect in A1 in the second season, probably because of differences between the years in the prey populations in the mud (Goss-Custard 1969).

FEEDING RATE AND PREY DENSITY

It was expected that at temperatures above 6°C the feeding rate would depend on the density of the main prey *Corophium*. This was not the case (Goss-Custard 1970). In five areas, peck rate was independent of the density of *Corophium* over 4 mm long (i.e. the size range taken by the birds). Feeding success was independent of prey density in area B1, and increased

TABLE I

The feeding rate of redshank in six areas in relation to the mean size and density of their prey

Area	Feeding rate† (number per minute)	Range in the density of *Corophium* above 4 mm in length (number per sq m)	Range in the mean weight of *Corophium* above 4 mm in length (mg)
A1	60·9–68·8	2488–5736	0·25–0·36
B1	60·7	2468–5676	0·23–0·45
A2	61·1	1032–7075	0·21–0·37
A3	54·4	3044–7824	0·28–0·49
A4	50·8	3825–7631	0·32–0·47
B4	54·5	4113 and 8144	0·31 and 0·53

†Full data in Goss-Custard (1970). The dependence of feeding success on prey density was known in areas A1 and B1 but not in the remainder. In these areas all the available measures of feeding success irrespective of prey density were used to calculate feeding rate, so the values of feeding rate shown may be approximate. In area B4, data were available from only two study periods.

only slightly as prey density increased in A1. The little information on feeding success from the other areas corroborated that there was either no change or only a slight increase in feeding success with increasing prey density. Thus feeding rate either remained constant or increased only slightly over large increases in prey density (Table 1).

One interpretation of this result is that the birds took more of an alternative prey when *Corophium* density was low. However, this was not so. In area A1 an average of 89% of the prey was *Corophium* during the five

study periods, the rest consisting almost entirely of *Hydrobia*. The proportion of *Hydrobia* in the diet varied independently of *Corophium* density
(Table 2) so, like the overall feeding rate, the number of *Corophium* taken
increased only slightly as prey density increased. As 97% of the prey in
B1 were *Corophium*, any variation in the numbers of *Hydrobia* or *Nereis*
taken would have had little effect on the rate of feeding on *Corophium*. Thus
the number taken per minute was independent of prey density.

TABLE 2

The proportion of *Hydrobia* in the diet in relation to the density of *Corophium*
in the mud in area A1

Number of gizzards examined	Proportion of *Hydrobia* in the diet (%)		Density of *Corophium* in the mud (number per sq m)
	Mean	S.E.	
7	0·7	0·3	4031
4	3·8	2·9	5736
13	7·2	5·3	3470
3	22·6	17·2	2488
7	25·7	9·7	5104

There were differences in feeding rate between areas, which were not
related to prey density (Table 1). Furthermore, it seems that the rate of
feeding on *Corophium* also varied between areas independently of prey
density. The diet in areas A3, A4 and B4 was not described fully but the
average overall feeding rates (54·4, 50·8 ,54·5 prey per minute respectively)
were generally less than the rate of feeding on just *Corophium* in area A1
(59·6, 58·3, 54·9, 52·8 and 64·1 per minute during periods 1 to 5 respectively)
and area B1 (58·6 per minute). Even if all the prey taken in areas A3, A4 and
B4 were *Corophium*, which is unlikely, they would still have been taken at a
slower rate than in areas A1 and B1.

FIELD WORK—DISCUSSION

The results at temperatures above 6°C raised the question of why the
differences in feeding rate between areas were not related to prey density
and why within areas the feeding rate was largely unaffected by wide
variations in prey density. It would be expected that the feeding rate in
short-term bouts of feeding would depend on the predator's rate of contact

with the prey. This rate would itself be a function of the density of available prey, searching speed, the time required to deal with one prey (handling time) and the effective visual range of the predator (Murton 1968).

Assuming that redshank responded to all the prey they found, the problem is what determined their rate of contact with the prey. Areas A1 to A4 were situated along a transect from the top to the bottom of the beach, A1 being nearest to high water mark. The results suggest a trend for feeding rate to decrease from the top to the bottom of the beach; the correlation between feeding rate and distance from high water mark was strong and significant despite there being only four observations ($r = 0.993$, $P < 0.01$). This trend was perhaps due to some physical variation of the substrate affecting one or more of the components which determined rate of contact. The handling time with such small prey probably did not vary much between areas, and in any case prey size was often similar in different zones (Table 1). However, substrate features may have influenced the remaining three determinants of feeding rate. Thus they may have affected prey behaviour and thence the density of available prey; the distance from which prey were visible; and the searching speed by simply affecting, for example, the rate at which the birds could walk. None of these possibilities has yet been studied, however.

Area differences in substrate do not account for the feeding rate within areas being so little affected by change in prey density. As prey density increased, one would expect the number of detectable prey, and thence feeding rate, to increase also. Possibly the proportion of prey available decreased as prey density increased, so that the density of detectable prey remained either constant (area B1) or increased only slightly (A1). However, there is no evidence for this at present.

An alternative assumption to the one adopted so far is that redshank did not take all the prey they encountered. *Hydrobia* often occurred in very high densities clearly visible on the mud but many were ignored, perhaps due to their small size making them an uneconomical prey or to their thick shell which may hinder digestion. Since small *Corophium* were much less likely to be taken than big ones (Table 3), perhaps the birds were tending to select the large ones and to ignore the small. Alternatively, the small ones may have been unavailable, either because their small size made them difficult to detect or because of some difference in behaviour. It is difficult in the field to determine whether this selection for larger size reflects a size difference in availability or preference. In the laboratory, however, it is possible to study size preferences and some early findings are discussed below.

TABLE 3

The ratio between the frequency of each size class of *Corophium* in the diet of redshanks and their frequency in the mud

Area	Mean length of *Corophium* in the mud (mm)	Size group of *Corophium* in mm								Numbers of *Corophium* measured		P†
		1	2	3	4	5	6	7	8	In birds	In mud	
A1	4·11	0	0	0·24‡	1·25	1·77	2·83	3·00	—	199	85	<0·001
A1	3·67	0	0·01	0·18	0·80	4·64	8·28	6·19	—	317	157	<0·001
A1	4·22	0	0	0·30	1·00	1·39	5·96	2·47	3·20	63	186	<0·001
A4	6·16	—	0	0·26	0·43	1·21	1·11	0·98	2·02	55	142	<0·20 >0·10

†P values show the significance of the difference between the frequencies of each size group taken by the birds and the frequencies they would have been expected to take had they no size preference. The non-significant difference in area A4 probably reflects the fact that the mean prey size in the substrate was large so that there were few small ones present.

‡The figures in this part of the table are the ratios mentioned in the title; in this particular case the 3 mm size group was apparently selected against by the birds, and the 6 mm and 7 mm size groups in the same area were selected for.

FEEDING EXPERIMENTS IN THE LABORATORY

In preliminary choice experiments, five captive redshank reared from eggs or wild-caught chicks strongly preferred large prey. However, these birds were presented with a simultaneous choice whereas in nature the choice would probably be sequential (Tinbergen 1960). An experiment was therefore designed to test this.

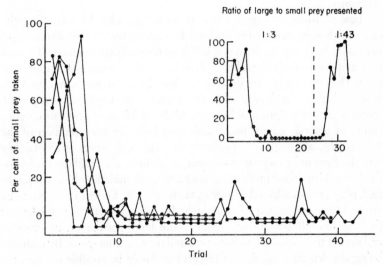

FIG. 1. The proportion of small prey ignored by redshank in a series of trials where they were presented with a sequential choice between large and small prey. These values contrast with the small proportion (1·9%) of the large prey missed. The inset shows the effect of a severe reduction in the density of the large prey on the number of small ones missed.

A row of holes was drilled down the centre of a 3 m length of wood and prey put in each. The small prey were single mealworm segments and the large prey were sections of eight segments. One large prey was put in each of several holes selected at random and between 1 and 4 small prey were then put in each of the remaining holes. There were often big gaps between adjacent large prey so that any preference for large ones would not be due to the birds being able to see the next one along. The birds were deprived of food for 6 hours and then allowed to collect a total number of segments that was constant between trials. Trials were repeated at regular intervals and in each the numbers of large and small prey taken and missed were noted.

In the first few trials the birds took both small and large prey but in later trials they ignored most of the small ones (Fig. 1). So far only two birds have been tested so the results should be regarded as tentative, but they indicate that the birds learn to take the large ones in a situation of sequential choice. Hence the predominance of large *Corophium* in the diet could have been due to a preference rather than to the small ones being undetectable. In other words, in nature the birds may not have taken all the prey they encountered.

These preliminary experiments are to be extended by increasing the range of choices presented to the birds by varying the relative densities and sizes of the large and small prey. It will be particularly interesting to discover how responsiveness to small prey depends on the density and size of large prey. The one experiment done so far (inset Fig. 1) suggests that the birds re-started taking small prey when the density of the large ones was greatly reduced. The description of how the birds exploit prey populations consisting of several sizes of individuals may help the interpretation of the field results, although it is probable that area differences in availability will also have to be taken into account. In addition, they will test the hypothesis that birds' food preferences are partly determined by how efficiently each prey can be collected (Hinde 1959; Kear 1962; Emlem 1966; Royama 1966). In the present experiments, whether it is most efficient to take just the large or both large and small prey would depend on the relative densities and sizes of the prey, the handling time and the searching speed. By manipulating the densities and sizes of the prey it should be possible to assess the extent to which the birds' performance tends to maximize the biomass of prey collected per unit time.

ACKNOWLEDGEMENTS

I should like to thank the Natural Environment Research Council for financial support.

SUMMARY

1. This paper discusses the effect of certain environmental factors and prey size on the winter diet and feeding rate of the redshank, an estuarine wading bird. Diet and feeding rate were independent of temperature

above 6°C but at lower temperatures considerable changes sometimes occurred in both. These changes were probably due to the effect of low temperature on the behaviour and thence availability of the prey.

2. *Corophium volutator* formed most of the diet (87%+) at temperatures above 6°C, and the feeding rate of redshank was largely unaffected by wide variations in the density of *Corophium*: in one study area it increased slightly as prey density increased and in another it remained constant. Furthermore, there were differences in feeding rate between areas where the prey density was similar. Small *Corophium* were rarely taken by the birds and the likelihood of a prey being eaten increased as its size increased.

3. Preliminary experiments on captive redshank indicate that the increase in prey risk with size could have been due to the preference of the birds for large prey. Current experiments to examine how the birds exploit a food supply consisting of numerous individuals of varying size are described.

REFERENCES

EMLEM J.M. (1966) The role of time and energy in food preference. *Am. Nat.* **100**, 611-7.

GOSS-CUSTARD J.D. (1969) The winter feeding ecology of the redshank *Tringa totanus*. *Ibis*, **111**, 338-56.

GOSS-CUSTARD J.D. (1970) The responses of redshank (*Tringa totanus* (L.)) to spatial variations in the density of their prey. *J. Anim. Ecol.* **39**, in press.

HASSELL M.P. (1966) Evaluation of parasite or predator responses. *J. Anim. Ecol.* **35**, 65-75.

HINDE R.A. (1959) Behaviour and speciation in birds and lower vertebrates. *Biol. Rev.* **34**, 85-128.

HOLLING C.S. (1965) The functional response of predators to prey density and its role in mimicry and population regulation. *Mem. ent. Soc. Can.* **45**, 1-60.

KEAR J. (1962) Food selection in finches with special reference to interspecific differences. *Proc. zool. Soc. Lond.* **138**, 163-204.

MURTON R.K. (1968) Some predator-prey relationships in bird damage and population control. *The Problems of Birds as Pests* (Ed. by R. K. Murton and E. N. Wright), pp. 157-69. London.

ROYAMA T. (1966) Factors governing feeding rate, food requirement and brood size of nestling great tits *Parus major. Ibis*, **108**, 313-47.

SOLOMON M.E. (1949) The natural control of animal populations. *J. Anim. Ecol.* **18**, 1-35.

SOLOMON M.E. (1964) Analysis of processes involved in the natural control of insects. *Adv. ecol. Res.* **2**, 1-58.

TINBERGEN L. (1960) The natural control of insects in pinewoods. I. Factors influencing the intensity of predation by songbirds. *Arch. néerl. Zool.* **13**, 265-343.

9

DISCUSSION

I. Prestt: First, as the prey species include the ragworm, molluscs and a crustacean, does movement of prey play a part in the attraction of the redshank to them? Secondly, do they select a part of the beach which is rich in prey, and if so, how do they find it? Thirdly, does one successfully-feeding bird attract others?

J. Goss-Custard: In answer to the first point, movement of prey may well attract redshank, but as I do not know the cues the birds respond to, I cannot really say for certain. It is important to determine these cues since I am assuming in the experiments that the birds in nature can differentiate between large and small *Corophium*; whether or not they can do this may depend on the cues they respond to. In answer to your last two questions, the birds were scattered over most of the estuary but tended to concentrate in certain areas and at certain parts of the beach within the study areas. The density of feeding redshank was positively correlated with *Corophium* density over the estuary as a whole and in transect A but not in transect B. There is good evidence that the birds concentrated where their feeding efficiency was greatest. Presumably they can assess the suitability of the various areas they hunt over and may tend to stay in the best areas. High densities of birds may attract others. I doubt, however, if a bird could tell the feeding efficiency of an another as it could not tell the value of the prey taken.

D. E. Glue: Some wader species feed by night. Have you noted behavioural changes then?

J. Goss-Custard: Yes, they fed at night in winter when they could not collect all their food during the day. They fed by rapidly moving their open bill from side to side through the mud. This method was used infrequently during the day.

D. C. Seel: How does the feeding activity of the redshank vary with the state of the tide, particularly with the depth of water over the mud? Your illustration of redshank feeding appeared to indicate that the bird was feeding in shallow water.

J. Goss-Custard: Most observations were made over the period 2–3 hours on either side of low water, and during that time there seemed to be no constant trend for feeding activity to change with time. The few observations I made on birds feeding at the tide edge suggested they were feeding at rates similar to those at low water, but I made no systematic observations as to the effect of water depth on feeding.

THE USE OF THE HERB LAYER
BY GRAZING UNGULATES IN
THE SERENGETI

By R. H. V. Bell

Serengeti Research Institute, Arusha, Tanzania

THE VEGETATION

The pattern of pasture use by African grazing ungulates has developed in response to the characteristic distributions within the herb layer of physical structure, mechanical properties, and chemical composition.

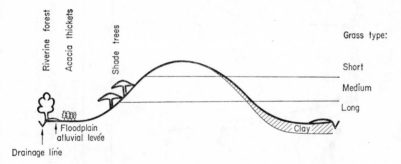

Fig. 1. The relationship of vegetation to topography on the catena.

These are largely determined by the growth form of the grass plants that are the largest component of the herb layer. Briefly, the concentration of the cytoplasmic constituents (protein and soluble carbohydrates) is highest in leaves, while that of structural constituents (mainly cellulose) is highest in stems. The proportion of fibre to protein increases in all tissues as their size increases, so that the protein/fibre ratio is highest in young plants, and in those species that never attain great size (French 1957). During grass growth, therefore, components with high proportions of protein

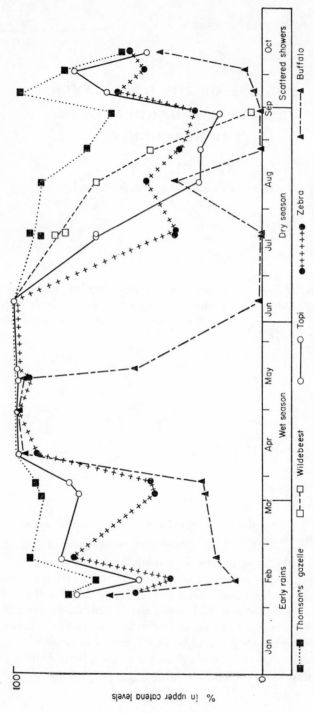

FIG. 2. Seasonal use of the catena by herbivores. This figure shows, at intervals through 1967, the percentage of each species occupying the upper catena levels (grass types short and medium in Fig. 1).

become rarer, and are progressively interspersed by taller, tougher structural components. This process continues farthest in those grass types with tall dense growth-forms.

The herb layer of the Western Serengeti study area was divided into five structural types, from short mat-forming to tall-robust. These types are distributed regularly in relation to the topography. The typical topography of the area is a series of undulations: rain falling on these gravitates from the peaks, across the slopes, to the sumps, resulting in the catenary soil sequence from leached sandy peaks to heavy clay sumps (Morison, Hoyle & Hope-Simpson 1948). The effective growing season, which is determined by availability of soil water, is therefore longer in the sumps than on the peaks. The grass types are arranged in relation to the catena levels, the shortest on the peaks, the longest in the sumps (Fig. 1).

THE DISTRIBUTION OF ANIMALS

In Fig. 2, the seasonal use of the pasture by herbivores is analysed in relation to these catena levels. The study area, chosen for its simple domed topography and hence its clear zonation of vegetation, was a minor granitic shield of about 12×6 km. The distributions of animals were mapped at frequent intervals throughout 1967. In the figure, the five grass types are lumped under two broader headings, short to medium, and long; at each interval recorded, the percentage of each grazing species occupying the short to medium grass types (i.e. on the upper levels of the catena) is shown. The five most numerous grazing species are shown: these are, in order of size, and with figures from Sachs (1967) for average weights of adult females:

		kg
Buffalo	*Syncerus caffer* (Sparrman)	447
Zebra	*Equus burchelli bohmi* (Matschie)	219
Wildebeest	*Connochaetes taurinus albojubatus* (Thomas)	163
Topi	*Damaliscus korrigum jimela* (Matschie)	108
Thomson's gazelle	*Gazella thomsoni* (Gunther)	16

Points to note about the seasonal use of the catena are the following:

1. *Wet season.* All species concentrate on the shorter grass types on the upper levels of the catena. This concentration lags about 3 weeks behind the rainfall, which reflects the lag in the vegetation's response to the rainfall. The exception is buffalo, which leave the peaks after the heaviest rainfall of

April and early May. (Wildebeest are migratory and were absent from this study area during the wet season.)

2. *Dry season.* The herbivores disperse from the peaks towards the longer grass of the sumps, in a clearly defined sequence. With the exception of Topi, preceding wildebeest, the succession is in order of size, and the order of descent is the reverse of the order of ascent at the beginning of the wet season. The descent of wildebeest and Topi is more complete than that of zebra at the end of the dry season. This partial inversion of the zonation of herbivores at the end of the dry season is confirmed by aerial surveys of the whole of the western Serengeti (Bell 1969).

3. *New growth of grass.* Following the first scattered dry season showers, herbivores concentrate on the resulting green flush, which, because of accidents of rainfall, is not distributed in relation to the catena.

4. *Early rains.* During the period of scattered rainfall preceding the main wet season, the herb layer fluctuates in intermediate stages of growth, and the herbivores (except wildebeest which are absent) are stratified in order of size on the catena levels.

The conclusions which can be drawn from these distributions are as follows:

(a) At the time when the herb layer is tallest and densest (the wet season), all species concentrate on the upper levels of the catena where the vegetation structure is least robust, and the protein content highest.

(b) As the herb layer is removed from the upper catena levels during the dry season, the herbivores are forced into the longer grasses of the sumps. The order of the grazing succession is evidence of different tolerances for long grass by different herbivores.

ASSOCIATION BETWEEN SPECIES

During the wet season, all species occupy the same catena levels; even during the dry season the zonation shown in Fig. 2 is a difference in emphasis rather than a spatial separation. In fact, mixed groups of several species are commonly seen at all seasons in the Serengeti. This situation is quantified by means of Cole's (1949) coefficient of interspecific association, using the same data as in the last section. The study area is divided by a grid with cells of 65 ha ($\frac{1}{4}$ sq mile), and the coefficient gives a measure of whether pairs of species occur in the same cells more or less often than they would if their relative distributions were due to chance.

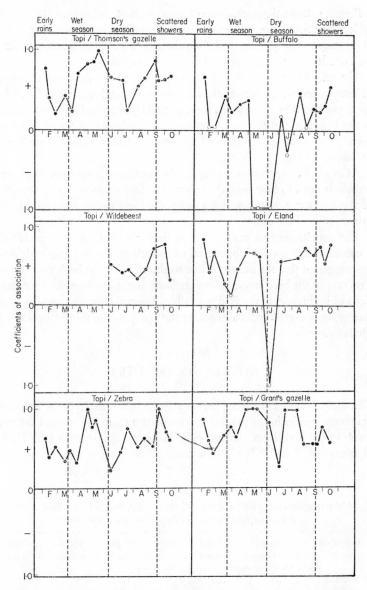

FIG. 3. Coefficients of association of Topi with six other species at intervals through 1967. Closed points, $P < 0.05$; open points $P > 0.05$, i.e. showing that the difference is due to chance.

Figure 3 shows the coefficients of association of one species (Topi) with each of the four other species mentioned on page 113; two others, eland (*Taurotragus oryx* (Pallas) and Grant's gazelle (*Gazella granti* (Brooke)) are included to show the generality of the situation.

The principal point is that all these pairs show a similar seasonal pattern of association, with a peak in the late wet season and another at the end of the dry season. These two peaks are separated by periods of lower association at the beginning of the wet season and the early dry season. These lows correspond to the two periods of movement relative to the catena, which are shown in Fig. 2.

As might be expected from Fig. 2, the association pattern of Topi with buffalo is out of phase with the others, the low coefficients of association of the late wet season coinciding with the early return of buffalo to the lower levels of the catena.

The conclusion that can be drawn from these associations is that the presence of one species is beneficial to the others. Vesey-Fitzgerald (1960) has suggested that the benefit is due to the effect of the herbivores on the structure of the herb layer. He pointed out that the wet-season concentrations of herbivores in the Rukwa valley, Tanzania trap the herb layer, by trampling and grazing, at a structure and composition more favourable to the herbivores.

ANALYSIS OF DIET

The analysis of diet described here was made to test the hypothesis that the herbivore species in the grazing succession shown in Fig. 2 select different levels of the herb layer. The work was done in conjunction with Dr M. Gwynne of E.A.A.F.R.O. (Gwynne & Bell 1968). Freshly-ingested

TABLE I

Mean proportions of different parts of plants occurring in the herbage intake of four ungulate species in the dry season, Sept. 1967

Ungulate species	Total dicotyledon	Total grass	Grass leaf	Grass sheath	Grass stem
Wildebeest	0	100	17·2	52·7	30·1
Topi	0	100	9·4	53·4	37·2
Zebra	0·1	99·9	0·2	48·7	57·0
Thomson's Gazelle	38·75	61·25	2·88	37·5	20·88

stomach contents were taken from four of the five species (buffalo was not
included), and analysed by Dr Gwynne according to the vegetation components present. The categories used were dicotyledon and monocotyledon,
subdividing these into leaf, stem, sheath, seed and fruit. Table 1 shows the
results for the collection in the dry season (Sept. 1967). The first point to

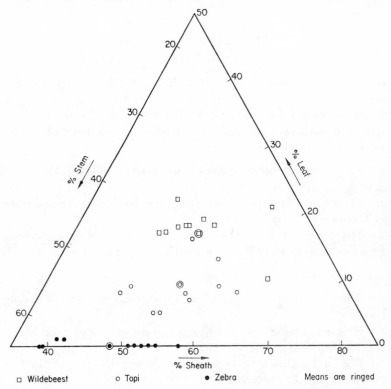

FIG. 4. Analysis of stomach contents of three grass-eating herbivores, showing
the frequency of the three principal grass components (after Gwynne & Bell
1968).

notice is that three of the species eat almost exclusively grass, while Thomson's gazelle selects 40% dicotyledon material, much of which is fruit. This
agrees with the observations of Talbot (1962) and Lamprey (1963). The
differences between the diets of the grass-eating species are shown in Fig. 4.
The diets do not differ significantly in their content of sheath, but the
contents of stem and leaf do differ, with $P < 0.05$ (Gwynne & Bell 1968).

Wildebeest select most leaf, zebra least, and Topi intermediate; the order of stem content is the reverse of this.

Comparable data have not been obtained for buffalo, but the evidence available suggests that, during the dry season, this species selects a high proportion of leaf from the very large riverine grasses that are not heavily used by the other grazing species mentioned here (Sinclair, pers. commun.).

SELECTIVITY AND ITS CONSEQUENCES

In the second section it was shown that the order of the grazing succession corresponds (except for Topi) with the order of decreasing body weight. It can now be seen from the analysis of diet that the order of the grazing succession also corresponds with some trends in diet selectivity of the four species analysed. These are:

1. The protein content of the diet—lowest in zebra, highest in Thomson's gazelle.
2. The mechanical properties of the diet—toughest in zebra, most delicate in Thomson's gazelle.
3. The spacing of the preferred diet—commonest and most accessible for zebra, rarest and least accessible for Thomson's gazelle.

These trends raise three questions.

1. Why does the protein content of the diet increase with decreasing body size?
2. How does a non-ruminant, the zebra, maintain itself on a diet higher in crude fibre than that of the ruminants?
3. What is the relationship between diet selectivity, interspecific associations and the grazing succession?

Review of the agricultural literature in conjunction with the field data reported here provides some understanding of these questions.

SELECTIVITY FOR PROTEIN

The relationship between body size and energy turnover is well known (i.e. Hungate, Phillips, McGregor, Hungate & Buechner 1959), the gradient being given by $H = K + W^{0.73}$. Brody, Procter & Ashworth (1934) have shown that the relationship between body weight and the rate of protein turnover is very similar (endogenous nitrogen excretion $= K + W^{0.74}$, and a functional link between the two has been suggested (Wahbi 1966)). The significance of this is that the maintenance requirement of protein per unit of body weight increases with decreasing body size in the same way as the requirement for energy increases. Given a series of animals, such as the ruminants, with comparable digestive systems, the maintenance of body weight by smaller animals requires an intake of diets with higher proportions of protein and soluble carbohydrates, at the expense of fibre.

THE DIGESTIVE SYSTEM OF HORSES

The problem of how the non-ruminant zebra maintains itself on the diet which is highest in crude fibre is less easy to understand. A review of available figures by Glover & Duthie (1958) shows that efficiency in the extraction of protein from foods of different compositions is similar for cattle and horses, and can be described by curves of the type:

$$D.C. = a + b \log C.P.$$

(D.C. = digestibility coefficient, C.P. = crude protein in feed.)

Alexander (1963), however, reviewing the mechanisms of digestion in horses, pointed out important differences from ruminants:

1. The sites of fermentation of structural carbohydrates in the horse are the enlarged colon and large intestine. Alexander (1952) has shown that the rate of breakdown of a cotton (cellulose) thread suspended in the ventral colon of the horse is twice as rapid as that of a similar thread suspended in the rumen of a cow (Balch & Johnson 1950).

2. The rate of passage of ingesta through the horse's alimentary canal is much faster than through that of the ruminant. Figures available for 95% excretion of markers are 30 hours (Alexander 1952) and 45 hours (Hintz & Loy 1966) for the horse, as opposed to 60–100 hours for the cow (Balch & Campling 1965). Given a similar efficiency of extraction of protein from food, these figures imply that the horse is capable of a higher rate of assimilation of protein than the ruminant, so that it can maintain its body weight on a diet (i.e. grass stems) too low in protein to constitute a maintenance diet for ruminants (Bredon & Juko 1961).

The problem of why the ruminant is relatively more efficient when using diets with high protein contents is discussed elsewhere (Bell 1969).

DIET SELECTIVITY,
INTERSPECIFIC ASSOCIATIONS AND
THE GRAZING SUCCESSION

Considering the basic grazing succession of zebra, wildebeest and Thomson's gazelle, one can now understand the increased selectivity by the later members for vegetation components that are high in protein. The question remains: what is the relationship of diet selectivity with the grazing succession and with the interspecific associations?

Arnold (1962) showed that the rate of intake by a selective grazer (in his case, sheep, selecting for green leaf) is reduced by the presence of unwanted components of the vegetation. This principle, that the rate of intake by an animal depends on the structure of the vegetation in relation to the animal's selectivity, can be extended to provide an interpretation of the present observations. It can be supposed that the interference by unwanted components is least when the selected component is common and near the surface of the herb layer (i.e. grass stems) and greatest when it is sparse and low down (i.e. dicotyledon material).

Thus, during the wet season, the intake of a maintenance diet is easiest for all species from the shorter grass types on the catena peaks, where there are high ratios of leaf to stem and of protein to fibre. In addition, as suggested by Vesey-Fitzgerald (1960), the mixed concentrations of species trap the herb layer in a sub-mature state, of which the physical structure and chemical composition both facilitate the intake of protein.

When growth ceases at the beginning of the dry season, the relatively small volume of dry matter on the catena peaks is rapidly removed, forcing the grazers to descend into the longer grasses in the lower levels of the catena. Zebra, with their high tolerance of grass stems, lead the succession; in so doing, they open up the herb layer by trampling and they increase the relative frequency of leaf by selecting stems, thereby increasing the suitability of the vegetation structure for the wildebeest that follow them. These in turn reduce the vegetation in such a way as to facilitate the use of dicotyledon material by gazelles. During this phase the grazers are still positively associated, but the coefficients of association tend to have lower values since the associations are progressive rather than static.

Finally, at the onset of the next wet season, the leafy mat of early grass growth provides an ideal structure for the intake of protein, so that the grazers become independent of each other's effect on the herb layer; this season of superabundance thus coincides with the period when coefficients of association are lowest.

ALTERNATIVE GRAZING SUCCESSIONS

The last point to be considered here is that of the two species which resemble each other most closely—wildebeest and Topi. These are adjacent in diet, body size, and position in the grazing succession as well as being members of the same tribe (Alcelaphini) (Simpson 1945). Comparison of the world distributions of the two genera (Sidney 1965) shows that they overlap in only two areas, one in the north-western Kalahari, and the other in Lake Province, Tanzania, which includes the Serengeti. Further, comparison of their distributions in the western Serengeti shows a quantitative difference, wildebeest being commoner in the north-west of the area, Topi in the south-east (Bell 1969). The significance of this difference is that the short mat-forming grass types predominate in the north-west; whereas in the south-east, the medium, upright grass types, dominated by *Themeda triandra*, reach the catena peaks, largely eliminating the shorter grass types (Bell 1969). The suggestion here, then, is that both these Alcelaphines select for grass leaf, but because of differences in structure, particularly of their jaws and teeth (Bell 1969), wildebeest are the more successful in a herb layer with a low frequency of stem and whose leaves incline to the horizontal, Topi in one with a higher frequency of stem and in which the leaves incline to the vertical. The result is that these two species become alternative members of the grazing succession according to the detailed structure of the herb layer. It is possible that the great variety of faunal composition found in African grazing ecosystems can be looked at in terms of such substitutions in the grazing succession in relation to minor differences of vegetation structure. Such a system has great flexibility and could prove a basis for the high efficiency of pasture utilization of these ecosystems.

ACKNOWLEDGEMENTS

This work was supported by a grant from the Nuffield Foundation. In addition, I would like to thank the Director of Tanzania National Parks and the Director of the Serengeti Research Institute for allowing the work to be carried out in the Serengeti National Park and for assistance during the project.

SUMMARY

1.   There is a characteristic relationship between the structure, mechanical properties and chemical composition of the Serengeti herb layer resulting from the growth form of the plant types in it. These vary in relation to topographic undulations in a catenary sequence, the peaks supporting a short, the sumps a tall herb layer. The use of these catena levels by ungulates was analysed.

2.   At the time of maximum growth in the wet season, all grazing species concentrate in mixed associations on the catena peaks. During the dry season, they descend in order of size into the sumps. Here distributions are more dispersed but still positively associated.

3.   Stomach contents of four species were analysed according to plant parts (monocotyledon or dicotyledon; stem, sheath, leaf or fruit). These analyses showed:

(a) The level of the herb layer selected corresponds to the position in the grazing succession: this suggests that feeding by the later members of the succession is facilitated by the activity of the earlier members, hence the positive associations.

(b) The chemical composition of the diet changes in sequence with the animal's position in the grazing succession, and in inverse sequence of body size.

4.   The reason for these differences in selectivity are discussed in terms of the maintenance requirements and digestive systems of the two main taxa concerned (ruminants and horses). The effect of the differences in selectivity on the pattern of use of the food supply is discussed in the light of feeding trials with sheep, from the agricultural literature.

REFERENCES

ALEXANDER F. (1952) Some functions of the large intestine of the horse. *Quart. Jl exp. Physiol.* **37**, 205–14.

ALEXANDER F. (1963) Digestion in the horse. *Progress in Nutrition and Allied Sciences* (Ed. by D. P. Cuthbertson), pp. 259–68. Edinburgh.

ARNOLD G.W. (1962) Factors within plant associations affecting the behaviour and performance of grazing animals. *Grazing in Terrestrial and Marine Environments* (Ed. by D. J. Crisp), pp. 133–54. Oxford.

BALCH C.C. & CAMPLING R.C. (1965) Rate of passage of digesta through the ruminant digestive tract. *Physiology of Digestion in the Ruminant* (Ed. by R. W. Dougherty), pp. 108–24. Washington.

BALCH C.C. & JOHNSON V.W. (1950) Factors affecting the utilisation of food by dairy cows. *Br. J. Nutr.* **4**, 389–94.

BELL R.H.V. (1969) The use of the herb layer by grazing ungulates in the Serengeti National Park, Tanzania. Ph.D. thesis, Univ. of Manchester.

BREDON R.M. & JUKO C.D. (1961) The chemical composition of leaves and whole plant as an indicator of the range of available nutrients for selective grazing by cattle. *Trop. Agric., Trin.* **38**, 179.

BRODY S., PROCTER R.C. & ASHWORTH U.S. (1934) Growth and development with special reference to domestic animals, Part 34. *Res. Bull. Mo. agric. Exp. Stn,* **220**, 32.

COLE L.C. (1949) The measurement of interspecific association. *Ecology,* **30**, 411–24.

FRENCH M.H. (1957) Nutritional value of tropical grasses and fodders. *Herb. Abstr.* **27**, 1–9.

GLOVER J. & DUTHIE D.W. (1958) The nutritive ratio/crude protein relationships in ruminant and non-ruminant digestion. *J. agric. Sci. Camb.* **50**, 227–9.

GWYNNE M.D. & BELL R.H.V. (1968) Selection of vegetation components by grazing ungulates in the Serengeti National Park. *Nature, Lond.* **220**, 390–3.

HINTZ H.F. & LOY R.G. (1966) Effects of pelleting on the nutritive value of horse rations. *J. Anim. Sci.* **25**(4), 1059–62.

HUNGATE R.E., PHILLIPS G.D., MCGREGOR A., HUNGATE D.P. & BUECHNER H.K. (1959) Microbial fermentation in certain animals. *Science, N.Y.* **130**, 1192–4.

LAMPREY H.F. (1963) Ecological separation of the large mammal species in the Tarangire Game Reserve, Tanganyika. *E. Afr. Wildl. J.* **1**, 63–92.

MORISON C.G.T., HOYLE A.C. & HOPE-SIMPSON J.F. (1948) Tropical soil-vegetation catenas and mosaics. *J. Ecol.* **36**, 1–48

SACHS R. (1967) Liveweights and body measurements of Serengeti game animals. *E. Afr. Wildl. J.* **5**, 24–36.

SIDNEY J. (1965) The past and present distribution of some African ungulates. *Trans. zool. Soc. Lond.* **30**, 1–396.

SIMPSON G.G. (1945) Classification of Mammals. *Bull. Am. Mus. nat. Hist.* **85**, 1–350.

TALBOT L.M. (1962) Food preferences of some East African wild ungulates. *E. Afr. agric. For. J.* **27**(3), 131–8.

VESEY-FITZGERALD D.F. (1960) Grazing succession among East African game animals. *J. Mammal.* **41**, 161.

WAHBI S.D. (1966) Comparative studies on serine deaminase. Ph.D. thesis, Univ. of Manchester.

DISCUSSION

V. C. WYNNE-EDWARDS: Are the smaller herbivores like Thomson's gazelle dependent on the presence of the large ones such as buffalo and zebra? If the larger species were exterminated would you expect the small ones to die out?

R. H. V. BELL: In the Western Serengeti, Thomson's gazelle do occur in the absence of the other species, but only in areas where the herb layer is kept short by other influences, in particular by human activity, including burning and grazing by domestic stock. It is certainly possible that in the absence of these influences Thomson's gazelle would become extinct in this area.

In reply to an anonymous questioner asking whether perissodactyls were less efficient than artiodactyls, R.H.V. Bell replied that studies on the nutrition of horses suggest that they are more efficient in some respects, and on this evidence it is not easy to see why perissodactyls don't sweep the board. The problem could probably be resolved by feeding trials, measuring rates of intake and assimilation of the two groups.

R. C. BIGALKE: One should beware of thinking that because the zebra has been shown to take tall, coarse, cellulose-rich grass, this feeding behaviour need necessarily be related to the digestive physiology and efficiency of perissodactyla in general. The square-lipped rhinoceros is a short-grass feeder par excellence.

J. J. R. GRIMSDELL: What is the role of buffaloes in the grazing succession in the study area, and how do they affect the smaller herbivores?

R. H. V. BELL: There are few data on buffalo; the evidence suggests that buffalo at all phases of the seasonal cycle select leaves from larger species of grass, so that they tend to be rather separate from the other members of the succession in the Western Serengeti. In areas of higher productivity such as Lake Rukwa, buffalo are of the greatest importance to the smaller animals, by breaking down the very tall stands of *Echinochloa pyramidalis* (Vesey-Fitzgerald 1960).

A. PAGE: What size of grid did you use in calculating an 'association coefficient' between different species of animals? The size of grid will affect whether one obtains positive, negative or no association between species. Also, what led you to choose this grid size? Thirdly, did you find any correlation between the grid size that you used and the size of pattern in the vegetation in the area?

R. H. V. BELL: The grid used in this analysis had cells of 65 ha ($\frac{1}{4}$ sq mile). In general, the associations between species are so close that the mixed group falls well within such a cell, so that decrease of cell size results in an increase in coefficients of association. The study area was chosen because of its broad, simple topography, resulting in clear vegetation zonation on the catena, so that the degree of heterogeneity of vegetation within a cell is not great. The cells were made no smaller because I felt that this represented the limit of accuracy for mapping the distributions properly from aircraft.

FOOD NICHE AND CO-EXISTENCE
IN LAKE-DWELLING TRICLADS

By T. B. Reynoldson and R. W. Davies†

Department of Zoology, University College of North Wales,
Bangor, Caernarvonshire

Earlier work (for references see Reynoldson (1966)) has suggested that the key to understanding the distribution and abundance of lake-dwelling species of triclad in Britain is competition for food. This hypothesis was tested by examining the natural food of the four common species, *Polycelis nigra*, *P. tenuis*, *Dugesia polychroa* and *Dendrocoelum lacteum*. A serological technique was used and rabbit anti-sera were produced for those prey organisms which earlier studies had shown were eaten frequently, namely, oligochaetes, gastropods and *Asellus*. An anti-serum was also produced against *Gammarus* since this is a widespread, abundant freshwater organism. The chironomidae are the main potential food item omitted, but Pickavance (1968) has shown they are not eaten much by these triclads.

Twelve habitats covering a range from unproductive hill lakes to productive lowland lakes were sampled and 1843 positive antigen-antibody reactions, spread approximately equally among the four triclad species, were analysed. This showed that the food regimes of the triclads overlapped but the species in each genus fed more on a particular type of food than the others; the *Polycelis* species took more oligochaetes, *Dugesia lugubris* took more gastropods, and *Dendrocoelum lacteum* more *Asellus*. Such contrasts were found in all environments and also persisted during all seasons of the year.

Contrasts in the food niches of the species were quantified by measuring the differences for each prey item according to the simple formula $N_1 - N_2$ where N_1 and N_2 represent the percentages of the particular prey item in the total food of the two triclads being compared. Such measurements fell into two non-overlapping groups, those with a low value of 10·4% (95% limits 0–21·8), and those with a high value of 46·2% (95% limits

†Present address: Department of Zoology, University of British Columbia, Vancouver, British Columbia, Canada.

31·4–61·0). The prey items in the latter group are identical with those which the earlier work suggested were 'food refuges' of the triclad species, i.e. food items for which intraspecific competition is more severe than interspecific competition. It was also shown that the distribution of *Dendrocoelum lacteum* was significantly associated with that of its 'food refuge' *Asellus*. It was not associated with, for example, the distribution of *Gammarus* on which it also feeds but to a much lesser extent.

It is concluded that if two or more species of organism utilizing the same limiting food resource are to co-exist, without spatial or temporal separation, then food must be partitioned so that each species has at least a 30% superiority, as calculated here, in one area of the general food resource. Such partitioning of prey could be on a taxonomic basis as occurs for triclads or on a size basis as suggested for some birds (Schoener 1965). A theoretical explanation for this has been suggested by Bossert (in MacArthur & Wilson 1967).

The two *Polycelis* species had similar food niches as revealed by the serological technique, and yet competition is the most severe between them. This paradox may be explained if each feeds more extensively on a different family of oligochaetes. Limited evidence suggests that the Naididae may provide a 'food refuge' for *Polycelis nigra* and Tubificidae for P. *tenuis*.

REFERENCES

MacArthur R.H. & Wilson E.O. (1967) *The Theory of Island Biogeography*. Princeton.

Pickavance J.R. (1968) The ecology of *Dugesia tigrina* Girard, an American immigrant planarian. Ph.D. thesis, Univ. of Liverpool.

Reynoldson T.B. (1966) The distribution and abundance of lake-dwelling triclads—towards a hypothesis. *Adv. Ecol. Res.* **3**, 1–71.

Schoener T.W. (1965) The evolution of bill size differences among sympatric congeneric species of birds. *Evolution, Lancaster, Pa.* **19**, 189–213.

DISCUSSION

J. Phillipson: In view of the demonstrated food refuges of P. *nigra* and P. *tenuis*, could Dr Reynoldson tell me anything about the distribution of the two respective prey species? Am I correct in thinking that the distribution of Naididae versus Tubificidae is somewhat different and would thus give rise to spatial separation of the two *Polycelis* species?

T. B. REYNOLDSON: While the Naididae are capable of swimming, the triclads are not, so spatial separation in the manner suggested could not operate. Further, the amount of time Naididae spend free-swimming is very short, so they would be available both to *Polycelis nigra* and *P. tenuis* on the substratum for considerable periods.

J. H. LAWTON: Is it reasonable to compare the observed percentage differences in food composition of sympatric triclads, with the similar values for observed percentage differences in bill sizes of birds and gapes of lizards, found by MacArthur and others?

T. B. REYNOLDSON: I prefer the term co-existence to sympatry, since Cain has defined these as different situations. I think that the comparison is reasonable only in the sense that the fit of the triclad 'difference' values to the Bossort model is so striking that it was worth pointing out. This, I believe, does not invalidate the conclusions to be drawn regarding the contrasts in the 'difference' values for refuge and non-refuge prey of triclads.

E. R. B. OXLEY: I would like to ask Dr Reynoldson if the food refuges he has determined could be the consequence rather than the cause of co-existence. Can these differences in preference account for co-existence or is food preference highly correlated with the ability to catch and utilize the refuge prey in a situation with interspecific competition?

T. B. REYNOLDSON: It is my view that present-day food refuges are both the cause and the consequence of co-existence, in the sense that natural selection will have enlarged the initial difference between the prey of the various species. Such differences in preference can account for co-existence, since intraspecific competition for a particular 'food refuge' prey has been shown to be more severe than interspecific competition for the same type of food.

G. C. VARLEY: Can Dr. Reynoldson express the differences between flatworms by numerals representing search efficiencies for different prey? I feel that only when he has made a precise mathematical model can he determine whether his concept of refuges is enough to explain co-existence.

T. B. REYNOLDSON: It would be very useful to quantify differences between triclad feeding activities. However, we have carried out laboratory studies in which we have used three different food refuges in seven combinations using single, double and treble refuges. In six of these, predictions based on the hypothesis presented in the paper have been substantiated. The single failure was in the technically most complex situation since it involved the three food refuges which should have resulted in *Polycelis tenuis, Dugesia* and *Dendrocoelum* co-existing. Failure was due to the near simultaneous death of adult *Dendrocoelum* after breeding, which resulted in fouling of the water. This is regarded as a technical failure since it is unlikely to happen in nature and is therefore probably not a refutation of the hypothesis.

A STUDY OF VEGETATIONAL
DYNAMICS: SELECTION BY SHEEP
AND CATTLE IN *NARDUS* PASTURE

By I. A. Nicholson, I. S. Paterson

The Nature Conservancy, Edinburgh

and A. Currie

Hill Farming Research Organisation, Edinburgh

INTRODUCTION

This is a short report on an investigation still in progress. The research is concerned with the dynamics of *Nardus* pasture, the effects of grazing on vegetational succession and the characteristics of the food resource for sheep and cattle.

Nardeta of various types occur extensively in the Cheviot hills. The development of Nardeta at the higher elevations has been attributed to the redistribution of peat and the influence of acid drainage water from peaty summits (Smith 1918). Nevertheless, *Nardus stricta*, which is of limited nutritive value (Thomas & Fairbairn 1956), is said to have increased in upland pasture as a result of selective grazing by sheep on free range (Fenton 1937). A large area of *Nardus* grassland does, in fact, occur on rangeland that has been exposed to sheep grazing since mediaeval times.

Fenton (1933) has described a generalized succession to *Nardus* grassland and King (1960) has suggested a mechanism whereby grazed and burnt Callunetum may be replaced by a *Nardus*-dominant community. Although the grazing of sheep on open range may encourage *Nardus* dominance, experiments in Wales have shown that under controlled grazing the trend can be reversed. *Nardus* can be eliminated completely by a combination of grazing and fertilizer treatment (Jones 1967). In practice, there are many examples where cattle-grazing has promoted the retrogression of Nardetum, and the sequence has been described in general terms (Fenton 1937). The processes associated with these trends, however, have not been described. In particular there is little understanding of the nature of selection by grazing animals or of the significance of selective defoliation on vegetational succession.

The data in this account are taken mainly from three experiments. Close clipping is used in two experiments and animal grazing in the third, to bring about changes in botanical composition. In the grazing experiment all the areas are grazed by sheep except one which involves cattle for comparison. No attempt is made to review all the data, some of the salient features only are outlined, and a hypothesis concerning the interrelationships between animals and pasture is suggested.

THE EXPERIMENTAL AREA

This lies at 330 m at Sourhope, the Hill Farming Research Organisation's field station in southern Scotland at the eastern end of the Cheviot range. As the plots cover nearly 1 ha it was necessary to accept some variability in site conditions. The soil material is derived from andesitic lavas and tuffs of the Old Red Sandstone which give rise to soils of higher fertility than on many hill pastures. At the site of the clipping experiments the soil is a podsolic brown earth similar to those of the Sourhope association described by Muir (1956), and the vegetation is described as *Festuca-Deschampsia* (*Nardus*) (King 1962). The area for the grazing experiment lies on a peaty podsol of the same association, with a well-developed H horizon extending to 20 cm in parts. The vegetation is described as *Molinia–Festuca–Deschampsia* (King 1962), and though *Molinia* is well represented it is not abundant. *Nardus* is dominant at both sites. The mean annual rainfall is about 760 mm.

METHODS

(a) CLIPPING EXPERIMENTS

The herbage is cut with hand shears. One of the treatments involves cutting monthly throughout the year. Each of the others, except the control, comprises a series of four successive monthly cuts at different seasons of the year. Samples for determining yield and *Nardus* content are taken from two random quadrats measuring 25 cm × 25 cm, in each of three replicated plots measuring 1·5 m × 1 m. Botanical analysis is done with the point quadrat. Measurements of shoot weight are made on samples taken from a separate experiment.

(b) GRAZING EXPERIMENT

The grazing treatments are related to the growth of the ungrazed pasture which is measured in exclosures. The treatments consist of changing the seasonal curves of the standing crop of pasture on different plots. By manipulating sheep grazing, the curves are made to conform to pre-determined models to provide a range of conditions for studying selection and the effects of grazing on botanical composition. Only one of these treatments is considered in this account (Fig. 1). In each year the standing crop is permitted to rise almost to the seasonal peak giving, in the first 2 years of the experiment, an average value of approximately 2000 kg/ha, before grazing down to less than 500 kg/ha. All measurements of standing crop refer to living or 'green' dry matter. Cattle are put in an additional plot to provide a direct comparison with sheep under the same conditions of vegetation. The plot sizes are 0·058 ha for sheep and 0·4 ha for cattle. The point quadrat is used for botanical analysis.

Selection in grazing is assessed by examining twenty quadrats (20 cm × 20 cm) per plot at intervals during grazing (Hercus 1963). All species occurring in each quadrat are recorded and signs of grazing noted. Severity of grazing is recorded on a scale of 1 to 4, and the most severely grazed shoot of a species determines the score for that species in a quadrat. Score 1 represents grazing within the top one-third of a leaf or shoot, and score 4 indicates extremely close grazing down to the level of the shoot base.

RESULTS

SOME CHARACTERISTICS OF *Nardus* PASTURE

The 'rhizomes' of *Nardus stricta* (Kruijne 1965) spread slowly, but under light grazing repeated branching leads to the accumulation of a dense mass of shoots. The old shoots eventually die, leaving behind the dead shoot bases in rows or clustered together in dense groups. Examination of the pasture on the experimental area showed a spatial pattern of living shoots, dead remains and 'bare ground' sparsely covered with litter. Associated species, mainly *Festuca ovina*, *Deschampsia flexuosa* and *Agrostis canina*, were most abundant in the 'bare ground' areas and on the *Nardus* debris where decomposition was most advanced. The spatial pattern and examination of living 'rhizomes' suggested that a cyclical process of colonization and decomposition of the *Nardus* plants maintains a dynamic equilibrium in the pasture.

Systematic observations have been made on the unenclosed range in the vicinity of the plots (Currie & Nicholson, unpublished). *Nardus* begins to grow relatively early in the season, vegetative shoots emerging from the sheaths in March or April. Most of the inflorescences appear towards the end of May though new ones continue to develop until July or August.

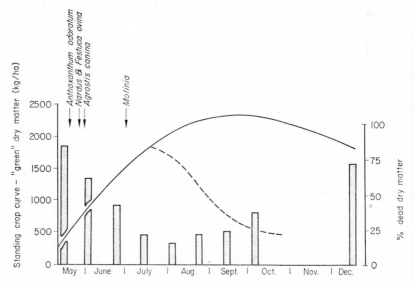

FIG. 1. The unbroken curve represents the standing crop of ungrazed *Nardus* pasture from May to December, expressed as 'green' dry matter. The histogram shows the percentage of dead material in the community at nine sampling dates. Arrows indicate the approximate dates of appearance of the inflorescences of species of five grasses. The figure also illustrates the actual curve (broken line) of standing crop which occurred as a result of the grazing treatment discussed in the text.

Figure 1 illustrates the seasonal growth-cycle of one of the *Nardus* swards on the experimental area and shows the changes in the proportion of accumulated dead material in the standing crop from May to December.

Nardus is grazed by sheep when the young shoots appear and grazing may be sustained for several weeks during spring. Grazing may continue within the *Nardus* community throughout the summer, but pressure on *Nardus* itself is light. In midsummer, especially with little rain, much of the herbage dies and though a succession of new shoots are produced until early autumn, a large quantity of dead and ungrazed material accumulates. In winter, sheep forage for persistent green shoots amongst this dead material and *Nardus* itself receives much of the grazing.

STUDIES ON PASTURE REACTIONS TO CLIPPING AND GRAZING

(a) DEFOLIATION BY CLOSE CLIPPING

1. *Nardus shoot weight.* In a clipping experiment, individual shoots of *Nardus* were weighed every month for 3 years. The level of cutting was much closer than that found possible under the heaviest experimental grazing. With no clipping the mean monthly weight of shoots sampled in the period June to August increased by 11% in the third year; while under monthly clipping the mean shoot weight declined by 54%. Although this represents a substantial reduction, *Nardus* shoots were still abundant and vigorous enough to compete with the associated species when cutting stopped, despite the severity of the treatment.

2. *Botanical composition.* Repeated close clipping resulted in a decline in the proportion by weight of *Nardus* in the sward (Table 1). In plots cut monthly for 2 years the percentage of *Nardus* fell by 70%. There was, however, a marked difference in response from clipping at different seasons.

TABLE I

Mean proportion by dry weight of *Nardus* (%) in the sward under different cutting treatments

	1965–66	1966–67	% change
Cut monthly all year	40	12	−70
Cut April, May, June, July	72	43	−40
Cut December, January, February, March	63	55	−13

Cutting in the four months April to July gave a decline of 40% in the second year while in a similar series of cuts beginning in December the proportion of *Nardus* was reduced by only 13%.

The response of the sward associated with the decline of *Nardus* between 1965 and 1967 is illustrated most clearly in the plots repeatedly defoliated (Table 2).

TABLE 2

Frequency (in %), as determined by the point quadrat†, of some sward
components of Nardetum cut monthly from May 1965

	September 1965	September 1967
Nardus stricta	46	23
Deschampsia flexuosa and *Festuca ovina* (combined)	78	86
Agrostis canina	17	21
Galium saxatile	12	34
Hypnum cupressiforme	11	32

Between April 1965 and April 1968 the frequency of litter-covered 'bare ground'
declined from 38% to zero.

†Based on seventy-five needles, each of which was stratified into 5 cm zones to take
into account the vertical structure of the sward.

Other species showed a general increase in frequency, particularly
Galium saxatile and *Hypnum cupressiforme*. Both species were prominent
in colonizing the areas between the original *Nardus* tussocks.

The dead shoot bases in Nardetum occupy a considerable area of ground.
With the decline of *Nardus* the proportion of this material is increased, but
the establishment and spread of plants upon it tends to be delayed until
decomposition of the debris is far enough advanced to provide a suitable
medium (Plate 1). This is one of the mechanisms responsible for the inertia
of *Nardus* sites to change in botanical composition. It also indicates that
although a trend towards a pasture type of higher quality for animal grazing
may begin soon after changing the management, there may be a considerable period of low production while the new sward develops.

(b) DEPLETION OF STANDING CROP BY
    HEAVY GRAZING

1.   *Grazing pressure.* The pasture is grazed down from just below the peak
yield of standing crop in each year (Fig. 1). Considering 1966 and 1967

PLATE I

1.   Ring quadrat (4·5 cm diam.) on the site of the clipping experiment, showing
typical cluster of dead *Nardus* shoots. Shoots may remain like this for several years
before they decompose and new plants colonize.

2.   Ring quadrat placed on ground between *Nardus* tussocks, showing colonization
by *Galium saxatile*, *Deschampsia flexuosa* and *Hypnum cupressiforme*. The quadrat was
placed in the plot at the beginning of the experiment in 1965. The photograph was
taken 18 months later, showing how slow colonization can be even where old *Nardus*
tussocks are absent. Once *Deschampsia* becomes established it rapidly covers the ground.

I

2

together, the median date for the beginning of grazing was 10 July and for the end of grazing 6 October. The mean number of sheep-grazing days for the 2 years was 2906 per ha. For some periods during the grazing-down process a sheep density of 69 per ha was required. For the same period on the unenclosed range with a stocking of 1 sheep per 0·4 ha, the mean figure was approximately 100 sheep-grazing days per ha for the range as a whole. As Nardetum is relatively neglected by sheep during the summer and early autumn (Hunter 1962), the comparable figure for this pasture type would be much lower.

These figures indicate the very high additional pressure required to deplete the standing crop of Nardetum to the level of a close-grazed pasture and thus to the point when a rapid differentiation of sward components may be expected to occur.

2. *Effect of grazing on botanical composition.* The main effects of heavy grazing on the composition of the sheep-grazed pasture were a decline in *Nardus* and *Molinia* and an increase of *Agrostis, Anthoxanthum* and miscellaneous species (Table 3). There were only two grazing seasons between the analyses but even with the heavy grazing pressure applied, the frequency

TABLE 3

Botanical composition, as determined by the point quadrat† (% frequency) of heavily grazed *Nardus* pasture 1966 to 1968

	1966	1968
Nardus stricta	30	26
Molinia caerulea	14	6
Festuca ovina	25	24
Deschampsia flexuosa	2	2
Agrostis spp.	13	16
Anthoxanthum odoratum	2	5
Carex spp.	5	4
Luzula spp.	4	3
Miscellaneous spp.	5	14

†Based on 125 needles, each of which was stratified into 5 cm zones to take into account the vertical structure of the sward.

PLATE 2. 1, 2 and 3 show three stages of grazing-down on the same area. 1, on 8 August 1967, 26 days after grazing began, shows grazing amongst the *Nardus* tussocks but very little on the *Nardus* itself. 2, on 17 August shows heavy grazing on *Nardus* and very close-grazed patches between. In 3, on 5 October the *Nardus* is still more heavily grazed though the tussocks are still prominent. The tufts of *Festuca ovina* shoots, pulled out in grazing and then rejected, should be noted in 2 and 3.

changes were comparatively slight. However, as expected, *Molinia* showed the greatest sensitivity to continuous close grazing, declining from 14% to 6%. The trend suggested by the frequency figures is thought to

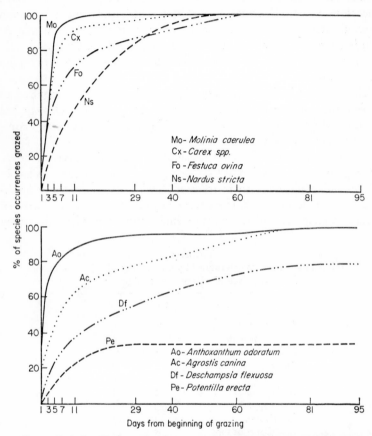

FIG. 2. Relationship between the percentage of species occurrences grazed by sheep and the number of days from the beginning of grazing, for different plant species. The data were derived by noting the presence of grazing on species within a quadrat, thrown twenty times per sampling date during a grazing period from 7 July to 13 October 1966.

be a real one because of supporting evidence from observations on the reactions of the sward as a whole, and by comparison with other treatments in the project where the rate of decline of *Nardus* was accelerated. Supporting evidence for the incipient development of another sward type included the production of short shoots by *Nardus* which, even where ungrazed for long periods, failed to develop to the normal size, an effect which was accentuated in the second grazing season; the establishment of seedlings of *Agrostis* spp. on the exposed areas amongst the tussocks; and the treading damage to the tussocks themselves. This damage was particularly apparent when it occurred after the protective growth had been removed. The

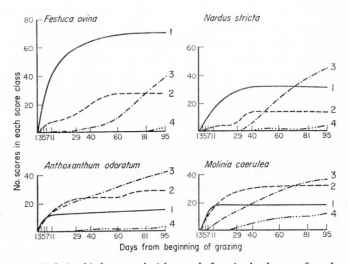

FIG. 3. Relationship between the 'closeness' of grazing by sheep on four plant species and the number of days since grazing began. Score 1 represents grazing within the upper one-third of a leaf or shoot, score 4 down to the base of the shoot close to the ground, and scores 2 and 3 intermediate stages. The data were derived from twenty quadrat throws at ten sampling dates during a grazing period from 7 July to 13 October. In each quadrat the species occurring were scored, and the most grazed individual of a species determined the score for that species in the quadrat.

herbage growing in urine patches also showed a relative increase in the vigour of the *Agrostis* and *Anthoxanthum* associated with the *Nardus*.

3. *Selection of sward components.* The observations on selection involved noting evidence of grazing and scoring its severity on individual plots in random quadrats. The plot was sampled ten times during grazing which lasted for 95 days in 1966. During this period the standing crop was reduced from approximately 2200 to 110 kg/ha of 'green' dry matter. Half the samples were taken in the first 11 days as rapid changes were expected to occur initially.

The results for 1966 are illustrated in Figs. 2 and 3 which show the trends in the percentage of the species occurrences grazed and the severity of grazing respectively.

In 5 days from the beginning of grazing the occurrence of grazed *Molinia* rose to about 90% (Fig. 2). *Anthoxanthum* showed a similar rise and both species reached 100% in 11 days. *Carex* spp. also showed a sharp initial rise though 100% occurrence was reached somewhat later. *Agrostis canina* and *Festuca ovina* both showed sharp initial rates of increase though neither reached 100% grazed occurrence until grazing was well advanced. *Nardus* and *Deschampsia flexuosa* showed the lowest rates of increase of all species except for *Potentilla* which was hardly grazed at all. The mean frequency in the pasture of all these species, based on the 20 cm × 20 cm quadrat, were in the range 63–91%. *Molinia*, *Anthoxanthum* and *Carex* spp., whose grazed occurrences increased most rapidly, were in the range 55–63%.

In the first few days after the sheep were put into the plot, grazing was extremely selective. Young shoots or leaf tips of *Anthoxanthum* or *Agrostis canina* were taken and also leaf tips of *Festuca*. The herbage of *Festuca* was dense and widely dispersed throughout the sward. Though the data are inadequate to support the observation, it was clear in the field that at this early stage *Festuca* was often grazed inadvertently. Thus, the combination of size of mouth and density of herbage often resulted in the defoliation of material that was not deliberately selected.

Figure 3 shows relative changes in the severity of grazing for *Festuca*, *Nardus*, *Anthoxanthum* and *Molinia*. *Festuca* and *Nardus* show a similar sequence, with score 1 (light grazing) increasing to a peak over a period of 40 days or more. Scores 2 and 3 occur in succession and score 3 rises rapidly between 40 and 60 days after grazing commenced. Score 4 was recorded at 40 days for *Nardus* but in the case of *Festuca* this score was not recorded for a further 40 days.

In contrast *Anthoxanthum* and *Molinia* showed an entirely different sequence, the frequencies of scores 1 to 3 rising in rapid succession; in *Molinia* score 4 occurred comparatively early and rose steadily throughout the grazing period. The more prostrate habit of many *Anthoxanthum* shoots appeared to provide some protection from the extremely close grazing represented by score 4. There was evidence of a sequence of selection in grazing from one species to another as grazing-down progressed.

From the sequence illustrated in Figs. 2 and 3, the main pasture constituents in this experiment can be ranked into three categories of preference.

Molinia, Carex and *Anthoxanthum* are apparently the most highly preferred. *Festuca ovina, Agrostis canina, Nardus* and *Deschampsia flexuosa* can be roughly grouped together, while *Potentilla* showed no sign of being deliberately selected, in fact there is evidence that avoidance was deliberate. Except for a few plants, e.g. *Vaccinium myrtillus* which occurred at very low frequencies, *Potentilla* was the only species in this category.

4.   *Comparison of selection by sheep and cattle.* The effects of grazing by sheep and cattle in these experimental treatments are not compared fully here. However, several salient features about herbage selection are worth noting. Although the order of preference of plant species was approximately the same, the differences amongst them were much less clearly defined. *Nardus* was more closely grazed by cattle than by sheep at the beginning, but extremely close grazing of *Nardus* hardly occurred at all, in fact close-cropping was largely confined to *Molinia* and *Anthoxanthum*. The sparse occurrence of very close-grazed patches in the pasture grazed by cattle was one of the most striking differences.

(c) OBSERVATIONS ON THE GRAZING PROCESS

In grazing at the high stocking rates used in the experiment, essentially the same process of depletion in the standing crop was observed in all treatments with sheep. Four stages were recognized:

1.   Extremely selective grazing by animals at first, at many places scattered throughout the plot. Often, no more than the upper few millimetres of a leaf or shoot were grazed.

2.   Development of centres of concentration. Single plants or clumps of *Anthoxanthum* or young *Molinia* shoots usually provided the foci which were moderately grazed. As grazing increased around the edges of these patches, material tended to be taken less selectively.

3.   Close grazing of the initial centres and expansion outwards producing a patchwork of close-grazed areas. Simultaneously, the areas lacking the preferred plants received further pressure and sometimes secondary centres of grazing concentration developed within them. There was also repeated return to the initial centres as new shoots appeared.

4.   The close-grazed areas eventually joined, and pressure over the entire area became relatively uniform, except for isolated areas that were partly rejected—usually *Nardus* tussocks.

Three levels of selection, or rejection, could be recognized: namely, at the level of the species, of the individual plant and of the area. Selection for a plant or part of a plant was inevitably imprecise owing to the complexity of the sward, and growth form affected this precision. Discrete shoots were readily selected or avoided while a dense mass of interwoven leaves and shoots resulted in relative unselective grazing. In this way *Nardus* leaves were often grazed unless the tussocks were fairly isolated. This is clearly shown in the final phase of grazing down (Plate 2).

The development of similar harder- and lighter-grazed patches has also been noted in experimental studies on Callunetum (Hunter & Grant 1968).

DISCUSSION

Hunter (1962) found that the chemical composition of herbage containing *Nardus* was related to the proportion of *Nardus* in the sward. Relatively high values for percentage CaO, K_2O and crude protein and low values for crude fibre were associated with low frequencies of *Nardus*. He found grazing intensity to be more closely related to CaO and crude fibre content than to any other factor, and showed that *Nardus* communities on mixed range tended to be neglected in summer. On many ranges with extensive areas of Nardetum the structure could be improved by encouraging a succession towards pasture types in which *Agrostis*, *Anthoxanthum* and other species are better represented.

Nardetum is well adapted to maintaining itself under the usual conditions of grazing by sheep on free range. The inertia of Nardetum to successional change is associated both with the growth behaviour of *Nardus* and with the characteristics of the persistent, compact clusters of dead rhizomes and shoot bases left by the plant as it continually spreads within the sward. This may be regarded as a cyclical form of autogenic succession which, under the seasonal pressure in free-range grazing, maintains the stability of the plant community.

The vigour of *Nardus* can be considerably reduced by extremely close clipping in the growing season, but sustained pressure for several years may be necessary to eliminate the plant completely. Although the reduction of *Nardus* creates opportunities for a new competitive balance amongst the species, the remaining debris of *Nardus* retards the spread of plants.

Under controlled sheep grazing there is evidence of a similar though less pronounced trend towards a reduction of *Nardus*. When the seasonal development of *Nardus* is well advanced, close cropping of the shoots does not readily occur even with heavy and sustained grazing. A number of other species including *Molinia*, *Anthoxanthum* and *Agrostis canina* receive much of the initial pressure. However, at the high animal density used in the experiment, the tendency to avoid *Nardus* by preference may be compensated by other factors. These are the nature of selection itself and the structure of the sward combined with the pattern of depletion which develops during grazing.

The grazing-down process observed in *Nardus* grassland suggests a general hypothesis. In terms of grazing behaviour and pasture response the sequence reveals certain features at the sward level analogous to the process which has probably led to, and now maintains, the present differentiation of pasture types on mixed range. In this situation it has been shown (Hunter 1962) that a disproportionate concentration of grazing animals occurs on the vegetation associated with certain soils. Individuals of certain preferred species in the community appear to have the same relationship to the unit sward as the individual sward type bears to the whole mosaic on a unit of range. At high animal densities the more preferred species in the sward may have a similar functional significance in attracting concentration. The attraction of grazing over an expanding area round the preferred plants appears to be an important means by which the less preferred species of certain growth forms are subjected locally to intensive pressure. The regrowth of species within the patches attracts further grazing which helps to maintain the herbage digestibility at a relatively high level (Black 1967). Throughout a grazing season an important aspect of this localized pressure would be to increase the frequency of defoliation and cycles of regrowth, and hence accelerate the reduction of *Nardus* and promote a trend towards dominance by other species. These observations suggest that the occurrence of certain preferred species able to withstand heavy grazing may have a significance to vegetational succession out of all proportion to their abundance.

High densities of grazing animals occur on parts of the range naturally or they may be encouraged by management aimed at controlling the animals' behaviour. Unless the density reaches a critical level which will vary according to past grazing, type of soil, animal selectivity and other factors, the above mechanism of change in the structure of the pasture is unlikely to operate. The undesirable aspects of selective close-grazing are well known, but on certain soils the habit when suitably directed may have an essential role in creating a more balanced system on managed range. Under natural conditions selective close-grazing may be a beneficial factor inducing diversity in range structure and affecting a variety of inter-relationships amongst different animal species.

ACKNOWLEDGEMENTS

The authors appreciate the facilities provided by Dr R. L. Reid and latterly by Dr J. M. M. Cunningham, Director of the Hill Farming Research Organisation, and for the considerable assistance kindly given by Mr J. Eadie and his colleagues at Sourhope. This enabled the experiments to be continued after two of the authors were appointed to the staff of the Nature Conservancy. The authors are also indebted to several colleagues for helpful criticism of the draft.

SUMMARY

1. Clipping experiments show the inability of Nardus to withstand repeated close defoliation in the main period of growth. However, the plant has a remarkable capacity for survival; and this, combined with the effects of the persistent dead tussocks, makes Nardetum resistant to successional change.

2. Grazing the pasture down to the level of a close-grazed sward between July and October required a mean stocking with sheep thirty times greater than the average for the open range. General close-grazing of Nardus was delayed owing to a marked selection for other plants. Grazing tended to be concentrated round several preferred species, the close-grazed area expanding outwards from initial centres.

3. Cattle were less selective. They grazed Nardus earlier in the same July–October period but less closely than sheep.

4. It is suggested that selective close-grazing may accelerate pasture differentiation at a given animal density. By encouraging very high grazing pressures locally in expanding areas the habit may have, on certain soils, beneficial effects on the animal environment by improving food quality in both natural and managed situations.

REFERENCES

BLACK J.S. (1967) The disgestibility of indigenous hill pasture species. *Rep. Hill Fmg Res. Org.* **4**, 33–7.

FENTON E.W. (1933) The vegetation of an upland area (Boghall Glen, Midlothian). *Scott. geogr. Mag.* **49**, 331–54.

FENTON E.W. (1937) The influence of sheep on the hill grazings in Scotland. *J. Ecol.* **25**, 424–30.

HERCUS J.M. (1963) Botanical sampling as a means of identifying the components of sheep's diet in tussock grassland. *N.Z. Jl agric. Res.* **6**, 83–9.

HUNTER R.F. (1962) Hill sheep and their pasture: a study of sheep-grazing in south-east Scotland. *J. Ecol.* **50**, 651–80.

HUNTER R.F. & GRANT S.A. (1968) Interactions of grazing and burning on heather moors and their implications in heather management. *J. Br. Grassld Soc.* **23**, 285–93.

JONES L.T. (1967) Studies on hill land in Wales. *Tech. Bull. Welsh Pl. Breed. Stn Aberystwyth* **2**.

KING J. (1960) Observations on the seedling establishment and growth of *Nardus stricta* in burned Callunetum. *J. Ecol.* **48**, 667–77.

KING J. (1962) The *Festuca-Agrostis* grassland complex in S.E. Scotland. *J. Ecol.* **50**, 321–55.

KRUIJNE A.A. (1965) *Nardus stricta* L. as a grassland species in the Netherlands. *Neth. J. agric. Sci.* **13**, 171–7.

MUIR J.W. (1956) *The Soils of the Country round Jedburgh and Morebattle*. Edinburgh.

SMITH W.G. (1918) The distribution of *Nardus stricta* in relation to peat. *J. Ecol.* **6**, 1–13.

THOMAS B. & FAIRBAIRN C.B. (1956) The white bent (*Nardus stricta*): its composition, digestibility and probable nutritive value. *J. Br. Grassld Soc.* **11**, 230–4.

DISCUSSION

G. C. VARLEY: Is *Nardus* weight for weight as good a diet as other grasses, and is any chemical growth inhibitor present which might suppress growth in the neighbourhood of *Nardus*?

I. NICHOLSON: *Nardus* is of comparatively low value throughout the year. In answer to the second question, experiments on leaching failed to support the suggestion that an inhibitor might be present.

J. EADIE: *Nardus* is less digestible. At any given level of digestibility, voluntary intake is lower. *Nardus* rapidly senesces in winter and therefore adds to the fund of dead material, making it substantially less desirable.

SOIL AMOEBAE: THEIR FOOD AND THEIR REACTION TO MICROFLORA EXUDATES

By O. W. Heal and Margaret J. Felton

The Nature Conservancy, Merlewood Research Station,
Grange-over-Sands, Lancashire

INTRODUCTION

The object of this study was to define which of the soil microflora are eaten by soil amoebae. Many previous investigations have examined the relationship between amoebae and bacteria, but other microflora have been largely ignored (Stout & Heal 1968). In most animal feeding studies, observations and experiments can be made under field or laboratory conditions. Field studies are technically very difficult with micro-organisms, but laboratory studies are relatively easy owing to the availability of suitable techniques for isolation, cultivation and observation. In the following laboratory study a wide range of soil microflora was offered to amoebae to test their edibility. As a result of feeding tests, some indication of factors influencing edibility were obtained and further tests were made on the effect of microflora exudates on amoeba activity.

MATERIALS AND METHODS

AMOEBAE

The amoebae studied were *Acanthamoeba* sp. (Neff 1957), *Hartmannella astronyxis* Ray and Hayes (Ray & Hayes 1954) and *Mayorella palestinensis* Reich (Reich 1933). Adam (1964) has produced good evidence for including these three amoebae in a single genus, *Hartmannella*. The amoebae were obtained from the Culture Collection of Algæ and Protozoa, Cambridge, and maintained axenically in 4% mycological peptone.

MICROFLORA

Details of the test microflora are given in Appendix 1.

FEEDING EXPERIMENTS

These were made on soil extract agar (S.E.A.) in 10 cm petri dishes. S.E.A. was prepared by autoclaving equal volumes of soil and water for 30 minutes at about 1 kg/sq. cm. The filtrate was diluted to 25% with distilled water, and 1% agar added. In each experiment seven streaks, 0·5–1·0 cm long, of the test microflora plus a control streak of *Aerobacter aerogenes* were placed around the perimeter of the plate and a drop of 4 to 6-day-old amoeba culture was added to each streak. The plates were incubated at 25°C and examined microscopically at frequent intervals until all the amoebae encysted. Edibility of microflora was assessed by reproduction of amoebae which was classified as 0—none, 1—very slight, 2—moderate, 3—abundant but encystment before all the food was consumed, 4—abundant and food supply completely consumed.

EXUDATE EXPERIMENTS

Exudate experiments were made on *Streptomyces* agar (St.A) composed of: maltose 10%, tryptone 0·5%, KH_2PO_4 0·05%, NaCl 0·05%, $FeSO_4$ 0·01%, Oxoid Ionagar No. 2 1%, in distilled water. The test microorganism was streaked across the centre of 10 cm petri dishes containing 30 ml St.A and incubated at 25°C for 7 days to allow exudates to diffuse into the agar. The test colony was then removed, and cores of 1 cm diameter were transferred from the remaining agar to plates of non-nutrient agar. Control cores were prepared from sterile St.A. Drops of 3 to 6-day-old amoeba cultures were added to test and control cores, and incubated at 25°C. A random sample of 100 amoebae was examined daily on each core for 3–4 days. The amoebae were classed as active, rounded, encysted or disintegrated. Contaminated cores were rejected but 6–14 replicate counts were made on each occasion. Means of the test and control counts of active amoebae were compared using a *t*-test for small samples.

RESULTS

FEEDING EXPERIMENTS

For most of the microflora, tests were made with all three species of amoeba, but because no marked differences in feeding habits of these amoebae were observed, only the results for *Acanthamoeba* sp. are given. The results are summarized in Table 1 and details for individual microflora given in Appendix 1.

(a) BACTERIA

Previous observations on bacteria as food for soil amoebae have been concerned with selected members of the bacterial flora. To find the proportion of edible bacteria in a given soil, thirty-seven isolates of the predominant aerobic bacteria from two soils in Roudsea Wood National Nature Reserve were tested. These were isolated on soil extract-yeast extract agar and characterized by Dr A.J. Holding. Of the thirty-seven isolates from Roudsea 43% supported abundant reproduction until the food was completely consumed and the amoebae encysted. The remaining 57% of isolates supported reproduction similar to the control *Aerobacter*, but the amoebae encysted before the bacterial streaks were completely consumed.

A further eighteen strains of bacteria were examined to provide a wider range of types. The results for all bacteria showed that gram-negative rods gave most growth of amoebae. The bacilli supported good growth, but amoebae usually encysted before all the food was consumed. Spores of anaerobic *Clostridium*, tested under aerobic conditions, were egested apparently undamaged after the rest of the cell was digested.

With a number of *Bacillus* spp. and *Chromobacterium* spp., reproduction of amoebae was slight in areas where long filaments were formed, but abundant reproduction occurred in the older parts of the colonies where filaments fragmented. Some feeding on the tips of filaments was observed.

The isolate of *Serratia* produced red pigment when grown at 25°C, but no pigment at 35°C. Reproduction of amoebae was recorded only on the latter. Approximately similar reproduction was recorded on three strains of *Chromobacterium* which produced dark purple, pale purple and no pigment, but there were indications that more reproduction occurred in the pale areas of colonies of the dark purple strain.

It was confirmed that amoebae reproduced less rapidly on *Aerobacter* when the latter was grown on sugar-rich media causing the production of large capsules (Wilkinson 1958). Poor growth of amoebae was also recorded on three strains of *Cytophaga* which were 2–10 μ long with orange pigmentation. On one strain there was antagonism, some of the amoebae becoming rounded and highly vacuolate.

(b) ACTINOMYCETALES

The distinction between *Mycobacterium*, *Nocardia* and *Streptomyces* is difficult and still under critical review (Küster 1968). *Streptomyces* are common soil forms, but *Nocardia* and *Mycobacterium* have not been recorded in large numbers, possibly because of the lack of selective techniques. However they probably form a normal part of the soil microflora (Jensen 1963a; Stout 1958, 1961). Most of the present isolates were from soil or are known to occur there.

The three strains of *Mycobacterium* which supported most reproduction of *Acanthamoeba* had rods < 10 μ long, and the less edible strains had

TABLE I

Reproduction of *Acanthamoeba* on various microflora. Details of the microflora are given in Appendix 1. The categories of reproduction are given on p. 146

Microflora	Number tested	Reproduction of *Acanthamoeba*				
		0	1	2	3	4
Pseudomonas	12				1	11
Achromobacter	4					4
Aerobacter (= *Klebsiella*)	2				1	1
Chromobacterium	4				3	1
Serratia	2	1			1	
Arthrobacter	13			2	5	6
Bacillus	15			1	13	1
Clostridium	3		1	1	1	
Cytophaga	3		3			
Nocardia	11	3	2	3	3	
Mycobacterium	6	1	1	1	3	
Streptomyces	20	18			2	
Algae	14	11	3			
TOTAL	109	34	10	8	33	24

filaments < 100 μ long. The *Nocardia* spp. which supported good reproduction had smooth moist colonies composed mainly of rods < 10 μ long and usually lacked aerial mycelium. Those giving poor or no reproduction of amoebae formed dry colonies with aerial mycelium and vegetative hyphae with little or no fragmentation. The inedible strains are closely related to *Streptomyces* (Gordon & Mihm 1957), 90% of which were inedible (Table 1). The two edible *Streptomyces* were strains of *S. rimosus* which formed soft colonies fragmenting into rods and short filaments (Gordon & Mihm 1962, Plate 2). Strains of *S. rimosus* forming dry colonies with persistent hyphae were inedible.

The non-wettable spores of *Streptomyces* were confined to the surface film, thus preventing contact with amoebae, and none were seen to be ingested. When spores were wetted with 2% 'Teepol', amoebae ingested 20–30 of those within 3 hours, but no reproduction of amoebae was observed over 48 hours. In controls of *Aerobacter* in 2% 'Teepol' the amoebae did reproduce.

In the presence of *Streptomyces* and *Streptomyces*-like *Nocardia*, amoebae often encysted rapidly, or developed large vacuoles and disintegrated. This reaction was not observed with the edible strains of *S. rimosus*.

(c) MYCELIAL MICROFUNGI AND YEASTS

Feeding studies on these groups (Heal 1963) were not extended. Fungal mycelium was inedible but spores of two species supported moderate growth of amoebae. Spores of other species were ingested but not digested. Nineteen species of soil yeasts supported reproduction of amoebae, *Saccharomyces* and *Kloeckera* being most edible.

(d) ALGAE

A wide range of algae occur in soil in sufficient numbers to provide a potential food source for protozoa (Tchan & Whitehouse 1953; Rosa 1957). A representative group of fourteen soil algae was grown on plates by a north-facing window at room temperature (12–16°C). Nine species were ingested by *Acanthamoeba* but only three supported reproduction, which was slight. Those which were not ingested had capsules about 15 × 20 μ (*Navicula pelliculosa*, *Nostoc muscorum*), formed long chains or filaments (*Hormidium flaccidum*, *Anabaena cylindrica*) or were large and motile (*Euglena viridis* v. *terricola*). The tests confirm Oehler's (1916) findings that algae are not suitable as food for small (10–20 μ diameter) soil amoebae.

(e) OTHER ORGANIC MATTER

Apart from the microflora, organic particles such as humus may be eaten. Amoebae are known to contain cellulase and chitinase (Tracey 1955) and testate amoebae have been grown on sterilized organic matter (Schönborn 1965). Samples were collected from litter, fermentation and humus horizons in four soils: a woodland with moder humus on slate and with mull humus on limestone (Roudsea Wood National Nature Reserve), and on acid peat under *Juncetum squarrosii* and a brown earth under *Festucetum-Agrosti* (Moor House National Nature Reserve). Samples were sterilized by autoclaving after removal of large particles and homogenizing the litter and fermentation samples. *Acanthamoeba* and *Hartmannella astronyxis* encysted without reproducing when added to this organic matter which included dead microflora.

CONCLUSIONS ON FOOD

Bacteria have long been regarded as the main, or only, food of soil amoebae and the present and other surveys have shown that a wide range of bacteria are consumed, and support varying amounts of reproduction. There is also increasing laboratory evidence of the edibility of live yeasts (Drozanski & Drozanska 1961; Soneda 1962; Heal 1963; Nero, Tarver & Hedrick 1964) and field evidence of their availability (Lund 1954; di Menna 1960; Jensen 1963b).

The results show that congeneric species of the microflora tend to support similar amounts of reproduction and different species of small soil amoebae have similar food preferences (Cutler & Crump 1927; Singh 1941, 1942; Chang 1960; Heal 1963; Groscop & Brent 1964). Therefore the soil microflora can be listed in descending order of edibility:

1. *Pseudomonas, Achromobacter, Aerobacter* (= *Klebsiella*), *Saccharomyces* and *Kloeckera*
2. *Chromobacterium* and *Arthrobacter*
3. *Bacillus, Nocardia* (bacteria-like forms), *Hansenula, Pichia, Cryptococcus* and *Candida*
4. *Mycobacterium, Cytophaga* and a few mycelial fungi
5. *Nocardia* (*Streptomyces*-like forms), some algae
6. *Streptomyces*, most mycelial fungi and some algae.

Other surveys of the food of free-living amoebae have largely been confined to bacteria and the present results are in good agreement with these (Sewertzova 1928; Nikoljuk 1956; Kunicki-Goldfinger, Drozanski, Blaszczak, Mazur & Skibinska 1957; Chang 1960; Groscop & Brent 1964). Unfortunately, in a major survey, Singh (1942, 1945) gave no identifications of the bacterial strains tested, so preventing detailed comparisons. *Nocardia* and *Mycobacterium* supported better growth in the present study than reported elsewhere (Rhines 1935; Kunicki-Goldfinger *et al.* 1957; Drozanski & Drozanska 1961; Dudziak 1962). There are however problems of nomenclature, as with *Arthrobacter*, which may be represented in other studies under a different name.

The sparse literature on non-bacterial groups also suggests that mycelial fungi, algae and *Streptomyces* are rarely eaten by amoebae (Oehler 1916; Sewertzova 1928; Sandon 1932). Pfennig (1958) observed reproduction of amoebae on *Streptomyces* spores produced on Rossi-Cholodny slides but he also observed *Streptomyces* spores germinating on dead amoebae.

EXUDATE EXPERIMENTS

During the feeding experiments amoebae were observed to become rounded, develop large vacuoles and eventually distintegrate in the presence of certain microflora. The reaction of *Acanthamoeba* to exudates from eighteen species of microflora was studied in detail, compared in each case with a control (p. 146). In the absence of food the normal reaction of the amoebae population is to encyst over a period of 3 or 4 days. The results (Table 2) show the differences in numbers of active amoebae between test

TABLE 2

The influence of microflora exudates on activity of *Acanthamoeba*. Figures show the difference (+ or −) in number of active amoebae on exudates compared with those on control agar. One hundred amoebae were examined on each occasion. Differences significant at $P < 0.05$ are shown in heavy type

Species	Day			
	1	2	3	4
Aerobacter aerogenes	+ **9·5**	+ **9·1**	+ 6·9	+ 2·3
Arthrobacter simplex	− 2·2	+ 4·4	− 2·9	
Nocardia globerula	+ **7·2**	+ **5·3**	+**12·9**	
N. cellulans	+ **5·7**	+**11·5**	+ 5·9	+ 2·2
N. coeliaca	− 2·6	−**21·1**	−**23·0**	
Streptomyces rimosus	−**70·6**	−**51·3**	−**56·1**	−**41·3**
S. griseus	−**23·6**	− **9·3**	−**12·8**	
S. noursei	− 4·4	− 0·8	−**16·4**	− 2·1
Saccharomyces cerevisiae	+**12·7**	+**15·6**	+**16·4**	+**16·7**
Candida humicola	+**14·4**	+**23·9**	+**26·5**	+**16·4**
Kloeckera africana	+**23·0**	+**28·9**	+**37·6**	+**19·0**
Paecilomyces elegans	+ 5·5	− 6·7	− 9·3	
P. farinosus	+**16·6**	+**12·1**	+ 3·7	
Trichoderma viride	− 6·2	− **6·8**	+ 5·8	
Penicillium frequentans	+ 0·6	− 0·4	− 8·4	− 7·2
Bumilleria exilis	+ 6·1	+ 4·4	+ 7·6	
Monodus subterraneus	+ 7·7	+ 3·8	+ 7·8	
Hormidium flaccidum	+ 5·7	− 3·0	+ 6·9	

and control. The antagonism or antibiosis noted in feeding studies was recorded in tests with *Streptomyces*, and some *Nocardia* and mycelial fungi. In these cases amoebae encysted more rapidly than on the control; and especially with *Streptomyces*, they showed vacuolation and disintegration

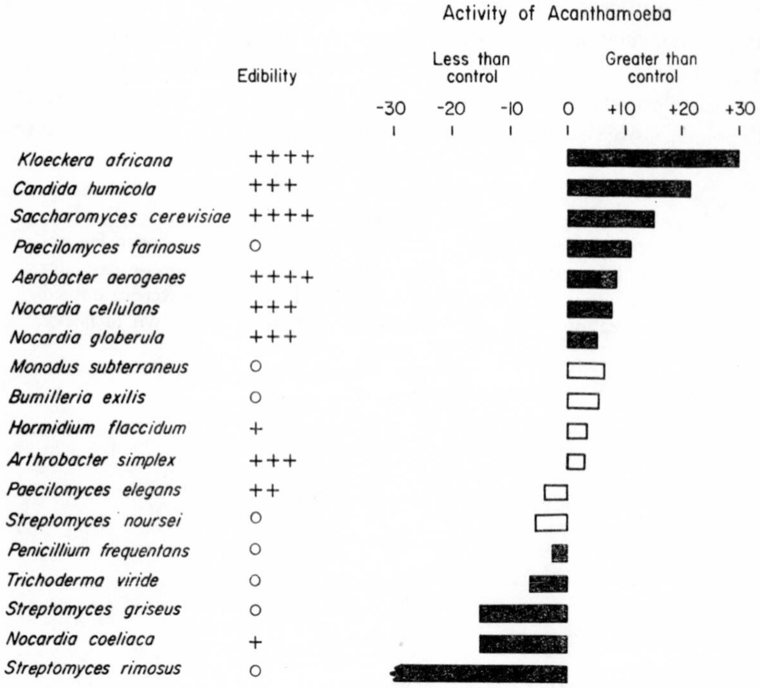

FIG. 1. Relationship between the edibility of microflora and the effect of their exudates on amoeba activity. Edibility is expressed as the reproduction of amoebae, in categories 0–4 (p. 146). Effect of exudate is expressed as the difference in number of active amoebae on test and control, □ = difference not significant, ■ = difference significant $P < 0.05$. Exudate data are from Table 2, day 2.)

An unexpected result was the stimulatory or probiotic effect produced by exudates from yeasts, *Aerobacter*, some *Nocardia* and *Paecilomyces farinosus*. In these, encystment was retarded contrary to the findings of Cutler & Crump (1935).

Comparison of the results from the exudate and feeding experiments (Fig. 1) show that microflora which support abundant reproduction of

Acanthamoeba tend to produce exudates which stimulate activity. Those which are inedible or support little reproduction tend to produce exudates which inhibit or do not effect amoeba activity. This was supported by further tests in which exudates from a number of microflora were rapidly screened. No counts of amoebae were made and only gross differences between test and control amoebae were recorded.

The only genera in which a wide range of edibility was recorded were *Nocardia* and *Mycobacterium*; sixteen species and strains were screened. The eight which supported moderate reproduction (*Nocardia globerula*, *N. lutea*, *N. calcarea*, *N. cellulans*, *Mycobacterium phlei*, three strains of *M. rhodochrous*) produced no obvious differences from the control, whereas the six which supported little or no reproduction (*N. caviae*, *N. coeliaca*, two strains of *N. asteroides*, *M. smegmatis*, *M. fortuitum*) caused amoebae to encyst more rapidly than on the control.

Similar tests on eleven strains of fungi showed that *Penicillium claviforme* and *Aspergillus flavipes* were strongly antagonistic; *P. thomii*, *P. chrysogenum*, *P. nigricans*, *A. fumigatus* and *Trichoderma viride* were weakly antagonistic. Exudates of *P. terlikowski*, *Gliocladium catenulatum*, *Scopulariopsis brevicaulis* and *Stachybotrys* sp. showed no effect.

The reaction of amoebae to *Streptomyces* exudates was very marked especially with *S. rimosus* which caused disintegration within a few hours. All of fourteen *Streptomyces* spp. screened showed antagonism in the following order of effect—*S. rimosus* > *S. lavendulae*, *S. griseus*, *S. aureofaciens*, *S. fradiae*, *S. polychromogenes*, *S. phaeochromogenes*, *S. antibioticus*, > *S. venezualae*, *S. erythreus*, *S. olivaceus*, > *S. noursei*, *S. vinaceus*, *S.* sp. Vacuolation, a pre-death characteristic, was observed in each case.

Other tests were carried out with *Streptomyces* to answer specific questions as follows:

1. Are *Streptomyces* antagonistic to a wide range of amoebae and is amoebae sensitivity an artifact resulting from prolonged culture? To answer these questions four amoebae which had been in axenic or monoxenic culture for 2–30 years were used—an unidentified hartmannellid amoeba, *Hartmannella astronyxis*, *Acanthamoeba* sp. and *Mayorella palestinnensis*. In addition seven unidentified small amoebae were isolated from a woodland soil and were tested within 18 days of isolation. All the amoebae showed vacuolation within 18 hours of contact with exudates of *S. rimosus*, showing that the reaction is widespread and unrelated to culture history.

2. Is the reaction related to starvation of the amoebae? Addition of *Aerobacter aerogenes* or 4% mycological peptone at the start of tests with *S. rimosus* did not influence the reaction of *Acanthamoeba*.

3. Are amoeba cysts resistant to toxic exudates? An unidentified hart-mannellid amoeba was used because it produced cysts more readily than axenic *Acanthamoeba*. Cysts were incubated in the presence and absence of *S. rimosus* exudate for 3 days. Samples of thirty cysts were removed from test agar and control agar and single cysts placed on streaks of *Aerobacter aerogenes* to test viability. A further thirty cysts from the test agar were washed for two periods of an hour in sterile distilled water before placing with *Aerobacter*. The results (Table 3A) showed that pre-formed cysts were viable but washing was necessary, probably to remove adhering exudates. Viability was lower than that of control cysts ($\chi^2 = 10 \cdot 7$, $P < 0 \cdot 01$).

TABLE 3

Viability after exposure to exudates of *Streptomyces rimosus* of A—pre-formed cysts, and B—cysts formed in the presence of *S. rimosus* exudates

	A		B	
	No. of cysts tested	Proportion excysted after 48 hours (%)	No. of cysts tested	Proportion excysted after 48 hours (%)
Unwashed cysts	30	0	20	0
Washed cysts	30	43	20	55
Control (unwashed)	30	100	20	85

4. Are cysts viable when formed in the presence of toxic exudates? Some amoebae manage to encyst in the presence of *S. rimosus* exudates. Tests similar to those described above showed that these cysts are viable (Table 3B), but again less than control cysts.

5. Can *Streptomyces* produce toxic exudates when growing on natural substrates? Previous tests had used organisms grown on nutrient-rich media, so radiation-sterilized oak leaf litter was used as a more natural substrate. *S. rimosus* and *S. aureofaciens* were grown on the leaves for 3 weeks at 25°C. *Acanthamoeba* placed on these test leaves for 5 hours were reduced in number and showed vacuolation, compared with amoebae on leaves without *Streptomyces*. Although growth of *Streptomyces* was sparse, enough exudates were produced on a natural substrate to kill amoebae.

6. What is the nature of the toxic exudates? This remains unanswered but the toxin(s) of *S. rimosus* were shown to be water-soluble, colourless, and resistant to heating for 15 minutes at 100°C. In tests with solutions of different osmotic pressures, surface active agents, chelating agents and

antibiotics, only streptomycin sulphate and terramycin hydrochloride caused the characteristic vacuolation and disintegration. These antibiotics are produced by *S. griseus* and *S. rimosus*. However as similar reactions were caused by *Penicillium claviforme*, and to a small extent *Cytophaga* as well as a wide range of *Streptomyces*, a wide range of toxins must be involved. Some bacterial toxins cause similar vacuolation in amoebae (Singh 1945; Knorr 1960).

FACTORS INFLUENCING EDIBILITY

There has been much discussion about the properties of different micro-organisms which influence their suitability as food for amoebae and these can be grouped as follows.

(a) PHYSICAL PROPERTIES PREVENTING INGESTION

The size and shape of micro-organisms are regarded as important factors influencing their value as food (Sewertzova 1928; Sandon 1932; Kunicki-Goldfinger *et al.* 1957; Groscop & Brent 1964). The present survey emphasizes the importance of large size and filamentous habit in restricting ingestion; many filamentous forms when fragmented were edible. The non-wettability of many fungal and *Streptomyces* spores also prevents ingestion, but even when wet, these spores are apparently indigestible.

(b) CHEMICAL PROPERTIES PREVENTING INGESTION

Many microflora are known to produce exudates antagonistic to other microflora. In the present study, antagonistic exudates recorded from fungi, *Nocardia*, *Mycobacterium* and *Streptomyces* were mainly inedible (p. 152). It is not known if the relationship is causal because the production of antagonistic exudates is associated with non-fragmentory filamentous growth which also restricts feeding.

Singh (1945) and others have recorded toxic effects of pigmented exudates from bacteria, especially *Pseudomonas*, *Chromobacterium* and *Flavobacterium* (Imszeniecki 1947; Gräf 1958; Chang 1960; Knorr 1960; Groscop & Brent 1964). Removal of pigmented exudates by washing of *Pseudomonas pyocyanea* (Chang 1960) and cultivation of *Serratia* free from pigment

(p. 147) allowed amoebae to feed on both these inedible bacteria. However in the present study *Serratia* was the only organism in which edibility was related to pigment production. Purple *Chromobacterium* and pink, yellow or cream colonies of *Arthrobacter*, *Nocardia* and *Mycobacterium* were eaten as readily as related non-pigmented strains. Pigmentation is therefore an unimportant factor influencing feeding on the majority of the microbial population.

(c) PHYSICAL/CHEMICAL PROPERTIES PREVENTING DIGESTION

Spores of fungi may be ingested but not digested (Sewertzova 1928; Heal 1963). When egested, these spores germinate, thus indicating a spore wall resistant to digestion. Bacterial spores may also resist digestion (Oehler 1916; Sewertzova 1928; p. 147) and be capable of development when egested (Knorr 1960). Similarly *Acanthamoeba* was observed to egest some algae apparently undamaged.

(d) CHEMICAL PROPERTIES PREVENTING DIGESTION OR ABSORPTION

Little information is available on intracellular toxins released when the micro-organism is digested, but Knorr (1960) recorded one unidentified substance of this type in *Pseudomonas fluorescens*.

(e) NUTRITIONAL DIFFERENCES OF DIGESTED ORGANISMS

It is widely recorded that some edible bacteria support less reproduction, or a lower reproductive rate, of amoebae than others. (Cutler & Crump 1927; Sewertzova 1928; Sandon 1932; Singh 1941, 1942; Nikoljuk 1956; Kunicki-Goldfinger *et al.* 1957; Chang 1960). This may result from differences in the nutritional value of protoplasm of different bacterial species or from differences in their ease of digestion by amoebae. However, no detailed work on this appears to have been done.

DISCUSSION

This study clearly indicates that bacteria, bacteria-like actinomycetes, and yeasts are the main food of soil amoebae but there are dangers in extrapolating from laboratory to field conditions. Some of the more obvious problems are that: culture history may alter feeding habits (Stout & Heal 1968), feeding behaviour may be different when mixed populations of micro-flora are available (Singh 1941), potentially edible microflora may live in unavailable micro-habitats (Stout & Heal 1968), the morphology of the microflora under cultivation differs from that in the field (Gray, Baxby, Hill & Goodfellow 1968), the microflora tested in the laboratory have been isolated by selective culture methods and may not be truly representa-tive of the soil population (Casida 1968), amoebae may feed on organic matter in the soil solution by pinocytosis. Despite these problems, the limited field data tend to confirm the general relationships shown in the laboratory. For example, a number of studies have shown a relationship between numbers of amoebae and bacteria in soil (Stout & Heal 1968), and Geltzer (1963) in examining the distribution of amoebae in the rhizo-sphere noted that they were concentrated near bacterial colonies, but their growth was suppressed by *Streptomyces* and fungi.

Laboratory studies allow close observation of interactions between amoebae and microflora. Antagonism between microflora and protozoa has been recorded (Zaher, Isenberg, Rosenfeld & Schatz 1953; Stout & Heal 1968), but the stimulation of amoebae by exudates of edible micro-organisms suggests that exudates can act as a 'token stimulus' indicating the presence of food and also retarding encystment. This stimulation is probably comparable with the observations of various workers that bacteria and bacterial extracts can stimulate excystment of amoebae (Singh 1941; Crump 1950). Further, Kunicki-Goldfinger *et al.* (1957) and Dudziak (1955) showed that the rate of excystment was related to the edibility of the bacteria, and amino-acids isolated from the bacterial extracts have been shown to induce excystment (Singh, Mathew & Anand 1958; Drozanski 1961). Drozanski & Drozanska (1961) also found that when amoebae were cultured on heat-killed bacteria, the addition of disinte-grated *Aerobacter aerogenes* or *Saccharomyces cerevisiae* increased reproduction and ingestion of food by the amoebae. This may result from the addition of a food source, an essential growth substance or a stimulant.

ACKNOWLEDGEMENTS

The authors wish to express their sincere thanks to the following people and organisations for their generous advice and their help in supplying organisms and materials: Drs J. B. Cragg, J. C. Frankland, R. E. Gordon, T. F. Hering, A. J. Holding, J. W. G. Lund, J. D. Stout, the present staff at Merlewood Research Station, Glaxo Laboratories Ltd, the Culture

Collection of Algae and Protozoa, the National Collection of Industrial Bacteria and the National Collection of Type Cultures.

SUMMARY

1.   To define the food of small soil amoebae, 109 isolates of bacteria, actinomycetes, fungi, yeasts and algae were offered individually to *Acanthamoeba* sp., *Hartmannella astronyxis* and *Mayorella palestinensis* in the laboratory.

2.   Bacteria, yeasts and some *Nocardia* and *Mycobacterium* strains were eaten and supported abundant reproduction of amoebae. *Streptomyces*, some *Nocardia* and *Mycobacterium* strains, mycelial fungi and algae were inedible or supported little growth of amoebae.

3.   Microflora which support abundant reproduction tend to produce exudates which stimulate amoeba activity. Those which are inedible or support little reproduction tend to produce exudates which inhibit or do not affect activity.

4.   *Streptomyces* spp. produce exudates which are antagonistic to a wide range of soil amoebae. At least 50% of amoeba cysts are viable after exposure to toxic exudates. Production of toxins occurs when *Streptomyces* are grown on natural substrates.

5.   Large cell size ($> 20 \mu$) and filamentous habit appear to be important features of inedible microflora. Production of antagonistic exudates by microflora is associated with a filamentous habit but the exudates may directly prevent feeding. Pigment production is probably unimportant in preventing feeding on the majority of microflora.

REFERENCES

ADAM K.M.G. (1964) A comparative study of hartmannellid amoebae. *J. Protozool.* **11**, 423–30.

CASIDA L.E. (1968) Methods for the isolation and estimation of activity of soil bacteria. *The Ecology of Soil Bacteria* (Ed. by T. R. G. Gray and D. Parkinson), pp. 97–122. Liverpool.

CHANG S.L. (1960) Growth of small free-living amoebae in various bacterial and in bacteria-free cultures. *Can. J. Microbiol.* **6**, 397–405.

CRUMP L.M. (1950) The influence of the bacterial environment on the excystment of amoebae from soil. *J. gen. Microbiol.* **4**, 16–21.

CUTLER D.W. & CRUMP L.M. (1927) The qualitative and quantitative effects of food on the growth of a soil amoeba (*Hartmanella hyalina*). *Br. J. exp. Biol.* **5**, 155–65.

CUTLER D.W. & CRUMP L.M. (1935) The effect of bacterial products on amoebic growth. *J. exp. Biol.* **12**, 52–8.

DROZANSKI W. (1961) The influence of bacteria on the excystment of soil amoebae. *Acta microbiol. pol.* **10**, 147–53.

DROZANSKI W. & DROZANSKA D. (1961) Dead bacteria as food for amoebae. *Acta microbiol. pol.* **10**, 379–88.

DUDZIAK B. (1955) Kielkowanie i wzorst ameb w zaleznosci od pokarmu bakteryjnego. *Acta microbiol. pol.* **4**, 115–25.

DUDZIAK B. (1962) Wplyw pratkow kwasoopornych na pelzaki glebowe. *Acta microbiol. pol.* **11**, 223–44.

GELTZER J.G. (1963) On the behaviour of soil amoebae in the rhizospheres of plants. *Pedobiologia*, **2**, 249–51.

GORDON R.E. & MIHM J.M. (1957) A comparative study of some strains received as Nocardiae. *J. Bact.* **73**, 15–27.

GORDON R.E. & MIHM J.M. (1962) The type species of the genus *Nocardia. J. gen. Microbiol.* **27**, 1–10.

GRÄF W. (1958) Über Gewinnung und Beschaffenheit protostatisch-protocider Stoffe aus Kulturen von *Pseudomonas flourescens. Arch. Hyg. Bakt.* **142**, 267–75.

GRAY T.R.G., BAXBY P., HILL I.R. & GOODFELLOW M. (1968) Direct observation of bacteria in soil. *The Ecology of Soil Bacteria* (Ed. by T. R. G. Gray and D. Parkinson), pp. 171–92. Liverpool.

GROSCOP J.A. & BRENT M.M. (1964) The effects of selected strains of pigmented micro-organisms on small free-living amoebae. *Can. J. Microbiol.* **10**, 579–84.

HEAL O.W. (1963) Soil fungi as food for amoebae. *Soil Organisms* (Ed. J. Doeksen and J. van der Drift), pp. 289–96. Amsterdam.

IMSZENIECKI A.A. (1947) Cited by Kunicki-Goldfinger *et al.* (1957).

JENSEN V. (1963a) Studies on the microflora of Danish beech forest soils. III. Properties and composition of the bacterial flora. *Zbl. Bakt. ParasitKde*, II, **116**, 593–611.

JENSEN V. (1963b) Studies on the microflora of Danish beech forest soils. IV. Yeasts and yeast-like fungi. *Zbl. Bakt. ParasitKde*, II, **117**, 41–65.

KNORR M. (1960) Verusche über die biologische Sperre gegen Bakterien und Viren bei vertikaler Bodeninfiltration. *Schweiz. Z. Hydrol.* **22**, 493–502.

KUNICKI-GOLDFINGER W., DROZANSKI W., BLASZCZAK D., MAZUR J. & SKIBINSKA J. (1957) Bakterie jako pokarm pelzakow. *Acta microbiol. pol.* **6**, 331–44.

KÜSTER E. (1968) Taxonomy of soil actinomycetes and related organisms. *The Ecology of Soil Bacteria* (Ed. by T. R. G. Gray and D. Parkinson), pp. 322–36, Liverpool.

LUND A. (1954) *Studies on the Ecology of Yeasts*, pp. 50–69. Copenhagen.

MENNA M.E. di (1960) Yeasts from soils under forest and under pasture. *N.Z. Jl agric. Res.* **3**, 623–32.

NEFF R.J. (1957) Purification, axenic cultivation and description of a soil amoeba, *Acanthamoeba* sp. *J. Protozool.* **4**, 176–82.

NERO L.C., TARVER M.G. & HEDRICK L.R. (1964) Growth of *Acanthamoeba castellani* with the yeast *Torulopsis famata. J. Bact.* **87**, 220–5.

NIKOLJUK V.F. (1956) Pochvonnye Prosteishie I Ikh Rol' V Kil'turnykh Pochvakh Uzbekistana. *Izdatel'stvo Akademii Nauk Uzbekskoi SSR.* 1–144. Tashkent.

OEHLER R. (1916) Amöbenzuckt suf reinem Boden. *Arch. Protistenk.* **37**, 175–90.

PFENNIG N. (1958) Beobachtungen des Wachstumsverhaltens von Streptomyceten auf Rossi-Cholodny-Aufwuchsplatten im Boden. *Arch. Mikrobiol.* **31**, 206–16.

RAY D.L. & HAYES R.E. (1954) *Hartmanella astronyxis*: a new species of free-living amoeba. Cytology and life cycle. *J. Morph.* **95**, 159–88.

REICH K. (1933) Studien über die Bodenprotozoen Palästinas. *Arch. Protistenk.* **79**, 76–98.

RHINES C. (1935) The relationship of soil protozoa to tubercle bacilli. *J. Bact.* **29**, 369–81.

ROSA K. (1957) Vyzkum midroedafonu ve smrkovem porostu na Pradedu. *Přírodov. Sb. ostrav. Kraje.* **18**, 17–75.

SANDON H. (1932) The food of protozoa. *Publ. Fac. Sci. Egypt Univ. Cairo*, **1**, 1–187.

SCHÖNBORN W. (1965) Untersuchungen über die Ernährung bodenbewohnender Testacean. *Pedobiologia*, **5**, 205–10.

SEWERTZOVA L.B. (1928) The food requirements of soil amoebae with reference to their inter-relation with soil bacteria and fungi. *Zbl. Bakt. ParasitKde* II, **73**, 162–79.

SINGH B.N. (1941) Selectivity in bacterial food by soil amoebae in pure mixed cultures and in sterilized soils. *Ann. appl. Biol.* **28**, 52–64.

SINGH B.N. (1942) Selection of bacterial food by soil flagellates and amoebae. *Ann. appl. Biol.* **29**, 18–22.

SINGH B.N. (1945) The selection of bacterial food by soil amoebae, and the toxic effects of bacterial pigments and other products on soil protozoa. *Br. J. exp. Path.* **26**, 316–25.

SINGH B.N., MATHEW S. & ANAND N. (1958) The role of *Aerobacter* sp., *Escherichia coli* and certain amino acids in the excystment of *Schizopyrenus russelli*. *J. gen. Microbiol.* **19**, 104–11.

SONEDA M. (1962) An additional paper on animal-dung inhabiting yeasts and on the symbiosis with an amoeba. *Trans. Mycol. Soc. Japan*, **3**, 36–42.

STOUT J.D. (1958) Biological studies of some tussock-grassland soils. IV. Bacteria. *N.Z. Jl agric. Res.* **1**, 943–57.

STOUT J.D. (1961) A bacterial survey of some New Zealand forest lands, grasslands, and peats. *N.Z. Jl agric. Res.* **4**, 1–30.

STOUT J.D. & HEAL O.W. (1968) Protozoa. *Soil. Biology* (Ed. by A. Burges and F. Raw), pp. 149–95. London.

TCHAN Y.T. & WHITEHOUSE J.A. (1953) Study of Soil Algae. II. The variation of the algal population in sandy soils. *Proc. Linn. Soc. N.S.W.* **78**, 160–70.

TRACEY M.V. (1955) Cellulase and chitinase in soil amoebae. *Nature, Lond.* **175**, 815

WILKINSON J.F. (1958) The extracellular polysaccharides of bacteria. *Bact. Rev.* **22**, 46.

ZAHER F., ISENBERG H.D., ROSENFELD M. & SCHATZ A. (1953) The distribution of soil Actinomycetes antagonistic to Protozoa. *J. Parasit.* **39**, 33–7.

APPENDIX 1

The species or genus of all the microflora used in feeding tests are listed below. After each are given the source, number of isolates tested and category of reproduction of *Acanthamoeba* which the isolate(s) supported. The latter are 0–4, see p. 146. Sources are abbreviated as follows: AJH—isolated from Roudsea Wood N.N.R. and identified by Dr A. J. Holding (University of Edinburgh); REG—supplied by Dr Ruth Gordon

(Rutgers University, U.S.A.); JDS—isolated and identified by Dr J. D. Stout (Soil Bureau, Lower Hutt, New Zealand); CCAP—Culture Collection of Algae and Protozoa, Cambridge; NCIB—National Collection of Industrial Bacteria, Aberdeen; NCTC—National Collection of Type Cultures, London. Mycelial fungi and yeasts are listed in Heal (1963).

Bacteria

Pseudomonas spp. AJH, 11, 4
Pseudomonas spp. AJH, 1, 3
Achromobacter spp. AJH, 4, 4
Aerobacter aerogenes NCIB 418, 1, 4
Klebsiella sp. AJH, 1, 3
Chromobacterium sp. AJH, 1, 4
Chromobacterium sp. AJH, 3, 3
Serratia sp. JDS, 1, 0–3 (p. 147)
Arthrobacter pascens NCIB 8910, 1, 4
A. globeformis NCIB 8602, 1, 4
A. spp. AJH, 4, 4
A. citreus NCIB 8915, 1, 3
A. ureafaciens NCIB 7811, 1, 3
A. spp. AJH, 1, 3
A. aurescens NCIB 8912, 1, 3
A. simplex NCIB 8929, 1, 3
A. terregens NCIB 8909, 1, 2
A. tumescens NCIB 8914, 1, 2
Bacillus circulans AJH, 1, 4
B. circulans AJH, 1, 3
B. subtilis NCIB 8533, 1, 3
B. cereus v. mycoides AJH, 1, 3
B. sphaericus AJH, 1, 3
B. cereus/megaterium AJH, 9, 3
B. cereus/megaterium AJH, 1, 2
Clostridium butyricum NCIB 7423, 1, 3
C. butyricum NCIB 9380, 1, 2
C. pasteurianum NCIB 9486, 1, 1
Cytophaga spp. AJH, 3, 1

Actinomycetales

Mycobacterium phlei NCTC 525, 1, 4
M. rhodochrous NCTC 8139 & 8154, 2, 4
M. rhodochrous NCTC 8036, 1, 3
M. smegmatis NCTC 8159, 1, 1
M. fortuitum NCTC 2006, 1, 0
Nocardia globerula NCIB 8862, 1, 3
N. calcarea NCIB 8863, 1, 3
N. cellulans NCIB 8868, 1, 3
N. lutea NCTC 576, 1, 2

N. pelletieri REG 408, 1, 2
N. spp. REG 711, 1, 2
N. coeliaca NCIB 8939, 1, 1
N. caviae NCTC 1934, 1, 1
N. asteroides NCTC 6761 & 8595, 2, 0
N. madurae REG 431, 1, 0
Streptomyces griseus NCIB 8237, REG 712, 2, 0
S. fradiae NCIB 8233, REG 3535 & 3535–7B, 3, 0
S. rimosus NCIB 8229, REG 3558, 2, 0
S. rimosus REG 3595 & 3699, 2, 3
S. lavendulae NCIB 6959, 1, 0
S. aureofaciens NCIB 8234, 1, 0
S. polychromogenes NCIB 8791, 1, 0
S. phaeochromogenes NCIB 8505, 1, 0
S. antibioticus NCIB 8504, 1, 0
S. venezualae NCIB 8231, 1, 0
S. erythreus NCIB 8594, 1, 0
S. olivaceus NCIB 8238, 1, 0
S. noursei NCIB 8593, 1, 0
S. vinaceus NCIB 8852, 1, 0
S. sp. NCIB 8592, 1, 0

Algae

Chlamydomonas moewusii CCAP 11/16, 1, 0
Coccomyxa subellipsoidea CCAP 216/13, 1, 1
Chlorella pyrenoidosa CCAP 211/8A, 1, 0
Chlorococcum minutum CCAP 213/7, 1, 0
Hormidium flaccidum CCAP 335/2A, 1, 0
Botrydiopsis arrhiza CCAP 806/2, 1, 1
Monodus subterraneus CCAP 848/1, 1, 1
Bumilleria exilis CCAP 808, 1, 0
Navicula pelliculosa CCAP 1050/3, 1, 0
Nitzschia palea CCAP 1052/3, 1, 0
Euglena viridis v. terricola CCAP 1224/21, 1, 0
Nostoc muscorum CCAP 1453/8, 1, 0
Anabaena cylindrica CCAP 1403/2, 1, 0
Gloeocapsa alpicola CCAP 1430/1, 1, 0

DISCUSSION

T. B. REYNOLDSON: Is anything known, qualitatively and quantitatively, about the relationship of soil amoebae to the various types of soil?

O. W. HEAL: Very little is known about variation in species composition in different soil types because of taxonomic problems. In general, the numbers of soil amoebae are directly related to soil fertility, as are numbers of bacteria. No detailed studies have been made which show variation in numbers between specific soil types.

GENERAL DISCUSSION

R. C. BIGALKE: As a member of a species rare in its own environment, meeting with a large population of a related, but much commoner species, I am tempted to assume that the numbers of ecologists are not limited by the supply of mental food available. Africa has many problems but has few ecologists to delve as deeply as you are able to do here.

A few general points which have struck me, may help to stimulate a general discussion.

Firstly, from what we have heard of the redshank and the triclads it is clear that some predator-prey relationships are susceptible to clear and precise description. It is far more difficult to deal with herbivores or decomposers in such a neat fashion. Perhaps therefore we may expect to arrive at the basic principles underlying the relationships between animal numbers and their food resources more quickly by further work on secondary and tertiary consumers.

Secondly, Dr Lack made the point that phytophagous insects and mammalian herbivores almost always have more food than they can use. It is, however, not simply how much but what kind of food that is important. How herbivores select what to eat from all that is available is an important question and one that comes within the ambit of students of animal behaviour. We shall not accept too readily a convenient label such as 'search image' and expect it to explain everything. Species with a wide distribution must, in different parts of their range, select many different foods from many different assemblages of food. The selection of food plants growing on particular kinds of soil is extremely important in some cases. Dr Klein made this point and our experience of several ungulates in the montane regions of Natal support it. Our antelopes very accurately select regions with short and apparently unattractive grasses. But these are growing on small, widely spaced areas of base-rich soils overlying basaltic rocks. Recent work near Aberdeen on the relation between numbers of red grouse and the base status of the moorland soils ties in exactly. I wish to emphasize then the importance, at least to many herbivorous mammals, of selecting areas of country in the first instance, and only then of selecting food plants.

Finally, a general point, also touched on by Dr Klein, seems to me to need more attention. It is that mechanisms for population regulation must be expected to differ quite widely in response to the degree of stability of the environment. In an area such as Great Britain, where climate and plant growth do not fluctuate very much, it will be advantageous for a species to regulate its numbers fairly accurately. By contrast, in an arid area, for example, where complete failure of the rains is not uncommon and severe droughts occur frequently but unpredictably, food supplies may disappear entirely and animals must be able to withstand and recover from very great fluctuations in numbers. They must be adapted to a great lack of constancy. Perhaps mechanisms as different as the production of twins and triplets, as against the production of single young more or less frequently, may have evolved in response to conditions of stability as against conditions of instability.

PART II · THE IMPORTANCE OF BEHAVIOUR MECHANISMS IN RELATING ANIMAL POPULATIONS TO THEIR FOOD RESOURCES

Populations may not interact directly with their food supply by a simple effect of food abundance on population processes such as breeding or mortality rates. The aim of Part II was to explore the problem whether behaviour or the social structure of the population may act as intermediary mechanisms, such that food supply affects behaviour which in turn influences the population.

The first paper (Watson & Moss) reviews the evidence on whether dominance and spacing behaviour limit numbers in wild vertebrates and are related to nutrition, and discusses the problem that more than one factor may simultaneously limit numbers. The second paper (Cowley) reviews laboratory studies with rats and other vertebrates on the effects of food intake on behaviour and describes one such study in detail. These aspects suggest relevant problems for ecologists studying behaviour populations in the wild.

The two remaining papers (Way & Cammell, Patterson) are examples of studies which examine behaviour to see how it is related to food supply and how it affects populations. The first of these reviews research on several species of aphids, which have self-regulatory mechanisms for limiting population growth, in relation to nutrition. The second examines fighting in flocks of rooks, to see whether this is related to food or population regulation.

I. J. Patterson and A. Watson

DOMINANCE, SPACING BEHAVIOUR AND AGGRESSION IN RELATION TO POPULATION LIMITATION IN VERTEBRATES

By Adam Watson and Robert Moss

Nature Conservancy, Blackhall, Banchory, Scotland

INTRODUCTION

This review assesses the importance of behavioural mechanisms involving aggression, in limiting the size of breeding populations of vertebrates. We have had to select only some of the examples available, because space is short.

The paper is confined to factors which limit the breeding population alone. The non-breeding part of a population never increases indefinitely independently of the number of breeders, as could theoretically happen. Thus, in practice, factors which limit the breeding population, also, in the medium term, limit the total population.

We shall consider the question: Does socially-induced mortality, or socially-induced depression of recruitment, set upper limits to observed breeding populations of vertebrates? If so, do variations in nutrition affect these physiological and behavioural mechanisms? It is desirable to begin with a definition of what we mean by the terms 'limits' and 'limiting', and a summary of the conditions which are both necessary and sufficient before these questions can be answered in the affirmative. The rest of this Introduction is therefore theoretical, and is printed in small type so that readers who wish to read only the examples in the review proper may move on to the next main section (p. 173).

LIMITING FACTORS

Historically, the concept of limiting factors arose from consideration of simple systems, e.g. physical and chemical reactions, which require precisely measurable inputs of various materials for optimum functioning. If any one or more of these materials is in short supply then the system functions more slowly. In practice, one material is always in shorter supply than the rest, and it is this 'first limiting' material which limits the rate at which the system functions. Only one material can be limiting at any one time. If adequate quantities of the first limiting material are supplied then the material in next shortest supply—the second limiting material— becomes the effective limit upon the rate of functioning of the system, and so on for the third limiting material, etc. However, this no longer holds true for more complex biological systems such as entire animals or populations of animals, because such systems are very adaptable and can adjust their requirements in relation to supplies. The difference here is that two or more factors can be limiting at the same time. For example, a man can suffer from rickets and kwashiorkor at the same time. Both protein and calcium are limiting in this case. Supply of extra protein may cure the kwashiorkor and increase his functional performance, without curing the rickets or the restriction of performance due to the rickets. Another example is that lack of food and lack of shelter may both reduce an animal's performance at the same time.

The classic experimental design used to identify the first limiting factor of a system involves supplying the material which is thought to be limiting: if an increase in performance is observed it is concluded that the limiting factor has been identified. However, in animal populations, such an observation means only that one of the limiting factors, and not necessarily the most important one, has been identified. Thus, increasing the animals' food supply may increase their numbers, but it cannot be concluded that food was the only factor limiting numbers to their previous level. One can conclude merely that it was one of the factors. A good example of several-factor limitation of numbers by food, space and genetic constitution is given by Ayala (1968) in a recent laboratory study of *Drosophila* spp. This follows the earlier conclusions of Robertson & Sang (1944) that crowding and food shortage both restricted the fecundity of *Drosophila melanogaster*, but not crowding alone. So, when we ask 'does behaviour limit a breeding population?', we do not mean 'is behaviour the only factor which keeps the population at the observed level?'; rather we mean 'is behaviour one of the factors which keeps . . . ?'

We shall assume that upper limits to animal populations occur in the long term because breeding numbers fluctuate within a narrower range than is theoretically possible.

However by 'limit' in the short term we mean the actual number of animals which breeds each year within each area, no matter what the density may be. This will apply in cases where a breeding population is stable between years; and equally where it fluctuates greatly, even though the limit in some years may be only a tenth or less what it is in other years. We use the term 'limit' in this sense because (1) each year, breeding numbers reach a certain level and (2) this level could obviously be higher in many cases because some animals fail to breed or die. Thus, the population is limited each year and we call the factors which cause some animals to die or fail to breed 'limiting'. When animals die, the limiting factors are frequently not the same as the proximate factors which immediately cause death. For example, an animal may die of starvation because it has been ousted from suitable habitat by territorial behaviour. In this case starvation is the proximate factor but territorial behaviour is the limiting factor.

Limits for any species also differ greatly between areas. In this review we shall discuss the problem of differences within areas at the same time as differences between areas, though aware of the possibility (Chitty 1967a) that the two may in some respects be fundamentally different. As far as possible, we have tried to keep them separate in the various examples discussed.

In summary, we shall call a factor limiting when it helps to set a limit to a breeding population. Obviously, factors such as food, behaviour and space do not all affect the animals at the same level of biological organization, nor do they necessarily all act at the same time. For example, it is conceivable that a change in food supply may change behaviour, which in turn sets a limit to a population. In this case, behaviour has been identified as the proximate limiting factor. Nevertheless, if behaviour had not been studied, it would be a reasonable though incomplete conclusion that food set the limit. Alternatively, food and behaviour may act as limiting factors additively but independently of each other. In the present state of vertebrate population ecology, it is usually still a matter of semantics to decide whether food, or behaviour, or both, should be classified as limiting factors. The difficulty is that most studies at present do not provide sufficient data to decide for certain one way or the other. In the absence of these appropriate data, the frequent attempts in the literature to conclude that it has to be either one single factor or another are bound to be unreliable. In the above example in this paragraph we are therefore forced to describe both of these factors as limiting. This is an operational decision for current convenience only, and when better data are available in future it should become possible to frame more precise definitions.

In this paper we refer to numbers of animals when discussing populations. However, when comparing populations with resources such as food or space, biomass and individual growth rate may be more important than the actual number of animals in some cases. An inverse relation between biomass and numbers when resources remain constant is well known in fish. Variations in mean body weight of small rodents also occur at different population densities or different phases of a fluctuation, but this is less likely to be important with birds which usually reach a more definitive adult weight. While recognizing this, we have generally used numbers here because nearly all the work on behaviour in relation to population dynamics in the literature is in terms of numbers.

CONDITIONS WHICH WILL SHOW THAT SOCIALLY-INDUCED MORTALITY (OR DECREASE IN RECRUITMENT) LIMITS BREEDING POPULATIONS

The conditions which are both necessary and sufficient to show that social behaviour is an important limiting factor in any one breeding cycle within one area are summarized in Table 1 for ease of reference. Condition (a) is that a substantial part of the population does not breed, even though (b) such non-breeders are physiologically capable of breeding if the more dominant or territorial (i.e. breeding) animals are removed. These 'surplus' animals may fail to breed either because they die (e.g. Meslow & Keith 1968); or because they attempt to breed but they and/or their young all die; or because they are inhibited from breeding even though they survive, and may breed in later years. It is also necessary (c) to show that the breeding animals are not completely using up some resource, such as food, space or nest sites. In this case the resource might be the major limiting factor and behaviour relatively unimportant. If the breeding animals are only preventing other animals from using the resource and not completely using it up themselves, then condition (c) is satisfied.

TABLE I

Conditions which will show that behaviour, via socially-induced mortality (or socially-induced depression of recruitment), limits a breeding population

---

A   (a) A substantial part of the population does not breed, either because animals die; or because they attempt to breed but they and/or their young all die; or because they are inhibited from breeding even though they survive, and may breed in later years.
     (b) Such non-breeders are physiologically capable of breeding if the more dominant or territorial (i.e. breeding) animals are removed.
     (c) The breeding animals are not completely using up some resource, such as food, space, or nest sites. If they are, the resource itself is limiting.
     (d) The mortality or depressed recruitment due to the limiting factor(s) changes, (d_i) in an opposite sense to, and (d_{ii}) at the same rate as, other causes of mortality or depressed recruitment.

B   In addition, condition (e), see p. 172, is that numbers change following changes in food. This will show that food and behaviour are both limiting, if (a), (b), (c) and (d) are also met.

If (a), (b) and (c) are satisfied in any one breeding cycle, then behaviour is one of the factors limiting breeding numbers in that cycle. We must however see whether factors which set the limit in any one breeding cycle also hold in setting limits over a long period of time. Even if a given factor is limiting (i.e. fulfils (a), (b) and (c) in each breeding cycle), it is still theoretically possible for the population to continue increasing indefinitely. For example, if behaviour prevents half the population from breeding, this would be an effective limit in any one year or breeding cycle, but would not be over a longer period if recruitment of young to the mature breeding population were continually in excess of mortality. However, if (a), (b), and (c) are satisfied in each breeding cycle in a long-term study in which the population does not continue to increase indefinitely, then it is reasonable to say that behaviour is one of the factors setting a long-term limit to population size. In addition, there must still be a long-term effect which we shall call (d). It is invariably true that mortality—or depression of recruitment rate—is caused by a variety of different factors. If one (or more) of these factors is limiting a population, then (d), the mortality or depressed recruitment due to this factor(s), must change at the same rate and in the opposite sense to other causes of mortality or depressed recruitment. Thus, in a stable population, if mortality (or depression of recruitment) due to bad weather, for example, decreases by 100 individuals a year, the mortality (etc.) due to the postulated limiting factor(s) must increase by 100 individuals a year. In a fluctuating population the situation is more complex but this generalization must still hold over a period of years covering one complete fluctuation. If it does not hold, then not all the limiting factors have been identified.

The state of the art of population ecology at the moment is that condition (a) is well established in some cases, (b) and (c) have been shown in a few instances and suggested in many cases, and (d) is well documented in a few laboratory studies but appears to be almost beyond our present capacities for direct observation and experiment in the field. There is some evidence in the field and laboratory that mortality (or depression of recruitment), due to a postulated limiting factor, changes in the opposite sense to other factors and we shall call this d_i, but we have come across no conclusive evidence in wild vertebrates that it changes *at the same rate* (d_{ii}).

DENSITY DEPENDENCE

One must not confuse (d) with density dependence. Density-dependent mortality or recruitment are characteristics which workers, particularly on invertebrates (review Richards & Southwood 1968) have used to identify limiting or 'regulating' (Solomon 1964) factors. This appears logical on the assumption that density-dependent factors are regulating a population about a particular 'equilibrium' or mean level, and that they tend to bring this population, which may have suffered so-called 'density-independent disturbances', back towards this level. The equilibrium level varies between areas, as one might expect, but also varies within areas over different sets of years. Clearly, density dependence and equilibrium levels are statistical concepts, not necessarily biological realities. Although the methods involved may be useful for predicting future trends in populations if the only data available are counts of animals, they may not necessarily detect all the limiting factors unless the studies are broadened to include, besides population totals at different times, other methods of approach, such as the work on behaviour of aphids described in this symposium, or the increasing emphasis of entomological research on genetics, the quality of animals, and behaviour, summarized by Wilson (1968). In any case, we are concerned in this review only with seeing whether behaviour can be identified as a limiting factor. We cannot make use of the concept of density dependence to help us do this, because so little work has been done on direct measurement of density-dependent changes in socially-induced mortality and changes in recruitment in vertebrates. Indeed, Lack (1966) wrote that it is still virtually the case that no instance of density-dependent mortality has been demonstrated for a natural population of any animal. This is probably still the case in vertebrates.

BEHAVIOUR AND NUTRITION

We now move on to the next question: Is the behaviour which causes changes in mortality or recruitment affected by changes in nutrition? For example, does a well-fed animal change its behaviour so that its territory size or home range is reduced, or does a well-fed colony of breeding animals change its behaviour to allow an increase in numbers; and vice versa? Evidence that animals are dispersed, within and amongst areas, according to food supplies (literature reviewed by Wynne-Edwards 1962) is insufficient in the absence of evidence on (a), (b), (c)—and preferably (d) also if one looks at the problem in the long term. Such an association between animal densities and food supplies may simply mean that some factor other than behaviour has limited breeding numbers in the first place and that the breeding animals that are left then disperse themselves according to food supplies (Lack 1966). Equally, evidence that animals change in numbers within any one area following a change in food supplies may simply indicate that the population is limited by food, unless conditions (a), (b) and (c) in Table 1 are also satisfied. Conversely, such evidence does *not* show that numbers are limited solely by food, unless it can also be shown *inter alia* that (a), (b) and (c) are *not* true. For example, a change in food may cause a change in territory size via changes in physiology and behaviour, so that food is only one of the limiting factors.

Let us assume that evidence on (a), (b), (c) and perhaps (d) is available, that behaviour is an important limiting factor and that numbers are not limited solely by food supplies. The question still remains: do changes in food supplies cause changes in behaviour; i.e. are food and behaviour both limiting? Theoretically this is fairly simple to test: the prediction would be that numbers should change with changes in food supplies (e), in natural circumstances and experimentally. Section 1 of this conference shows the practical difficulties in measuring food, e.g. even in defining what is the important food material. Apparent evidence against (e), in studies such as that of Krebs & DeLong (1955), where extra artificial food was supplied to a wild *Microtus* population which showed no response, may be criticized (irrespective of whether (a), (b), (c) and (d) were satisfied), because the food was an unnatural one which may not have been adequate nutritionally.

DOMINANCE, SPACING BEHAVIOUR, AGGRESSION AND STATUS

The types of behaviour we are concerned with involve the establishment of dominance by some animals over others by means of aggression; they also affect the patterns in which the animals space themselves out. The result of these processes is that the resources of food, space, nest sites, etc., are divided up unequally in relation to the social position or status of the individual. There is no convenient term which includes all these aspects in its meaning and we have therefore used the four terms: dominance, spacing, aggression and status wherever each seems most appropriate. Categories of dominance behaviour are: (a) a hierarchy with a peck order or peck rights as in a group of poultry (Guhl 1953; Wood-Gush 1965), or peck dominance as in captive canaries (Shoemaker 1939); (b) a system of home ranges (Burt 1943; Jewell 1966; Leyhausen 1965), and (c) a territorial system where each individual is exclusively dominant over all others of the same sex at a particular site or area, but not elsewhere (Tinbergen 1957). There can be group territories or home ranges as well as individual ones (Davis 1942; Mykytowycz 1960; Carrick 1963) in which case a hierarchical system may be involved within a group range or territory. Overt aggression or hostility may be involved in all three, with or without 'mutual avoidance' (Tinbergen 1957; Terman 1962), or avoidance by subordinates towards superiors (Calhoun 1963; Bronson 1964). Many species show two or even three of these types of dominance behaviour in different seasons, or at the same time in different social circumstances (Davis 1958; Keenleyside & Yamamoto 1962; McBride 1966), or at different times in the same day (Leyhausen 1965).

SUBDIVISION OF QUESTIONS ON POPULATION LIMITATION

For the sake of clarity, we have organized this review into three main questions. We must emphasize that we do not think each species can be neatly pigeon-holed into one of the categories implied by the questions. All the questions might possibly be answered in the affirmative for any one species under different sets of conditions, for example whether the species is at the edge of its geographical range or not, whether the habitat is more or less homogeneous, and so on.

The first question is to ask if the breeding population is limited by dominance and spacing behaviour (A). The second is whether it is limited by some different factor (B) such as disease, starvation due to food shortage, habitat change, or predation. As a corollary, a third question (C) is to ask if spacing behaviour within populations, which may be limited in either of the above ways, is related to the distribution of food; in other words, do animals disperse on their territories or home ranges in relation to food supplies? Question C arises irrespective of whether A and B have been answered, and irrespective of whether these answers are yes or no. Lack (1966) clearly stated this important difference between A and C, which in some cases the original authors had not mentioned even implicitly.

For shortage of space we have had to exclude the large amount of literature on the question if and how, once animals have dispersed or spaced themselves out to breed, nutrition subsequently affects their breeding success (total number of young weaned or fledged per litter or brood). The important thing for the present review is the recruitment of young to the breeding population, not the number of young reared.

DO DOMINANCE AND SPACING BEHAVIOUR LIMIT BREEDING POPULATIONS?

Lack (1954, 1966) argued that territorial behaviour does not limit breeding populations, and Wynne-Edwards (1962), following Tinbergen (1957), Errington (1956), and earlier writers back to Moffat (1903), that it does. Hinde (1956), in a review of the functions of territorial behaviour in birds, could find no unequivocal field evidence that it does limit numbers, though he thought the evidence strongly suggestive of this. However, it is certainly not a well-established fact on the basis of which one can safely speculate, and is really no more than a likely idea which is still largely untested in the field. The conditions which are necessary to show that dominance behaviour—or more exactly (Tinbergen 1968) the presence of dominant individuals—limits a population are listed in Table 1, and examples where these conditions are fulfilled are now reviewed.

DOES A SUBSTANTIAL PROPORTION OF THE POPULATION NOT BREED ?

This is condition (a) in Table 1. Many species of birds have immature plumages for one or more years. Excluding such cases, there are descriptions of 'non-breeding adults' in many colonial sea birds during the breeding season and in various other birds (reviews by Wynne-Edwards 1962; Carrick & Murray 1964). Richdale (1957) found that a small number of male yellow-eyed penguins (*Megadyptes antipodes*) did not breed even though they had bred before, but he did not find this in females. The evidence that 'unemployed birds' in these colonies really are non-breeding adults which have bred previously is very poor and at best circumstantial in most studies, and the alternative proposition—that they are failed breeders or sub-adults which have not yet bred even once—cannot be effectively refuted. Nelson (1966), who also reviewed the evidence for and against these ideas in the Sulidae, showed in his study of gannets (*Sula bassana*) on the Bass Rock that adults, which had once bred, thereafter attempted to do so each year. He suggested that the many 'unemployed adults' were probably younger birds which had not yet bred.

Lockie (1966) found a substantial population of non-breeding transient male but not female weasels (*Mustela nivalis*), defining transients on the basis of trapping locations and gonad condition of individually marked animals. Errington (1943, 1963) also described a surplus of non-breeding musk rats (*Ondatra zibethica*) which were more vulnerable to predation than resident animals; this evidence was suggestive, though largely anecdotal. However it was not conclusive, because it was not based on quantitative analytical data and because the surplus 'transients' were not individually marked or independently recognizable for certain by any other characteristics. He also described (Errington 1945, 1957, 1963) 'thresholds of security' in muskrats and bobwhite quail (*Colinus virginianus*). Above these thresholds, the 'surplus' animals were said to be very vulnerable to predators, and the level of the threshold or 'carrying capacity' was said to vary greatly in different years or areas according to conditions in the environment and changes in the animals' behaviour. These threshold levels or carrying capacities, however likely the idea may appear, were not shown to be based on any objective evidence, and have already been criticized in more detail by Andrewartha & Birch (1954) and Lack (1954, 1966). On a largely intuitive basis, Errington may well have been aware of a genuine phenomenon, which may yet be shown conclusively by future workers doing critical experimental research, but it is not possible to conclude that he was right from the evidence presented.

Cases where non-breeding has been shown in adult females as well as males, and where there are data on the size of the non-breeding surplus in both sexes, are the shelduck (*Tadorna tadorna*) studied by Young (see Dunnet 1967)—but see an alternative view by Hori (1964)—the Australian magpie, *Gymnorhina tibicen* (Carrick 1963), and the red grouse, *Lagopus l. scoticus* (Jenkins, Watson & Miller 1963, 1967; Watson & Jenkins 1968; Watson & Moss 1970). Some evidence about a surplus in red grouse was based (Jenkins, Watson & Miller 1963, 1967) on our interpretations of differences in the total number of grouse counted at different times, and we do not now accept this suggestive but circumstantial type of evidence as good enough. However, the best evidence about a surplus in Jenkins *et al.* (1963) was based on direct observations of marked birds of known social status; this still stands and has been amply confirmed in later years (Watson & Jenkins 1968; Watson & Moss 1970).

The evidence is insufficient to generalize from, but shows that condition (a) in Table 1 is satisfied in some cases. More examples are given in the next section.

ARE SURPLUS ANIMALS CAPABLE OF BREEDING?

Condition (b) in Table I is that the surplus non-breeding animals are capable of breeding if more dominant individuals are removed.

Evidence on this rests partly on the widespread observation that territorial animals are replaced when killed naturally or experimentally. There are anecdotal records of this, scattered incidentally here and there among numerous studies about other aspects. A few documented examples are the snowy owl, *Nyctea scandiaca* (Schaanning 1907); raven, *Corvus corax* and peregrine, *Falco peregrinus* (Ratcliffe 1962); golden eagle, *Aquila chrysaetos* (Dixon 1937; Brown & Watson 1964); and the skylark, *Alauda arvensis* (Delius 1965). Removal experiments showing the same thing on a large scale with hundreds of birds which had settled in the breeding season and were therefore presumed to be territorial, were done by the important pioneering work of Stewart & Aldrich (1951) and Hensley & Cope (1951). Experiments where previously-studied, individual territorial birds were removed, were done by Orians (1961) with the red-winged blackbird (*Agelaius phoenicius*), Watson (1965) with the ptarmigan (*Lagopus mutus*), Holmes (1966, 1970 in this symposium) with the dunlin (*Calidris alpina*), and Lockie (1966) with the weasel. None of these examples fulfils condition (b) completely, however, because in no case was it proved that the colonizing individuals might not have bred elsewhere, although particularly in Holmes' work with the dunlin this possibility is very unlikely for reasons given in the Discussion after his paper in this symposium (p. 318). The study of blackbirds (*Turdus merula*) by Snow (1958) illustrates the above drawback about point (b): he found that when a gardener inadvertently left poison in one territory a succession of cock blackbirds occupied it and died in turn. Most incomers had previously had other territories nearby, and so would almost certainly have bred elsewhere if the first territory owner had not been removed.

At South Georgia, the average breeding age of male elephant seals (*Mirounga leonina*) was reduced following heavy culling of older animals, mostly males, by sealers (Laws 1956). In this colony the average breeding age of females is also low compared with unexploited populations (Carrick, Csordas & Ingham 1962). However, there were no measurements of the seals' food and behaviour was not studied, so that the cause of inhibition of breeding in the unexploited populations is not known. Wynne-Edwards (1962) and Lack (1966) give other examples of deferred maturity in birds. However, there is little quantitative evidence, food was not measured and there is no evidence that dominance behaviour is involved. One cannot therefore refute the possibility that food shortage, and not behaviour, directly inhibits breeding at high numbers.

Tompa's (1964) study of song sparrows (*Melospiza melodia*) near Vancouver provided better evidence; he found that first-winter males, which were previously non-territorial, replaced territorial males that died in autumn and winter. In the roe deer (*Capreolus capreolus*), dominant bucks occupy territories, and they evict younger bucks, which do not breed, survive worse, and emigrate more (Cumming 1966; Bramley in press); however, when dominant bucks are removed, younger bucks take territories. Marshall (1954) found in a few cases that young satin bower birds (*Ptilonorhynchus violaceus*), which do not usually mate or show vigorous territory defence, will do so if dominant older individuals are removed experimentally. Bendell & Elliott (1967) showed the same thing with a much larger sample of male blue grouse (*Dendragapus obscurus*). A weakness of many of the examples in the last three paragraphs is that they concern young males and mostly concern species which are polygamous. It might well be argued that adult females are the more important members of the breeding population.

Two cases where quantitative evidence from descriptive studies with large numbers of marked individuals has shown that birds of both sexes have social categories or classes whose spacing behaviour and contributions to the breeding stock are different, are the Australian magpie (Carrick 1963) and the red grouse (Jenkins *et al.* 1963; Watson & Moss 1970). The lower classes tend to occupy poorer habitat, survive worse, and do not breed. Several examples showing the same thing with rodents are mentioned in the next section. In the magpie and red grouse, occasional 'natural' experiments, where a high-class individual died and was replaced by a lower-class bird, showed that (b) was fulfilled. The research with red grouse has also shown by thirteen removal experiments on areas up to 40 ha (Watson & Jenkins 1968) that individually-marked lower-class grouse of both sexes, which usually all die before the breeding season, will readily take territories, survive well, and breed, when more dominant territorial birds are removed. This result fulfils condition (b). Such removal experiments are extremely time-consuming to do in the field, but there is a need for them to be attempted in a variety of species, to determine whether these findings are of general application or not.

POPULATION LIMITATION BY BEHAVIOUR, WHERE RESOURCES ARE NOT BEING COMPLETELY USED UP BY THE MORE DOMINANT, BREEDING ANIMALS

This is condition (c) in Table 1. We now review the importance of dominance and spacing behaviour, in cases where the breeding animals are not eating all the food, which is the most obvious resource. Because of the lack of examples, we shall not confine ourselves entirely to cases where condition (b) is shown to be satisfied in addition to (a) in the same study, nor to adult animals alone.

In confinement

There is much evidence in fresh-water animals that products excreted by individuals at high density inhibit the growth and survival of younger animals, even though food is in excess (see review by Rose & Rose 1965;

Licht 1967). This has been well demonstrated in tadpoles of several species of frog. A similar process in crowded guppies (*Lebistes reticulatus*) and other fish prevents spawning of individuals, which will spawn if the old 'crowded' water is replaced by a new supply. Products which have similar effect on populations of insects at high densities are also well known (references in Wynne-Edwards 1962).

Many experimental studies with confined animals show that dominance behaviour and spacing limit population density and affect population processes such as breeding success and survival, even though food, nests etc. are in excess. Usually but not always, population densities were artificially made higher than in the wild. Nevertheless the importance of these studies is that they show things which should certainly be looked for and not ignored in the field.

One of the best experimental demonstrations of population limitation by spacing behaviour and aggression is Van den Assem's (1967) study with sticklebacks at fairly natural densities in tanks. There was a fairly stable number of territories—and thus a stable population ceiling—for any one tank, when different batches of fish were introduced on different occasions. Below a certain mean minimum territory size, additional males could not get territories. Kalleberg (1958) clearly showed the importance of territorial behaviour with young trout (*Salmo trutta*) and salmon (*S. salar*) confined in a stream tank. At very high densities, there was a great amount of strife; many fish were unable to defend a territory, and were continually chased by the territory-holders. Keenleyside & Yamamoto (1962) also found many non-territorial parr of salmon at high densities, even though food was in excess. The frequency of agonistic encounters increased with the number of fish, but above a certain number of parr it decreased greatly; the non-territorial individuals had changed their spacing behaviour to form a closely packed school, and seldom went near the territory owners.

Toads (*Bufo bufo*) survive years in captivity if kept singly or not allowed to see each other, but suffer a high mortality if they are crowded, when they will starve to death in the presence of abundant food (Elkan 1960).

One example with birds is the study by Collias & Jahn (1959) of Canada geese (*Branta canadensis*) which were confined in semi-natural conditions, with plenty of food. Breeding success was greatly depressed by failure to pair, lay, and incubate, due to strife, domination and territorial behaviour in crowded conditions. Another is Shoemaker's (1939) study, where less dominant territorial canaries, in this case extremely close together in confinement, reared no young because of interference from more dominant territory owners.

An important study was done by Myers (1966) and Myers, Hale, Mykytowycz & Hughes (1970) with rabbits (*Oryctolagus cuniculus*) in unnaturally dense confined populations in Australia (see p. 183 for his studies of confined rabbits in more natural conditions). Again, there were profound effects of spacing behaviour, dominance and aggression in restricting population growth, although food was in excess. Increase in density greatly reduced fecundity and also caused ill health via upsets in the endocrine system and increases in strife. The effects of density on health were more marked with decreases in the amount of living space, which varies with the size of the enclosure, and living space became more important than density in very highly crowded conditions. Rabbits that were more aggressive and more dominant had larger territories. Rabbits whose mothers had been in crowded populations, where strife was common, were transferred to and reared in a different population, but they were then more aggressive there than the young of mothers from the less crowded populations. This also provides some evidence of a maternal effect on the aggression of the offspring, showing that changes in behaviour can be induced by effects more subtle than direct response to a shortage of some resource (see Cowley's review (1970) in this symposium).

Spacing behaviour, aggression and dominance have profound effects on populations of confined rodents. In the brown rat (*Rattus norvegicus*), aggression and spacing behaviour have been found to limit population growth in the presence of abundant food. Calhoun (1949, 1950), especially in his comprehensive and intensive study of brown rats confined in outdoor enclosures (1963), documented some of these aspects and made a detailed examination of factors affecting social rank or status. The rats were given an excess of food and boxes for nests or storing food. Spacing behaviour and aggression clearly had deleterious effects on subordinate animals, which emigrated more and survived worse, particularly at high densities. Food and behaviour were both occasionally involved, because food was made less available to some subordinate rats by the behaviour of other rats (p. 188). However, only a few of the effects of behaviour in Calhoun's 1963 study involved rats near food sources, and even there the conflict was territorial or social, and was due to a struggle over food only exceptionally when food ran out temporarily. A great deal of strife was territorial defence of burrows, passages and harborage boxes. In most cases, inhibition of body growth and low adult weight were associated with stress during early life, due particularly to strife between juveniles, attacks from adults, frequency of contacts with associates, type of associates, and social and geographical origin of associates. Furthermore, qualitative observations showed that most rats were able to eat and store plenty of food, yet marked differences in reproductive rate occurred among groups of different social status. Certain groups of high social class reproduced well even at high density, whereas others at the same density failed to give birth or to rear any young due to the extra stress arising from unstable social organizations. Calhoun concluded that spacing behaviour and aggression limited population growth. He suggested that the mechanism for this was a direct effect of strife on reproduction and space requirements. The evidence for this mechanism was criticized by Clarke (1955) on the grounds that food may have been relatively less available for the rats which reproduced poorly. Calhoun himself recognized the problem, which is partly a matter of semantics (p. 188).

Barnett (1958, 1964) also found that a class system occurred in rats in more closely confined conditions. Lower-class animals soon died following attacks by more dominant individuals when they were introduced into the home cage of the dominant. The spacing behaviour of rats is complex in that they tend to attack strangers, low-status adults, and young, but may be quite amicable and show hardly any strife among adults within an established local group or colony.

Clarke (1955) found three fairly recognizable classes among short-tailed voles (*Microtus agrestis*) in captivity, in which the frequency of wounding and persecution were related to size of home range, weight, and body condition. At high density, the breeding season became short, fertility declined, and infant survival worsened, even with food in excess. These effects were attributed to dominance behaviour and aggression, and the classes recognized were thought to be social classes, but these were assumptions because behaviour was not quantified.

Where food was unlimited but space and cover were restricted, Brown (1953) and Southwick (1955a, 1955b) found a high litter mortality of house mice (*Mus musculus*), due to lack of parental care associated with much fighting. As fighting increased at high densities, population growth slowed down or stopped completely in house mice (Southwick 1955b) and white mice (Petrusewicz 1963). In other cases in confined conditions with food in excess, population growth of house mice at high densities stopped due to lack of breeding, without much fighting or litter mortality (Crowcroft & Rowe 1957). Crowcroft & Rowe attributed some of these differences to differences in the layout of pens etc. in the experiments and differences in interpretation of the data on mortality. Where a ceiling population of adults which had had no litters for 6 months was allowed to disperse into a larger pen, reproduction soon began and numbers quickly doubled (Crowcroft & Rowe 1958). The frequency of strife or aggression was not directly measured in the above studies, except by Southwick.

In the semi-natural conditions of unthreshed corn ricks where food is in excess, Rowe, Taylor & Chudley (1964) found evidence of pysiological—and by assumption probably behavioural—factors tending to restrict population growth of house mice, particularly a failure of females to become fecund in oat ricks and more litter mortality in wheat ricks.

Rowe et al. (1964) and DeLong (1967) give useful reviews comparing studies of house mice in the laboratory and in the field. The main difference is that dispersal or emigration is not possible in the laboratory, although the studies of Strecker (1954) and Crowcroft & Rowe (1958) are important exceptions to this. The main similarity is that interference with the oestrous cycle is the final factor preventing further population growth.

In the wild

Many species of salmonids and other fish are territorial as parr or fry (references in Kalleberg 1958; Keenleyside & Yamamoto 1962; Le Cren 1965), and a general conclusion of workers in this field is that populations of fry are limited to a large extent by spacing behaviour and dominance. For example, Le Cren (1965) and Chapman (1962) found that aggressive behaviour by territorial juvenile trout and coho salmon (*Oncorhynchus kisutch*) caused emigration, and also death in the trout when densities were very high. Lindroth (1965) stated that a low food supply may limit the abundance of slow-growing parr in unfertile rivers, whereas in other rivers territorial behaviour or shelter determine abundance even if food is in excess or at least not deficient. In the wild, the evidence for the hypothesis that territorial behaviour may limit numbers is suggestive, but largely circumstantial and sometimes anecdotal; nevertheless much more interest is being taken in behaviour and we may therefore expect major advances in this field with freshwater-fish in future.

The foregoing section and the last paragraph give examples of populations limited by dominance, aggression and spacing behaviour. But most of them leave open the question of the limitation of wild breeding populations by behaviour, because the data refer either to confined conditions or immature stages.

Some of the effects described above, on the breeding rate, breeding success, juvenile survival and recruitment of rodents at high densities in confinement, have been clearly shown in the wild at high densities or during declines in brown rats (Davis 1951), *Clethrionomys rufocanus* in Lapland

(Kalela 1957), *Microtus* spp. in Germany (Frank 1957) and North America (Hoffman 1958; Krebs 1966), brown lemmings *Lemmus trimucronatus* (Krebs 1964) and house mice (DeLong 1967; Newsome 1969a, 1969b). Calhoun (1963), summarizing research on brown rats by several authors in Baltimore, stated that reduced populations within areas rose to the same ceiling level and thus were limited even though abundant unused food and cover were still in excess (see also Davis 1953). However, quantitative documentary evidence—as distinct from anecdotal evidence—about food supplies is seldom given in studies of wild populations, unlike controlled experiments in the laboratory (exceptions are Davis, Kalela, Krebs, DeLong and also Newsome by an indirect method). Despite these difficulties, the evidence indicates that, although food shortage is often involved in population limitation of rodents, this is by no means always the case. On the other hand, there is general agreement in these studies that changes in social behaviour invariably accompany the changes in reproductive physiology which are among the factors limiting populations. Although much more quantitative research on behaviour is needed in studies of wild rodents, and although some authors have come to conclusions about behaviour without directly studying it, nevertheless the tentative but overall indication of work in this field is that behaviour is a major limiting factor, as in the studies in confinement.

A widely-quoted illustration of the effects of behavioural competition on population limitation are various examples of poorer breeding success in great tits (*Parus major*) and other birds at high population densities (Kluyver 1951; review Wynne-Edwards 1962). However, in these cases it was not known directly whether dominance behaviour was involved, nor whether food supplies were in excess or not. Further, the critical measurement when discussing population limitation is recruitment of young, not breeding success. The importance of dominance behaviour is suggested in all the studies in the last two paragraphs, though it was not quantitatively measured in any of them, and therefore was assumed from circumstantial evidence only.

Jenkins (1961a, 1961b), who did measure behaviour, stated that differences in 'interaction' among adult partridges in two winters were associated with differences in breeding success during the following summer. However, differences in agonistic behaviour were not in fact measured; the measurement of 'interaction' was a combination of preening, resting, and alert or watchful behaviour, as well as overt agonistic behaviour. There was therefore no evidence that dominance behaviour was associated with the differences in reproduction. Furthermore, Blank & Ash (1962) and Southwood & Cross (1969) came to different conclusions about factors which limit breeding success in partridges (they suggested food and weather). In any case, the factors which limit the breeding population may not be the same as those limiting breeding success, and there is not yet conclusive evidence with this species showing how the breeding population is determined.

Some of the most comprehensive quantitative studies of behaviour yet attempted in relation to populations of vertebrates have been done with rabbits in Australia by Mykytowycz, Myers and Poole. Admittedly, the work was done in fenced enclosures, but population densities were similar to or not much higher than those in the wild, and food supplies were natural; the only big difference was that the fences stopped rabbits from immigrating or emigrating, and prevented mammalian predators getting in. In such conditions, fecundity decreases greatly as density increases (Mykytowycz 1960, 1961; Lockley 1961; Myers & Poole 1961). Mykytowycz concluded that behaviour restricted population growth; the less dominant females survived worse and lost more young by intra-uterine mortality. While agreeing that the decrease in fecundity with increase in density was sufficient to restrict population growth, Myers & Poole (1963) pointed out that it was not sufficient to stop growth. The mechanism for natural limitation suggested by them was emigration, combined with predation and sometimes starvation (p. 189). This was based on the facts that, at higher densities, more subordinate young adults of both sexes were forced by attacks from more dominant individuals to move about or lie in the open, and sometimes were seen trying to get out of the fences (Mykytowycz 1960; Lockley 1961; Myers & Poole 1963).

If behaviour does limit populations, we might reasonably expect changes in behaviour to precede changes in numbers in the wild. Many birds and mammals show territorial or other aggressive behaviour in autumn, when food is apparently plentiful, and a frequent speculative suggestion (e.g. Kluyver 1951; Kluyver & Tinbergen 1953; Kalela 1954) is that some juveniles fail to get a territory and are forced to emigrate at this time by the hostility of more dominant individuals. In some cases there is direct evidence of emigration, particularly of late young, rather than death, e.g. in the great tit (Dhondt & Hublé 1968). In other cases the seasonal reduction of population is in spring or summer. Again, young animals are usually, but not always, involved more than old ones, where this has been studied. Seasonal reductions due to dispersal, and in some cases due to dispersal and mortality combined, occur following a rise of hostility or strife in musk rats (Errington 1943), bobwhite quail (Errington 1945), prairie dogs *Cynomys ludovicianus* (King 1955), woodchucks *Marmota monax* (Snyder 1961), partridges (Jenkins 1961a, 1961b), song sparrows (Tompa 1964), ptarmigan (Watson 1965), the short-tailed voles (*Microtus agrestis*) studied by Chitty & Phipps (1966, but no direct evidence on behaviour in this case), the tree sparrows (*Passer montanus*) studied by Pinowski (1965), the black-capped chickadees (*Parus atricapillus*) studied by Smith (1967), and current research on red grouse (Watson & Moss 1970; Watson unpublished).

Only the last three studies have documented this by measuring changes in both the population size and the frequency of acts involving dominance, spacing behaviour or aggression. Nevertheless, the anecdotal evidence in some of the other cases strongly suggests that changes in spacing behaviour cause changes in numbers, because of the dramatic and sudden nature of the changes (e.g. the sudden appearance of many wounded animals (Errington 1943), and the sudden break-up of flocks with the formation of pairs (Errington 1945; Jenkins 1961a, 1961b; Watson 1965; Smith 1967)).

The possibility that the immediate cause of the change in behaviour is a sudden reduction in the amount of food available can be ruled out in the red grouse; the autumn decrease often occurs in September–October (Jenkins *et al.* 1967; Watson & Moss 1970), yet grazing of the main food plant, *Calluna vulgaris*, during the entire period of summer growth in April–October, is negligible (Moss 1969). This possibility can also be ruled out in the short-tailed vole by the quantitative data published by Chitty, Pimentel & Krebs (1968), and probably in the other species mentioned above. A sudden decrease in the quality of the food seems unlikely in many of these cases. For example, in the song sparrow, Tompa (1964) found that defence of territories is most vigorous just prior to, or at the beginning of breeding—a time when insect foods and newly growing vegetation begin to become available, so that the quality of the bird's diet is presumably improving. However the point has not been directly studied by specifically measuring the nutritive value of the food in relation to a seasonal reduction of a population.

In species with a short life and a rapid rate of reproduction, any deferment of maturity or inhibition of body growth (i.e. a special case of conditions (a) and (b) in Table 1) can contribute largely to bringing on a population crash, in lemmings, voles and other rodents (Kalela 1957; Krebs 1964). Sadleir (1965) found that the survival and recruitment of juvenile deer mice (*Peromyscus maniculatus*) in summer decreased at the same time as a seasonal rise in the aggressiveness of territorial adult males. In a maze in the laboratory, resident adults attacked intruding juveniles which died or were forced to live in small parts of the maze. In the field, juvenile survival was better on an area from which adults had been removed; this was later confirmed by Healey (1967), but was not found by Smyth (1968) in the bank vole, *Clethrionomys glareolus*. Sadleir's other conclusions were confirmed experimentally in the laboratory and in the field by Healey (1967). Healey also showed by laboratory experiments that inhibition of juvenile body growth varied according to the behavioural type (aggressive or docile) of the adult, and its location (at home or away from home). In the field, he tested these effects on inhibition of the body growth and survival of juveniles, by adding them to areas containing aggressive or docile adults. This ability to set up artificial populations of different behavioural types in the field will probably prove to be a most useful technique. Further suggestive evidence comes from other important population experiments in the wild (i) where rodents were removed, and animals then bred more successfully and/or immigration occurred, (ii) where rodents were added to an existing population and heavier mortality and/or emigration followed, and (iii) where the proportion of sub-adults reaching maturity was increased by removing other individuals. These experiments were done in the brown rat (Emlen, Stokes & Winsor 1948; Calhoun 1948; Davis 1953; Barnett 1964), the woodchuck (Snyder 1961; Davis 1962; Davis, Christian & Bronson 1964), in *Microtus californicus* (Krebs 1966) and in the bank vole (Smyth 1968). Terman (1962) gives other references involving migration.

Bronson (1964) was able to measure the rate of agonistic encounters of woodchucks in the field, as well as in the laboratory as Sadleir and Healey had done. The interaction rate increased rapidly as home ranges were established after the winter hibernation, and was at a peak about May–June, soon after breeding occurred. Young woodchucks tended to avoid the more dominant holders of home range, and tended to emigrate more in late summer. Krebs (1964) measured an effect of agonistic behaviour—frequency of wounds—in the wild with *Lemmus trimucronatus* in arctic Canada, and noted that this was associated with a period when numbers were declining and young males were failing to mature and thus failing to be recruited. A few direct measurements of agonistic behaviour in relation to population density and emigration were made by Clough (1968) in his exploratory study of *Lemmus lemmus* in Norway. However, much more intensive work on behaviour needs to be done with lemmings and voles, of the sort pioneered by Bronson, Sadleir, and Healey. This is now being done with *Microtus* by Dr C. J. Krebs and his co-workers in Indiana.

Less dominant individuals in the Australian magpie (Carrick 1963) and red grouse (Watson & Moss 1970)—which are both species that fulfil conditions (a) and (b) in Table 1—have access to food, and therefore fulfil condition (c) also. There is no evidence that groups of other species of birds in autumn or winter or mammals in late summer (when much seasonal mortality and emigration of less dominant individuals occurs) or of apparently non-breeding shelduck, birds or prey, and colonial sea birds in summer, are completely excluded from certain food sources, far less that the breeders are using up all the food resources.

This summary of evidence from the wild tends to confirm the observations on confined animals that dominance and spacing behaviour may limit populations in the presence of an excess of food, but the measurements of food in any one study in the wild are usually inadequate to establish the point fully. Nevertheless, in a few studies there is good evidence that seasonal reductions in breeding population due to changes in behaviour are not preceded by a reduction in the quantity of food. A reduction in the quality of the food seems unlikely in many cases, especially where the seasonal reduction is sudden, though the possibility of it has not been directly refuted by a specific quantitative study. In general, the evidence suggests that shortage of food may, in some cases, limit population growth, but that this is not a necessary factor. On the other hand, in research where workers directly studied behaviour and did not simply speculate about it, all the studies that we have read indicate that spacing behaviour or aggression (usually followed by changes in reproduction) is invariably involved in population limitation and is therefore likely to be a necessary factor and often the major limiting factor.

The discussion is now continued to include consideration of living space, as well as food and behaviour, as a factor limiting animal populations.

Both space and food limiting
There is evidence that space and food may both limit populations at the same time. In some cases, only space was studied and was found to be limiting, but in other cases where dominance behaviour was also studied, the effect of space restriction was found to operate through behavioural mechanisms. In the following three examples, behaviour was not directly studied. In the study of Johnson (1965), detailed quantitative evidence on differences in population densities and body size within and between lakes allows us to conclude that space and food both combined to limit wild populations of the fry of sockeye salmon (*Oncorhynchus nerka*). Lindroth (Discussion

after Johnson's paper), summarized some observations, from studies of the fry of Atlantic salmon in confined conditions, which can be interpreted in the same way. Silliman (1968) clearly showed a dual limitation by food and space by making quantitative measurements of the population growth of guppies in confinement. Some of these authors have made the reasonable tentative conclusion that dominance behaviour is probably involved in the apparent restriction of the populations by space. For example, in fresh-water salmonids, the fish which do not get territories probably get less food (Kalleberg 1958), grow more slowly, and so remain longer in the smaller size group which is known to be more vulnerable to predation (Larkin 1956; Kalleberg 1958; Johnson 1965). Population limitation thus involves space (and probably dominance behaviour), food and predation.

Newsome (1969a, 1969b) found that populations of feral house mice in the wild in Australia were limited by food supply, and by suitable cracks in the soil for making the burrows necessary for breeding. Burrows were probably the most important limiting factor in early summer, but by late summer we conclude from Newsome's evidence that both were apparently limiting. On the basis of circumstantial evidence about body weights, breeding and dispersal, he considered that dominance behaviour was a servo-mechanism governing the reactions of the mice to shortages of space and food. The references to a social order or to dominant and sub-ordinate animals were based on assumptions from this indirect evidence, as no work on behaviour was done.

Behaviour was studied by several workers using rodents in confined conditions. Where both food supply and space were limited, Strecker & Emlen (1953) found that young house mice died and the adults later stopped breeding but changed little in number. They concluded that space *per se* was not important in stopping breeding, as some 'cover units' were never used. Crowding, though important, was considered secondary to food shortage. Southwick (1955a) found that in some populations, where space but not food was restricted, fecundity remained high even at very high densities, provided voluntary food intake also remained high. Fecundity declined in populations where voluntary intake dropped below 3 g per head per day, despite food being in excess, and he found the same thing on examining the data of Strecker & Emlen. He attributed these differences in voluntary intake to the social structure of his different groups and gave some observations in support of this statement. Where food was limited but not space, house mice reached a ceiling of numbers, with a steady population and no food shortage (Strecker 1954). Reproduction was steady and high, but the excess was removed by a very high emigration rate once the population ceiling was reached, associated with increased strife. No animals were in poor condition. These studies illustrate the general point that shortage of a given resource can induce changes in behaviour before the resource becomes absolutely limiting (e.g. food shortage actually imposing starvation), so that behaviour may be the proximate factor limiting the population. In our terminology, both behaviour and food are regarded as limiting.

Petrusewicz (1966) also stated that food intake per head sometimes declined at high densities in white mice, due to social interaction. Calhoun's findings (1961, 1962) of a 'behavioural sink' with domesticated rats at high densities was another slightly different example. In this case, the positioning of food sources and other features resulted in rats being attracted to one food source in order to feed with other rats, and neglecting other food sources nearby. An unstable social system followed, with severe strife, pathological changes and inhibited reproductive rate, even though total food was greatly in excess.

Calhoun's (1963) intensive study of confined brown rats in outdoor enclosures raised another important point, already indicated by his earlier studies (1949, 1950). Even though food was in excess, rats with territories near the food source or the passages leading to it grew better, reproduced more successfully, and had young which also did better. Hence it could be argued that relatively poorer availability of food caused the observed effects on social status, body growth, and reproduction. Calhoun did find this association, but also noted that such an individual was unable, because of its low rank or status, to shift its residence to a better position nearer the food. He wrote that it is therefore difficult to separate cause and effect completely. Similarly, a rat, which is of low weight because of previous aggression from other rats during its early life, can less readily 'achieve priority in goals' during its adult life (e.g. getting a good territory near a feeding station). Another example is that certain unusual individuals or 'social outcasts' of very low rank showed abnormally inhibited feeding behaviour even though no other rats were at the food source, due to previous attacks and consequent timidity. In all the above cases, however, food shortage is no more than a proximate factor, with behaviour the important limiting factor. In our terminology, both are regarded as limiting. To put these examples in their proper perspective, only a few of the effects of behaviour on population growth in Calhoun's study involved rats near food sources, and most were associated with strife in early life (p. 179). Nevertheless, Calhoun freely admitted that he had no quantitative evidence to refute the possibility that most rats of lower social status, which grew more slowly, did so because they had a lower voluntary intake of food. Most people would provisionally accept his array of evidence that social factors did inhibit body growth. However, actual measurement of food intake in individuals of different social status will be an important and necessary future step to fill this gap in research on behaviour–nutrition–population relationships in confinement and in the wild.

Myers (1964) found that confined rabbits will breed successfully at very high densities provided food and living space are adequate. Food shortage causes poor breeding or stops breeding, and starvation kills more of the younger, less dominant animals; starved animals are also more vulnerable to predation (Mykytowycz 1961; Myers & Poole 1961, 1963). Mykytowycz found that the older and more dominant females tended to be buffered from food shortage in confined conditions by having larger and better feeding territories. Myers & Poole (1963) concluded that starvation was the main factor limiting populations in these fenced enclosures where emigration was prevented and mammalian predators were absent. Without fencing, they suggested that emigration of subordinates into unfavourable habitats, and their subsequent death there by predation (see Myers & Schneider 1964) and food shortage, would occur in the wild before catastrophic starvation set in. Nevertheless they speculated that, in much of Australia, where they stated that there are enormous areas of favourable habitat, this mechanism cannot operate and catastrophic plagues and starvation occur as in confinement. This seems analogous with the case mentioned by Klein (1970) in this symposium, where starvation apparently limits populations of North American deer on vast areas of 'uniform' suitable habitat, and where self-regulation is said to be unlikely because there are so few unfavourable areas for emigrants to disperse into. In the rabbit, it seems that social behaviour, starvation, and predation may all combine to limit the population. Possibly this might also be true with deer in parts of North America where predators are still numerous, but there has been hardly any research on behaviour in this context.

Hierarchies

Hierarchies are common among many wild vertebrates (Collias & Taber 1951; Wynne-Edwards 1962; Brown 1953), and also occur in many domestic animals (e.g. Guhl 1953; Wood-Gush 1965) and in man (e.g. Sommer 1970; and the work of Esser, Chamberlain, Chapple & Kline (1965) with patients in a mental hospital). Some research on hierarchies in confined conditions (references in Wynne-Edwards 1962) suggests that

14

this type of dominance behaviour might be potentially capable of limiting breeding stocks and recruitment of young in the wild.

In canaries kept very close together in a cage, males which were more dominant in the hierarchical flock were the only ones to rear young (Shoemaker 1939). However, the birds were also territorial around their nests, and much of the fighting was territorial. Also, social position happened to improve as birds came into breeding condition, so it is difficult to separate cause and effect here.

One case where a hierarchy in confined conditions might at first sight seem to limit numbers is the rabbit, but here the situation is more complicated (Myers & Poole 1963). During the breeding season, rabbits have group territories, and the effects of food shortage and mortality, in confined conditions at fairly natural densities, apparently vary according to quality of the territory rather than status of the individual (Mykytowycz 1960, 1961). Rather little strife usually occurs between the females within a group (Myers & Poole 1961); and Mykytowycz and Myers & Poole came to opposite conclusions about the relation between social dominance and the mortality of adults or young. Myers & Poole (1963), summarizing the Australian research on rabbits, state that territoriality breaks down during the non-breeding season and that the rabbits range widely in search of food, but rarely show any fighting or conflict over food. The more dominant individuals survive food shortage at this time better, because they are older and heavier, not because they show more dominance during this period.

Some studies of rodents in confinement (e.g. Clarke 1955; Southwick 1955b; Petrusewicz 1966) suggest that behaviour in hierarchies might limit numbers, but in most cases it is not exactly clear whether strife and fighting were within a hierarchical group or were also involved with other types of spacing behaviour. Barnett (1964) wrote that there is no evidence that dominance hierarchies ever occur in brown rats, even in confinement, and that probably all fighting of rats is territorial. Calhoun (1963, p. 259), who criticized Barnett's conclusion on semantic grounds, stated that both types of behaviour occurred among brown rats in outdoor enclosures, distinguishing them on the basis of sustained exclusion from a space (territory) or not (dominance). Low-class groups which showed aggression within hierarchies and not in territories were socially unstable, and animals in them reproduced less well than groups of high-class territorial animals, but the difference in behaviour was not quantitatively measured. In any case, the effects of strife within hierarchies cannot be fully separated from the other original reasons for the instability of the group or the low social status (e.g. social origin) of the animals composing it, or from the effects of territoriality which was apparently the commoner of the two types of behaviour. Brown (1953) wrote that fighting of confined house mice occurred within a social hierarchy, and caused litter mortality which limited the population. However, Crowcroft & Rowe (1963) who did a quantitative study of behaviour involving observations similar to Brown's, concluded that dominant-subordinate relationships resulted in territory formation if a suitable physical environment (nests and cover) is given. They indicated that the environments provided by Brown and Southwick were deficient in this respect.

We have therefore come across no conclusive direct evidence in the literature of populations in fairly natural conditions being limited by dominance or aggression solely within hierarchies. This seems surprising, and appears to be an obvious gap in research. It will be important to pay attention to the fact, suggested by Davis (1958) and commonly found and documented among many species since, that territorial behaviour and dominance hierarchies in any one species are two poles of a continuum varying with population density and type of environment, and hence varying in confined conditions with the circumstances of the experimental design used.

Turning now to the wild, Wynne-Edwards (1962) suggested that the function of hierarchies is to act as a social guillotine when food becomes scarce, by sacrificing the subordinate members of the hierarchy and ensuring that those at the top of the peck order get plenty of food. Similar suggestions were made by Lack (1966) to explain how spring breeding stocks of great tits and coal tits (*Parus ater*) are limited in the winter flocks. Ashmole (1963) and Carrick & Murray (1964) suggested that subordinate sea birds and elephant seals may be excluded from good feeding grounds within easy reach of their colonies, by the competition of more dominant individuals, and that this could limit the breeding population at the colony. In the literature we have found no direct evidence from the wild to confirm these speculative ideas, and this is a wide open field for research. Two apparent confirmations are criticized below.

Lockie (1956) found more fighting in flocks of rooks (*Corvus frugilegus*) when the weather was cold. Assuming that food was scarcer in cold weather, he speculated that submissive birds would often be displaced from food by the frequent fights, and would starve. However, Patterson (1970) in this symposium has shown by more intensive work that the frequency of winter 'food fighting' bears little relationship to the availability of food and this does not support Lockie's suggestion. Murton, Isaacson &

Westwood (1966) and Murton (1968) measured the feeding rate of wood pigeons (*Columba palumbus*) and the amount of food available. Birds at the front of the moving flock fed less. They suggested that these were individuals of low status, which would die of starvation when food was short. However they gave no evidence that these individuals were in fact of lower status, or that they starved and died later, so the conclusions are speculative.

To sum up so far, there is good evidence that conditions (a), (b) and (c) in Table 1 are met separately in fairly widespread examples, but only a few cases where all three have been met satisfactorily in any one species. This is probably because very few people have tried to satisfy all the conditions in one study, and not because the answer to question A is negative. In our opinion—and it is only a matter of opinion—the evidence suggests that population limitation by behaviour is widespread, but much more work is needed before this can be fully accepted.

Certainly, it must be admitted that many authors have concluded that populations are limited by territorial behaviour or by other forms of dominance and spacing behaviour, but on insufficient evidence and in some cases on no direct evidence (e.g. Kluyver & Tinbergen 1953; Gibb 1960).

COMPENSATION IN MORTALITY

Condition (d) in Table 1 was that limiting mortality factors—or limiting factors involved in depression of recruitment—should compensate for other mortality (etc.) factors in the long-term. There is good evidence in the short-term from populations of many invertebrates in confined conditions with food in excess that if one mortality factor—such as culling— increases, then breeding and recruitment rate subsequently increase or vice versa (review by Wynne-Edwards 1962). This is (d_i) in Table 1. There is also fairly good evidence of this from population experiments with rodents in the wild, already mentioned earlier (p. 185). However, there was little if any direct evidence on dominance or spacing behaviour in any of the above studies, in the wild or confinement, except those of Sadleir (1965) and Healey (1967). In the experiments in the wild there was no quantitative measurement of food, and thus no good evidence to refute the possibility that changes in food availability alone could have caused the observed changes in recruitment. Also, the studies on rodents in the wild were all short-term, for a few years only.

There has been a considerable literature on changes in recruitment in relation to stock size and exploitation rate in fishery research (e.g. Ricker

1954a, 1954b; Beverton & Holt 1957; Beverton 1962). These changes involve compensatory mortality, but we will not discuss this research any further here because there are practically no data on dominance or spacing behaviour in relation to the mortality of fish, nor on the actual contribution made by each mortality factor. Furthermore, it has been argued that food shortage—not behaviour—may cause the observed changes in recruitment (Beverton 1962). In any case, the population data are so variable that there is disagreement about the nature of the stock-recruitment curves, even by different observers using the same data (Gulland 1962).

There is a small amount of evidence in confined conditions that, if one mortality factor increases, then this has the effect of decreasing the limiting mortality factor (another variety of (d_i)), as distinct from merely affecting recruitment via changes in the reproductive rate. There is evidence of this from Silliman & Gutsell's (1958) study of confined guppies with food in excess, where the mortality from cannibalism of young fish by adults—which was this species' mechanism for limiting recruitment and the adult population—decreased when that due to culling increased, and vice versa.

There is good evidence from studies of rodents in the wild that different mortality factors compensate for one another. For example, several authors (p. 185) have found that on areas where many woodchucks and other rodents were removed, survival of the remainder improved and emigration decreased. Evidence that some of these effects can be due to behaviour was given by Bronson (1964) in the woodchuck and by Sadleir (1965) and Healey (1967) in the deer mouse. However in none of these cases was it conclusively shown which mortality factor(s) was limiting. Furthermore, all these studies were short-term, and although food was either ignored or considered to be in excess in all cases, it was not actually measured.

Evidence in wild vertebrates of mortality factors tending to compensate for each other in this way was given by Errington (1943, 1945) with the musk rat and bobwhite quail, but the evidence is suggestive and not conclusive, for reasons given earlier.

In the red grouse Jenkins, Watson & Miller (1963, 1964) suggested that the total number dying over the winter is pre-determined by territorial behaviour in autumn, when food is in excess (Moss 1969; Watson & Moss 1970). It follows that the number dying from any one cause varies inversely with the number dying from other causes within the same season. This has been the case in every year of a 12-year study, and no upward general trend in numbers has occurred; it therefore seems likely that behaviour operates as a limiting factor in the long-term but this has not yet been fully documented.

Another long-term example of compensatory mortality may possibly be the red deer (*Cervus elaphus*) studied by Lowe (1969). A year after the study began, the proportion killed by culling was greatly increased, and Lowe claimed that this largely (though not completely) replaced natural mortality. In North America it is implicit in much of the policy on deer management that shooting can replace natural losses (review by Klein (1970) in this symposium). It may be that, in practice, this does happen. However, we have not come across any conclusive demonstrations of the point by detailed research on mortality factors and populations on a given study area. In any case, dominance behaviour has not been measured in relation to population limitation in red deer and North American deer. Also, in contrast to the examples in rodents, food is claimed to be a limiting factor in deer, and the compensatory mortality described may therefore not be due to changes in behaviour.

In summary, there is good evidence that recruitment varies inversely with various mortality factors. There is evidence that the mortality factor acts in an opposite sense (d_i) to other mortality factors. However, both these effects are well documented only in the short term. There appears to be no good evidence that mortality factors and/or changes in recruitment rate compensate for each other exactly in the wild in the long term (d_{ii}). In only a few cases has it been shown that behaviour is a probable limiting factor in the long term.

CHANGES IN DOMINANCE AND SPACING BEHAVIOUR, FOLLOWING CHANGES IN NUTRITION

If we accept that populations are limited by dominance and spacing behaviour in some cases, the next question is whether these changes in behaviour—and thus numbers—are sometimes due to a change in nutrition. The examples in the preceding section, with rodents in confinement, show that changes in food supplies do cause changes in numbers and in dominance and spacing behaviour. This happens before food becomes absolutely limiting in the sense of causing a direct population decrease through starvation of mature animals.

Such an effect need not be direct. Cowley & Griesel (1966) and Cowley (1970) in this symposium showed that protein-poor diets given to domestic rats affected the behaviour of their offspring and that these maternal effects were cumulative over several generations. Although dominance and spacing behaviour were not studied, the exploratory activity of the young was affected, and this is known to be associated with aggression and spacing behaviour in several species of rodents (Calhoun 1963; Lagerspetz 1964; Healey 1967). An example from the wild is the red grouse (Watson 1967; Watson & Moss 1970), where the evidence at present indicates that differences in the quality of the food eaten by one generation affect the aggression and territorial behaviour of their offspring and thus population density in the next year, in the presence of abundant quantities of food. The hypothesis is that a high plane of nutrition for the parents improves egg quality, which leads to large broods of full grown young and subsequently to a set of territorial cocks that take small territories. In this case there is experimental evidence of a maternal effect, because the effects on chick survival still occur when samples of eggs are taken in from the field and hatched and reared in captivity (Jenkins, Watson & Picozzi 1965). Unless this is done, there may be difficulty, as Barnett & Evans (1965) point out, in distinguishing maternally-inherited effects from cultural effects learned from the mother or from other social experiences after birth.

The widespread phenomenon of differences in fertility amongst different areas shows that better food supplies are almost invariably associated with higher numbers of homoiothermic animals (e.g. reviews in Lack 1954; Wynne-Edwards 1962; Armstrong 1965; Schoener 1968). However, the majority of such examples do not include evidence relevant to points (a) (b) and (c) in Table 1, and are therefore open to the interpretation that food is the only major limiting factor. The discovery by Kluyver & Tinbergen (1953) that great tits in mixed woods have smaller territories than in pine woods, and also show a smaller amplitude of fluctuation from year to year and slower rates of increase, has been widely quoted and accepted as evidence, that populations are limited by territorial behaviour. In fact, it shows no such thing, as the conclusion is based on indirect, circumstantial evidence from populations, not behaviour. Nor are there data to prove Kluyver & Tinbergen's conclusion that great tits prefer mixed woods and fill up the good habitat there before moving into the poorer pine woods. Lack (1966) gives other criticisms of their conclusion about territorial behaviour.

Nevertheless, we can conclude from the evidence in previous paragraphs that condition (e) in Table 1 is fulfilled in a few cases, in confinement or in the wild, where (a), (b) and (c) are also fulfilled. In other words, there are some examples where animal populations are limited by nutrition and by spacing behaviour or dominance, even though the gross amount or total quantity of food is not limiting.

THE EFFECT OF FACTORS OTHER THAN FOOD UPON SPACING BEHAVIOUR

THE PHYSICAL STRUCTURE OF THE ENVIRONMENT

The physical structure of the environment affects spacing behaviour independently of nutrition in salmon and trout (Kalleberg 1958; Lindroth 1965). Saunders & Smith (1962) found that a population of brook trout (*Salvelinus fontinalis*) could be nearly doubled by adding physical structures which provided extra territorial stations. Boussu's (1954) study of trout populations in relation to cover showed the same point, as did McCrimmon (1954) with Atlantic salmon put into streams. Thus, 'living space' and the inverse of 'density' are not identical concepts, as also is clearly demonstrated by Myers, Hale, Mykytowycz & Hughes (1970) with closely-confined rabbits.

An example of the same effect with birds is given by Armstrong (1965). The size of the nighthawk's home ranges in Detroit is correlated inversely with the number of flat-roofed buildings per hectare of total area. Flat roofs provide nesting sites and suitable places for the birds' diving displays. Food was not measured, but the size of ranges was not correlated with the number of trees or shrubs that support the insects on which the birds feed.

Jenkins (1961b) found that home ranges in partridges tended to be larger in fields with poor ground cover; however this case is more complicated because poor cover was also associated with less food due to heavy grazing by domestic stock or to ploughing or other farming practices. In the red grouse, Watson (1964) found that territory size was inversely correlated with a measure of cover—or its converse, visibility. In this case, the variations in cover or visibility were mainly topographical, with small territories on hillocky ground and large ones on open flat ground, and were independent of the total quantity of food (Watson & Moss 1970, Miller & Watson unpublished).

Davis (1958) and Calhoun (1963) showed that the physical structure of the environment greatly affects spacing behaviour in confined house mice and brown rats, and also population density in wild brown rats (Davis 1953). Similar effects of variations in the physical environment on dispersion were found by McCabe & Blanchard (1950) with deer mice; on spacing behaviour by Terman (1963) with confined deer mice; on spacing and dispersion in the Uganda kob, *Adenota kob* (Buechner 1961); and on population dispersion of ringed seals (*Phoca hispida*) in relation to coastal topography and the physical structure of sea ice (McLaren 1962). H. Melchior (pers. commun.) has recently found that the physical structure of the environment greatly affects spacing behaviour and dispersion in arctic ground squirrels (*Citellus parryi*). Changes in dominance and/or spacing behaviour in relation to physical features in the environment have been described in man, e.g. with sailors in confined conditions (Altman & Haythorn 1967), students in college libraries (Sommer 1970), and blacks in the Chicago ghettoes (Hall 1970). An example of an experimental contribution in the wild is Snyder's (1967) study, where the number of scaled quail (*Callipepla squamata*) increased after piles of leafless brushwood were added to an area in Colorado.

An example where Bustard (1969) concluded that territorial behaviour limited the male population of the lizard *Gehyra variegata*, also included the contradictory conclusion that the upper limit was set by the number of suitable cracks in tree bark, which acted as 'home sites'. In this case, conditions (a) and (b) in Table 1 were satisfied, but not condition (c). Hence the physical structure of the environment, and not territorial behaviour, was limiting, because all the suitable cracks were utilized.

The evidence in this section so far shows that the physical structure of the environment greatly affects spacing behaviour and population dispersion. In addition, it has a marked effect on reproduction. This was clearly shown by Calhoun (1963) in the brown rat, and also in house mice in confined conditions (Southwick 1955a). When nests were made small and escape cover was poor, the confined mice fought more and reproduced less well even though food was in excess. In the semi-natural conditions of farm ricks, differences in the physical structure of oat ricks and wheat ricks were associated with differences in population size, proportion of females fecund, litter mortality, and wounding (Rowe *et al.* 1964). As Southwick's data also show evidence of limitation by food and space (p. 187) it seems that the nature of the physical environment may be a third limiting factor. Another example is McLaren's work on the ringed seal, where the nature of the sea ice affected breeding success.

Cover, visibility and other aspects of physical structure have seldom been quantitatively measured, but there are many anecdotal observations suggesting that they probably have important effects on spacing behaviour and dispersion in various other species. There is certainly enough evidence to say that it would be unwise to ignore the physical structure of the environment in any study of spacing behaviour or population dynamics.

BEHAVIOUR IN RELATION TO STRESS AND
GENETICS

(a) *Stress and behaviour*

The voluminous literature on stress in confined populations of mammals has been well reviewed (Barnett 1964; Christian, Lloyd & Davis 1965; Myers 1966; Christian 1968). Remarkably little has been found about stress in birds, and this is a wide open field for research. The general hypothesis is that crowding at high densities, and consequent increases in strife and persecution, produce pathological changes in the endocrine system of less dominant adults and their young. These changes cause mortality, lower densities, and thus less social pressure and stress. There is much evidence for this behaviour-endocrine feedback system at the very high densities which occur in most confined populations (but for an opposing view see Munday 1961). However, although there is some evidence for it in field studies, e.g. in brown lemmings (Andrews 1968) and in sika deer, *Cervus nippon* (Christian, Flyger & Davis 1960), it has not yet been well established in the field. For example, other field studies of the same species offer different explanations or do not agree with the expectations of the hypothesis, in house mice (Southwick 1958; DeLong 1967), voles (Chitty 1960; Krebs 1966), brown lemmings (Krebs 1964), and sika deer (references in Klein (1970) in this symposium).

(b) *Behaviour-genetic polymorphism*

Chitty's (1957, 1967a) alternative explanation is that declines are brought about by a genetic selection for aggressive individuals, which have an advantage in crowded populations but are less viable in other ways. This is balanced by selection for less aggressive individuals which are more fecund and rear more offspring during the subsequent increase from low densities. There is very little evidence for or against this hypothesis from the wild or in captivity (DeLong 1967), although there is much speculation about it in the literature. Krebs (1966) found that an introduced population of *Microtus californicus* increased greatly on an area where a decline had occurred, and it seemed reasonable to conclude that this was evidence of intrinsic differences between the expanding and declining phases, and that the area itself was not inherently unfavourable in that year simply because another stock had declined there. However, although this is an important first step in testing the hypothesis, it does not exclude the possibility of either maternally-inherited, non-genetic effects or else cultural effects on behaviour, rather than genetic effects. Nevertheless, this hypothesis is now being investigated in *Microtus* species (Semeonoff & Robertson 1968; Dr C. J. Krebs, pers. commun.). There is also evidence that tendency to aggression is partly an inherited character (see Moyer in press), and that it can be changed in a few generations by selective breeding, in birds (e.g. Craig, Ortman & Guhl 1965) and in mammals (e.g. Lagerspetz 1964).

(c) *Genetic stock and/or early social experience*

Spacing behaviour, activity, and the population asymptote or ceiling level may be greatly different in different stocks that are kept in the same laboratory conditions with food in excess. This has been shown in laboratory mice (Crew & Mirskaia 1931; Petrusewicz 1957, 1963, 1966), in house mice (Brown 1953; Southwick 1955a) and in deer mice (Terman, pers. commun.). Terman (1963) gives other references in *Mus musculus* and *Microtus* spp. These authors concluded that the original mice differed temperamentally, and that the differences in asymptote were due to inherent differences in behaviour. Calhoun (1963) gave brown rats, which were confined out of doors, several home ranges with equal opportunity for access to food and nest boxes. But rats in the groups in these home ranges grew and reproduced at different rates and showed differences in spacing behaviour, associated with differences in the cultural history and social origin of the groups.

Such differences in the effects of social behaviour on population growth can be inherited, as shown by Calhoun (1956) in inbred strains of house mice. An important study showing the effect of genetic constitution on differences in population size is Ayala's (1968) work on *Drosophila*. Changes in spacing behaviour can be induced by changes in the social environment. For example, Calhoun (1963) and Terman (1962, 1963) have shown the importance of early social experience in determining later spacing behaviour in brown rats and deer mice. The method of rearing and the early social experience of the lambs of Scottish blackface sheep affect their later spacing behaviour, and probably also their subsequent adult population density and performance in relation to the environment (Hunter & Davies 1963). Van den Assem (1967) found that simultaneous introduction of sticklebacks produced a higher asymptote (in this case a lower territory size), than does successive introduction of the same number of animals in similar environmental conditions. Some evidence confirming Van den Assem's conclusion in the wild is also suggested in red grouse that are colonising vacant areas (Watson & Jenkins 1968). Petrusewicz found that population growth of white mice from a stable asymptote could be induced by introducing a few newcomers or removing a few residents temporarily (1963), or by changing the cage in which the animals were kept (1957).

It is characteristic of the work with rodents that physiological changes and responses such as inhibition of reproduction, higher mortality, lower natality and lower food intake, occur at widely varying densities even if food is in excess, in house mice (e.g. Southwick 1955a; Southwick 1958; Petrusewicz & Andrzejewski 1962) and white mice (Petrusewicz 1957, 1963, 1966). The common characteristic is the change in dominance or spacing behaviour which is involved, either at the same time or shortly before the other responses (Southwick 1955a; Petrusewicz 1966). An example from the wild is Lockie's (1966) suggestion in the weasel that the population density and territory size of residents may depend on the cohesion of the social system; when this apparently breaks down because many die, transients do not occupy the vacant spaces which they do on other occasions when a resident population is in almost full occupation of all the ground. This seems analogous to the 'ghost towns' observed by Calhoun (1961) with domesticated rats in confinement, where spatial desertion of certain areas occurred due to 'togetherness' or to attraction to other places, and not due to exclusion by dominants.

(d) *Group selection and behaviour*
Wynne-Edwards (1962) drew together a large body of information on social behaviour, and discussed the idea that various types of social behaviour—some of which we have discussed here—limit populations. He also suggested that group selection is involved in maintaining these systems. At present there is not yet critical evidence from the wild either for or against this hypothesis about group selection. One of the initial difficulties here is to design suitable null hypotheses or experiments for testing in a wild population, and a second will be the formidable difficulties of the field work.

To conclude, this main section shows that spacing behaviour—and its effects in limiting population density—can be modified by various factors independently of food. One cannot regard animals as being constant in different studies, because the physical environment, the animals' genetic constitution or their early social experience may produce unexpected variations in spacing, dominance behaviour and aggression.

POPULATIONS APPARENTLY LIMITED
BY FACTORS OTHER THAN
BEHAVIOUR

Some workers have concluded that the populations they studied are limited by factors other than dominance behaviour and spacing (question B in Table 1). This paper is not primarily concerned with such cases and hence does not give a review of them. However, we have noticed that one factor which many such studies have in common is that dominance and spacing behaviour have not been examined at all. We will therefore discuss a few relevant examples of this. Authors naturally tend to explain their results in terms of the parameters they have measured and are tempted to ignore other factors, like dominance or spacing behaviour, which they have not studied. In such cases the proximate mortality factors, e.g. apparent starvation, may well be confused with initial or causal factors, e.g. spacing behaviour.

A good example is the conclusion of the Grouse Inquiry of 1911 that disease caused declines in the numbers of red grouse (Committee of Inquiry on Grouse Disease 1911). The evidence was that many birds were found dying in winter or spring in poor condition and containing many parasitic trichostrongyle worms. More recent work (Jenkins et al. 1963; Watson & Moss 1970) has shown that most of this mortality is socially induced and that the poor condition of the dying birds is not primarily due to disease or lack of food.

An outstanding current example where behaviour has not been studied is the large amount of work on deer in North America (reviewed in this symposium by Klein 1970). Starvation has been widely claimed to be a major cause of population limitation, because (a) apparently-starving animals are often found in hard winters, (b) deer range is often damaged by excessive numbers of deer, and (c) the deers' condition and reproductive performance decrease in parallel with this damage. No direct quantitative research on behaviour in relation to population dynamics has been done.

Work on roe deer, on the other hand, has shown the vital importance of behaviour, coupled with predation, in limiting breeding populations. Dominant roe deer expel subordinate individuals into less favourable habitats where they suffer a higher mortality rate (Andersen 1961, 1962, 1963; Cumming 1966; Bramley 1970). When a population was fenced (Andersen 1961), reproductive performance, body weight and antler size decreased. The fenced deer had apparently reached a higher population level than would have been the case if emigration followed by predation had occurred. Obviously, food is important, but behaviour and predation may also limit deer populations in a natural situation. Only research on dominance and spacing behaviour can test this suggestion, and we find it surprising that so little work has been done in this field where such large financial resources are available. Suitable questions for research are to ask if the recruitment of young deer to the breeding population, and their subsequent survival and performance within it, are affected by social status, and if the choice of feeding places—and thus nutrition—are affected by spacing behaviour.

Some cases of depressed reproductive rate at high densities in wild rodents have already been mentioned (p. 181). In the brown rat (Calhoun 1963) and the house mouse (Rowe *et al.* 1964) there was evidence that food was not limiting. However, in species which show 'cyclic' fluctuations, some workers have claimed that food shortage—not behaviour—causes the observed changes in reproduction and in the breeding population. An association of events was noticed in arctic Alaska between a decline of lemmings and a measured decline of almost the entire total food supply by lemming activities over the winter (Thompson 1955). Pitelka (1957) stated that the removal of food and cover on Thompson's area exposed the lemmings to predation, which was the proximate factor directly causing the population crash, but in our view the quantitative evidence for this conclusion was not sufficient (see also Watson 1957a, p. 458). Kalela (1962) associated declines of *Clethrionomys rufocanus* in Lapland with some measurements (1957) indicating a decline in the amount of certain favoured parts of the plants, due to grazing by the voles. He associated years of peak numbers of lemmings and voles with high summer temperatures in the year before, or the year of high numbers, but the data in fact show various exceptions to this and were not treated by statistical analysis, so there is not yet evidence of a real correlation. Other authors (Schultz 1966) consider that changes in the content of nutrients in the food, especially phosphorus, cause the cycles in abundance of lemmings, but documentary evidence from this work, which is partly experimental, has not yet been published. Other workers disagree that food shortage is a necessary prerequisite, even though in some cases it may be sufficient to cause declines (Chitty 1960, 1967a; Krebs 1964, 1966; DeLong 1967). Chitty, and Krebs (1964) considered food quantity but not its nutritive value. However, Krebs & DeLong (1965) found that fertilizing a natural grass pasture with a nitrogen and phosphorus fertilizer, or adding supplementary artificial food, was insufficient to prevent a decline in *Microtus californicus*. Another explanation, based on an association of events covering five seasons, is that lemming abundance is limited by the length of the breeding season (Mullen 1968). Young matured and reproduced better in long breeding seasons, which in turn followed periods of high minimum temperatures in spring (i.e. affecting lemmings directly, but still too cold for appreciable growth of vegetation). Fuller, Stebbins & Dyke (1969) give another explanation, based on an association between mortality rates and declines of *Clethrionomys rutilus*, and cold periods under the snow in winter. With this somewhat bewildering variety of explanations, the problem of population limitation in rodent species exhibiting 'cyclic' fluctuations clearly remains unresolved. They are good examples of the current confusion in population ecology.

An example from marine fish is that Beverton (1962) showed that the number of progeny of North Sea plaice (*Pleuronectes platessa*) surviving to become recruits (i.e. in this case recruits to the fishery) was not related to adult biomass, but that their rate of survival to recruitment was inversely related to adult biomass. From circumstantial evidence, predation and food shortage at the larval stage were suggested as the most likely causes of this compensatory mortality. Behaviour was not mentioned as a possibility, and was not studied. By analogy with freshwater salmonids and terrestrial vertebrates, we suggest that spacing behaviour and aggression in marine fish should be studied as a possible factor which might limit body growth and recruitment. Johnson (1965) has suggested some possible mechanisms for such an effect, and the problem is analogous to that of sea birds or seals feeding on or in the water, where some authors have speculated that domi-nance behaviour in relation to food may limit populations.

Another important example is the conclusion of Lack (1954, 1966) that territorial behaviour does not limit the breeding populations of tits. Kluyver (1951) had earlier suggested that it did, though without any direct evidence. Lack's view was based largely on studies of great tits in Marley Wood near Oxford (Lack 1964; Perrins 1965). The great tit occupies territories in spring (Kluyver & Tinbergen 1953; Gibb 1956) and these vary in size from year to year.

Although he did not study territorial behaviour, Lack concluded that it did not limit the breeding population. His main reason was that the data on breeding populations did not fit with a preconceived concept of terri-tory, which included the assumption that territory size should remain constant from year to year. We can see no reason against the alternative possibility that territory size may vary from year to year within areas and amongst areas, like any other biological parameter. Conversely, the fact that breeding populations in other species are fairly stable or that there is little turnover among territorial males has been used to conclude that these populations are limited by territorial behaviour (Bustard 1969; Lack 1969). This may seem likely, but cannot be fully established in the absence of evidence on conditions (a), (b) etc. in Table 1. Similarly, the conclusion of Lack that winter starvation limited breeding numbers of tits was arrived at by circumstantial evidence, not direct evidence or experiment (see Chitty 1967b, for a more detailed criticism.)

The measured data on great tits in Lack (1966) fit equally well into the concept that behaviour is an important limiting factor. If winter starvation were the main limiting factor (Lack 1966), variations in the production of young should have little if any influence on population changes. Lack says that this is so, but we have found that the data in his Fig. 14 show that breeding success is significantly correlated with population density in the next spring ($r = 0.545$, $P < 0.02$). This is the same correlation as in the red grouse, where the young produced in years of poor breeding take larger territories (Watson 1967; Watson & Moss 1970), and where territory size has been shown by experiment to be a mechanism limiting the breeding population (Watson & Jenkins 1968). Perhaps the same is true in great tits; at any rate, this is a hypothesis which should be falsifiable and which cannot be refuted until it is tested by experiment.

There is not space here to review the large amount of literature which contains conclusions that factors other than behaviour (e.g. weather, starvation, predation, disease, shortage of nest sites etc.) are important. We wish to stress, however, that dominance and spacing behaviour may well limit a population, even if these other factors are found to be the proximate cause of death or of lack of recruitment, and even if they are definitely shown to be amongst the limiting factors.

ONCE POPULATIONS HAVE BEEN LIMITED, HOW DO THEY DISPERSE?

A population may have been found to be limited by dominance behaviour and spacing (A), or by starvation or predation etc. (B), or by both. Question C now is—how do the animals, whose population has been limited by whatever mechanism, subsequently disperse themselves? This also arises with the many studies where the mechanism of limitation has not been identified. Subdividing C, the question is whether the pattern of spacing behaviour or dispersion is related to the amount or quality of food available (C1), or not (C2).

DISPERSION IN RELATION TO FOOD

Surveys of the literature, comparing territory size in many different species (Armstrong 1965; Schoener 1968) show that, in general, bird species with larger body weight occupy bigger territories, presumably because they need more food. This kind of extensive comparison is often quoted but does not get us very far, and is somewhat similar to the other general statement that there are usually more animals where their food seems to be more abundant.

Examples where dispersion is clearly related to food are mentioned in more detail under question A, and others are reviewed by Wynne-Edwards (1962) and Lack (1954, 1966). Holmes' work (1970) in this symposium, comparing the population density and territory size of dunlin between two different areas in Alaska, is a good example. Other useful examples are the studies of year-to-year differences within areas in the breeding populations of the snowy owl and pomarine skua (*Stercorarius pomarinus*) feeding on lemmings (*Lemmus trimucronatus*) in arctic Alaska (Pitelka, Tomich & Treichel 1955), and the short-eared owl (*Asio flammeus*) feeding on *Microtus* in Britain (Lockie 1955). These species all have strictly defended, exclusive, feeding territories. In some years the number of predators which settle in spring to breed cannot be explained in terms of even 100% survival of the previous summer's population of old and young on the same area, and therefore must be partly due to immigration. The territory size of the predators is affected in all three species, being bigger in springs when rodents are scarcer. This suggests a direct adjustment of territorial behaviour to food. Areas that are occupied one year may be deserted in the next winter and summer. There may even be modifications in territory size within an area in the same summer, in accordance with changes in food supply.

Crossbills (*Loxia* spp.), which feed almost entirely on seeds from coniferous trees, are another similar example (Svärdson 1957; Newton 1970, in this symposium). In Scotland, they feed on pine seeds, a source of food that fluctuates greatly. Nethersole-Thompson (p. 354) has found by an intensive population study in Scotland that breeding stocks disperse in relation to food, and that in years of scarcity they nearly all emigrate and populations suddenly appear elsewhere.

Within any one area in one year, territories of individual animals may differ in size. There is some evidence in the three rodent predators mentioned two paragraphs back that territory size varies inversely with prey abundance in the territory. A good example of this, apparently indicating a direct

15

adjustment, is Stenger's (1958) study of ovenbirds (*Seiurus aurocapillus*), where the density of invertebrates in the leaf litter per given area of ground in each pair's territory or home range was inversely correlated with the size of the territory.

These examples show that the dispersion of many animals is related to their food supply, sometimes even at the level of the individual and its territory size. How they do this is another question. Several speculative possibilities are: (a) the animal assesses the food supply fairly quickly on its arrival on the breeding ground, perhaps visually, and adjusts its behaviour accordingly; (b) dominance and spacing behaviour are affected indirectly via the animal's physiology, in turn affected by its nutrition; and (c) the overall food supply on areas of good and poor food may be similar, but its pattern of distribution quite different, so that spacing may be affected by the animal's relative ease in getting enough food (e.g. time spent searching, distance between good sources of food etc.). Almost nothing is known about these possible mechanisms, and there may well be others.

In comparing animal density with food intake and the food resources available, it may be dangerous to assume that different sets of *n* animals all have the same requirements and the same effects on the food supply. For example, Cowley (1970) in this symposium reviews cases where the nutritional requirements of different behavioural types of some domestic animals vary. Ayala (1968) gives another example with *Drosophila*, showing that different genetic strains show varying efficiency at using the food available.

DISPERSION IN RELATION TO FACTORS OTHER THAN NUTRITION

Examples of dispersion affected by the physical structure of the environment are given earlier (p. 196). There are a few cases where changes in food from year to year within an area, or variation from one part of an area to another in the same year, show that population density and dispersion are not related to food. Territory size in the rabbit has been found to bear no relation to food supply between years (Myers & Poole 1963), but this was based on work in confined conditions and food was not measured quantitatively; it needs to be tested in the wild. A better example is the study of tawny owls (*Strix aluco*) by Southern & Lowe (1968). There was a fairly stable number of territories from year to year, despite changes in the abundance of small rodents—which were the main food supply— either from year to year or from place to place within the study area in the same year. Another example is the golden eagle in Scotland, where the number of pairs within one area was remarkably stable over many years, in spite of very large annual changes of food supply, such as occurred after myxomatosis in rabbits, or after winters with severe mortality of deer and large amounts of carrion, or in years with big changes in numbers of other wild prey (Watson 1957b, but details not documented quantitatively). Equally, differences in the dispersion of eagles between areas were not related to quantitatively measured differences in the total amount of potential food (Brown & Watson 1964). Lockie's (1966) study of weasels showed evidence of a stable number of territories within areas in the weasel, despite five-fold variations in the numbers of *Microtus agrestis* which were the main food. He suggested (p. 160) that male 'weasels hold as much ground as they can and this gives sufficient available food in all situations except the most extreme. Such extremely low prey numbers may come only once or twice or perhaps not at all in the early history of a forest plantation'. This illustrates the point that spacing and dominance behaviour—and hence population density and dispersion—may in some cases bear little direct relation to food.

In tawny owls, golden eagles, and weasels, the territories or home ranges are so large that variations in food apparently have little or no effect on spacing behaviour. It may be asked: what is the point of such large territories? According to Wynne-Edwards (1962), it is to adjust numbers to a level below the potential limit set by food, so as to avoid the disaster of over-utilization and possible damage to both resources and population. Such a hypothesis appears to be almost beyond our present capacities to test by practical observation and experiment in the field; we will therefore not discuss it further here.

CONCLUSIONS AND SUMMARY

There are many papers dealing with territorial behaviour and with other forms of dominance and spacing behaviour in vertebrates, many on population studies, and much work on nutrition. Unfortunately, very few studies consider any two of these aspects together, and practically none consider all three. Consequently there is an inordinate amount of speculation in the literature—but relatively few data—about the relation between these aspects.

There are too few experimental studies of dominance or spacing be-haviour in relation to the nutrition and populations of vertebrates to justify drawing any firm conclusions. The main division of opinion is (i) spacing behaviour does limit breeding populations of vertebrates or (ii) this behaviour is simply the way in which animals, whose breeding populations have been limited by some other factor (usually starvation), space themselves subsequently. There is certainly good evidence that dominance and spacing behaviour limit populations of confined vertebrates, and some evidence from the wild. However, far more work needs to be done before this can be established as a widespread phenomenon; many authors have concluded that populations are limited by territorial behaviour and also by spacing behaviour within groups, but on inadequate evidence and sometimes on no direct evidence whatever. In some cases, populations appear to be limited by dominance or spacing behaviour independently of food, but in others food is involved either directly or indirectly in time. There are a few examples where limits are set by two or more factors apparently acting simultaneously (e.g. dominance or spacing behaviour, food, physical features in the environment) and this situation may well be fairly common, though seldom actually studied so far. Unfortunately, there is a tendency in the literature to emphasize the importance of either one factor or another, to the exclusion of others. Present evidence suggests that dominance and spacing behaviour—and thus population limitation— are usually affected by a change in the plane of nutrition (i.e. food shortage or a reduction of food quality) before animals are actually killed by starvation.

Many studies have concluded that factors other than dominance and spacing behaviour (e.g. disease, starvation, weather, predation, shortage of nest sites) set limits to populations. Starvation by food shortage has been claimed in many studies to limit populations of deer, great tits and other animals. A general characteristic of most of these studies is that dominance and spacing behaviour were not directly studied or were ignored. Behaviour may be one limiting factor in all these populations, even if starvation, predation etc. do prove to be limiting also. At any rate this is a possibility that should not be discounted without study, and should be tested experimentally in these cases.

A third question in this review was to see if spacing patterns within populations, which have been limited by whatever factors, are related to food supplies. Some good examples show they are, although how animals do this is uncertain. In some cases, factors other than food also affect spacing behaviour and dispersion.

Although a general conclusion of this review is that behaviour is frequently a major factor limiting populations of vertebrates, there is still a great variety of interpretations about the relative importance of the various different limiting factors in the wild. One important reason contributing to this uncertainty is that too few studies include observations on enough different aspects to enable the author to decide conclusively the relative importance of each. Another reason is that too many studies of vertebrates still rely purely on describing and measuring the natural situation and too few experiments have been done or are being done.

ACKNOWLEDGEMENTS

We are greatly indebted to Dr D. Jenkins, Mr H. N. Southern and particularly Prof. V. C. Wynne-Edwards for their valuable criticisms of the manuscript. This does not imply that they agree with the points of view that we have expressed.

REFERENCES

ALTMAN I. & HAYTHORN W.W. (1967) The ecology of isolated groups. *Behavl Sci.* **12**, 169–82.

ANDERSEN J. (1961) Biology and management of roe deer in Denmark. *Terre Vie*, **1**, 41–53.

ANDERSEN J. (1962) Roe deer census and population analysis by means of a modified marking-release technique. *The Exploitation of Natural Animal Populations* (Ed. by E. D. Le Cren and M. W. Holdgate), pp. 72–82. Oxford.

ANDERSEN J. (1963) Populations of hare and roe-deer in Denmark. *Proc. Int. Congr. Zool.* **16** (Specialized Symp.) 3, 347–51.

ANDREWARTHA H.G. & BIRCH L.C. (1954) *The Distribution and Abundance of Animals*. Chicago.

ANDREWS R.V. (1968) Daily and seasonal variation in adrenal metabolism of the brown lemming. *Physiol. Zoöl.* **41**, 86–94.

ARMSTRONG J.T. (1965) Breeding home range in the nighthawk and other birds: its evolutionary and ecological significance. *Ecology*, **46**, 619–29.

ASHMOLE N.P. (1963) The regulation of numbers of tropical oceanic birds. *Ibis*, **103b**, 458–73.

AYALA F.J. (1968) Genotype, environment, and population numbers. *Science, N.Y.* **162**, 1453–9.

BARNETT S.A. (1958) An analysis of social behaviour in wild rats. *Proc. zool. Soc. Lond.* **130**, 107–52.

BARNETT S.A. (1964) Social stress. *Viewpoints in Biology*, **3** (Ed. by J. D. Carthy & C. L. Duddington), pp. 170–218. London.

BARNETT S.A. & EVANS C.S. (1965) Questions on the social dynamics of rodents. *Symp. zool. Soc. Lond.* **14**, 233–48.

BENDELL J.F. & ELLIOTT P.W. (1967) Behaviour and the regulation of numbers in blue grouse. *Can. Wildl. Serv. Rep. Ser.* **4**.

BEVERTON R.J.H. (1962) Long-term dynamics of certain North Sea fish populations. *The Exploitation of Natural Animal Populations* (Ed. by E. D. Le Cren & M. W. Holdgate), pp. 242–59. Oxford.

BEVERTON R.J.H. & HOLT S.J. (1957) On the dynamics of exploited fish populations. *Fish. Invest., Lond.* ser. 2, **19**.

BLANK T.H. & ASH J.S. (1962) Fluctuations in a partridge population. *The Exploitation of Natural Animal Populations* (Ed. by E.D. Le Cren & M.W. Holdgate), pp. 118–30. Oxford.

BOUSSU M.F. (1954) Relationship between trout populations and cover in a small stream. *J. Wildl. Mgmt,* **18**, 229–39.

BRAMLEY P.S. (1970) Territorial and reproductive behaviour in roe deer. *J. Reprod. Fert.* (in press).

BRONSON F.H. (1964) Agonistic behaviour in woodchucks. *Anim. Behav.* **12**, 470–8.

BROWN R.Z. (1953) Social behavior, reproduction and population changes in the house mouse. *Ecol. Monogr.* **23**, 217–40.

BROWN L.H. & WATSON A. (1964) The golden eagle in relation to its food supply. *Ibis,* **106**, 78–100.

BUECHNER H.K. (1961) Territorial behavior in Uganda kob. *Science, N.Y.* **133**, 698–9.

BURT W.H. (1943) Territoriality and home range concepts as applied to mammals. *J. Mammal.* **24**, 346–52.

BUSTARD H.R. (1969) The population ecology of the gekkonid lizard (*Gehyra variegata* (Duméril & Bibron)) in exploited forests in northern New South Wales. *J. Anim. Ecol.* **38**, 35–51.

CALHOUN J.B. (1948) Mortality and movement of brown rats (*Rattus norvegicus*) in artificially supersaturated populations. *J. Wildl. Mgmt* **12**, 167–72.

CALHOUN J.B. (1949) A method for self control of population growth among mammals living in the wild. *Science, N.Y.* **109**, 333–5.

CALHOUN J.B. (1950) The study of wild animals under controlled conditions. *Ann. N.Y. Acad. Sci.* **51**, 1113–22.

CALHOUN J.B. (1956) A comparative study of the social behavior of two inbred strains of house mice. *Ecol. Monogr.* **26**, 81–103.

CALHOUN J.B. (1961) Determinants of social organization exemplified in a single population of domesticated rats. *Trans. N.Y. Acad. Sci.* **23**, 437–42.

CALHOUN J.B. (1962) A 'behavioral sink'. *Roots of Behavior* (Ed. by E. L. Bliss), pp. 295–315. New York.

CALHOUN J.B. (1963) *The Ecology and Sociology of the Norway Rat.* Bethesda, Maryland.

CARRICK R. (1963) Ecological significance of territory in the Australian magpie *Gymnorhina tibicen. Proc. Int. orn. Congr.* **13**, 740–53.

CARRICK R., CSORDAS S.E. & INGHAM S.E. (1962) Studies on the southern elephant seal, *Mirounga leonina* (L.). IV. Breeding and development. *C.S.I.R.O. Wildlife Res.* **7**, 161–97.

CARRICK R. & MURRAY M.D. (1964) Social factors in population regulation of the silver gull, *Larus novaehollandiae* Stephens. *C.S.I.R.O. Wildl. Res.* **9**, 189–99.

CHAPMAN D.W. (1962) Aggressive behaviour in juvenile coho salmon as a cause of emigration. *J. Fish. Res. Bd Can.* **19**, 1047–80.

CHITTY D. (1957) Self-regulation of numbers through changes in viability. *Cold Spring Harb. Symp. quant. Biol.* **22**, 277–80.

CHITTY D. (1960) Population processes in the vole and their relevance to general theory. *Can. J. Zool.* **38**, 99–113.

CHITTY D. (1967a) The natural selection of self-regulatory behaviour in animal populations. *Proc. ecol. Soc. Aust.* **2**, 51–78.

CHITTY D. (1967b) What regulates bird populations? *Ecology*, **48**, 698–701.

CHITTY D. & PHIPPS E. (1966) Seasonal changes in survival in mixed populations of two species of vole. *J. Anim. Ecol.* **35**, 313–31.

CHITTY D., PIMENTEL D. & KREBS C.J. (1968) Food supply of overwintered voles. *J. Anim. Ecol.* **37**, 113–20.

CHRISTIAN J.J. (1968) The potential role of the adrenal cortex as affected by social rank and population density on experimental epidemics. *Am. J. Epidem.* **87**, 255–66.

CHRISTIAN J.J., FLYGER V. & DAVIS D.E. (1960). Factors in the mass mortality of a herd of sika deer *Cervus nippon*. *Chesapeake Sci.* **1**, 79–95.

CHRISTIAN J.J., LLOYD J.A. & DAVIS D.E. (1965) The role of endocrines in the self-regulation of mammalian populations. *Recent Prog. Horm. Res.* **21**, 501–77.

CLARKE J.R. (1955) Influence of numbers on reproduction and survival in two experimental vole populations. *Proc. R. Soc.* B, **144**, 68–85.

CLOUGH G.C. (1968) Social behavior and ecology of Norwegian lemmings during a population peak and crash. *Meddr St. Viltunders.* ser. 2, **28**.

COLLIAS N.E. & JAHN L.R. (1959) Social behavior and breeding success in Canada geese (*Branta canadensis*) confined under semi-natural conditions. *Auk*, **76**, 478–509.

COLLIAS N.E. & TABER R.D. (1951) A field study of some grouping and dominance relations in ring-necked pheasants. *Condor*, **53**, 265–75.

COMMITTEE OF INQUIRY ON GROUSE DISEASE (1911) *The Grouse in Health and in Disease*. 2 vols. London.

COWLEY J.J. (1970) Food intake and the modification of behaviour. *Animal Populations in relation to their Food Resources* (Ed. by A. Watson), pp. 223–8. Oxford & Edinburgh.

COWLEY J.J. & GRIESEL R.D. (1966) The effect on growth and behaviour of rehabilitating first and second generation low protein rats. *Anim. Behav.* **14**, 506–17.

CRAIG J.V., ORTMAN L.L. & GUHL A.M. (1965) Genetic selection for social dominance ability in chickens. *Anim. Behav.* **13**, 114–33.

CREW F.A.E. & MIRSKAIA L. (1931) The effects of density on an adult mouse population. *Biologia gen.* **7**, 239–50.

CROWCROFT P. & ROWE F.P. (1957) The growth of confined colonies of the wild house mouse (*M. musculus* L.). *Proc. zool. Soc. Lond.* **129**, 359–70.

CROWCROFT P. & ROWE F.P. (1958) The growth of confined colonies of the wild house-mouse (*Mus musculus* L.): the effect of dispersal on female fecundity. *Proc. zool. Soc. Lond.* **131**, 357–65.

CROWCROFT P. & ROWE F.P. (1963) Social organization and territorial behaviour in the wild house mouse (*Mus musculus* L.). *Proc. zool. Soc. Lond.* **140**, 517–31.

CUMMING H.G. (1966) Behaviour and dispersion in roe deer (*Capreolus capreolus*). Ph.D. thesis, Univ. of Aberdeen.

DAVIS D.E. (1942) The phylogeny of social nesting habits in the Crotophaginae. *Q. Rev. Biol.* **17**, 115–34.

DAVIS D.E. (1951) The relation between level of population and pregnancy of normal rats. *Ecology*, **27**, 168–81.

DAVIS D.E. (1953) The characteristics of rat populations. *Q. Rev. Biol.* **28**, 373–401.

DAVIS D.E. (1958) The role of density in aggressive behaviour in house mice. *Anim. Behav.* **6**, 207–11.

DAVIS D.E. (1962) The potential harvest of woodchucks. *J. Wildl. Mgmt*, **26**, 144–9.

DAVIS D.E., CHRISTIAN J.J. & BRONSON F. (1964) Effect of exploitation on birth, mortality, and movement rates in a woodchuck population. *J. Wildl. Mgmt*, **28**, 1–9.

DELIUS J.D. (1965) A population study of skylarks *Alauda arvensis*. *Ibis*, **107**, 466–92.

DELONG K.T. (1967) Population ecology of feral house mice. *Ecology*, **48**, 611–34.

DHONDT A.A. & HUBLÉ J.H. (1968) Fledging date and sex in relation to dispersal in young great tits. *Bird Study*, **15**, 127–34.

DIXON J.B. (1937) The golden eagle in San Diego county, California. *Condor*, **39**, 49–56.

DUNNET G.M. (1967) Social organization in shelduck, *Tadorna tadorna* on the Ythan estuary. *Proc. R. Soc. Popul. Study Grp*, **2**, 45–8.

ELKAN E. (1960) The common toad (*Bufo bufo* L.) in the laboratory. *Br. J. Herpet.* **2**, 177–82 (cited by Wynne-Edwards 1962).

EMLEN J.T., STOKES A.W. & WINSOR C.P. (1948) The rate of recovery of decimated populations of brown rats in nature. *Ecology*, **29**, 133–45.

ERRINGTON P.L. (1943) An analysis of mink predation upon muskrats in north-central United States. *Res. Bull. Iowa agric. Exp. Stn*, **320**.

ERRINGTON P.L. (1945) Some contributions of a fifteen-year local study of the northern bobwhite to a knowledge of population phenomena. *Ecol. Monogr.* **15**, 1–34.

ERRINGTON P.L. (1956) Factors limiting higher vertebrate populations. *Science, N.Y.* **124**, 304–7.

ERRINGTON P.L. (1957) Of population cycles and unknowns. *Cold Spring Harb. Symp. quant. Biol.* **22**, 287–300.

ERRINGTON P.L. (1963) *Muskrat Populations*. Ames, Iowa.

ESSER A.H., CHAMBERLAIN A.S., CHAPPLE E.D. & KLINE N.S. (1965) Territoriality of patients on a research ward. *Recent Adv. Biol. Psychiat.* **7**, 37–44. New York.

FRANK F. (1957) The causality of microtine cycles in Germany. *J. Wildl. Mgmt*, **21**, 113–21.

FULLER W.A., STEBBINS L.L. & DYKE G.R. (1969) Overwintering of small mammals near Great Slave Lake Northern Canada. *Arctic*, **22**, 34–55.

GIBB J. (1956) Territory in the genus *Parus*. *Ibis*, **98**, 420–9.

GIBB J. (1960) Populations of tits and goldcrests and their food supply in pine plantations. *Ibis*, **102**, 163–208.

GUHL A.M. (1953) Social behaviour of domestic fowl. *Tech. Bull. Kans. agric. Exp. Stn*, **73**.

GULLAND J.A. (1962) The application of mathematical models to fish populations. *The Exploitation of Natural Animal Populations* (Ed. by E. D. Le Cren & M. W. Holdgate), pp. 204–17. Oxford.

HALL E.T. (1970) The facilitation of communication in a cultural environment. *The Use of Space by Animals and Men* (Ed. by A. H. Esser), in press.

HEALEY M.C. (1967) Aggression and self-regulation of population size in deermice. *Ecology*, **48**, 377–92.

HENSLEY M.M. & COPE J.B. (1951) Further data on removal and repopulation of the breeding birds in a spruce-fir forest community. *Auk*, **68**, 483–93.

HINDE R.A. (1956) The biological significance of the territories of birds. *Ibis*, **98**, 340–69.

HOFFMAN R.S. (1958) The role of reproduction and mortality in population fluctuations of voles (*Microtus*). *Ecol. Monogr.* **28**, 79–109.

HOLMES R.T. (1966) Breeding ecology and annual cycle adaptations of the red-backed sandpiper (*Calidris alpina*) in northern Alaska. *Condor*, **68**, 3–46.

HOLMES R.T. (1970) Differences in population density, territoriality and food supply of dunlin on arctic and subarctic tundra. *Animal Populations in relation to their Food Resources* (Ed. by A. Watson), pp. 303–22. Oxford & Edinburgh.

HORI J. (1964) The breeding biology of the shelduck *Tadorna tadorna*. *Ibis*, **106**, 333–60.

HUNTER R.F. & DAVIES G.E. (1963) The effect of method of rearing on the social behaviour of Scottish blackface hoggets. *Anim. Prod.* **5**, 183–94.

JENKINS D. (1961a) Population control in protected partridges (*Perdix perdix*). *J. Anim. Ecol.* **30**, 235–58.

JENKINS D. (1961b) Social behaviour in the partridge *Perdix perdix*. *Ibis*, **103a**, 155–88.

JENKINS D., WATSON A. & MILLER G.R. (1963) Population studies on red grouse, *Lagopus lagopus scoticus* (Lath.) in north-east Scotland. *J. Anim. Ecol.* **32**, 317–76.

JENKINS D., WATSON A. & MILLER G.R. (1964) Predation and red grouse populations. *J. appl. Ecol.* **1**, 183–95.

JENKINS D., WATSON A. & MILLER G.R. (1967) Population fluctuations in the red grouse *Lagopus lagopus scoticus*. *J. Anim. Ecol.* **36**, 97–122.

JENKINS D., WATSON A. & PICOZZI N. (1965) Red grouse chick survival in captivity and in the wild. *Trans. Int. Un. Game Biol.* **6**, 63–70.

JEWELL P.A. (1966) The concept of home range in mammals. *Symp. zool. Soc. Lond.* **18**, 85–109.

JOHNSON W.E. (1965) On mechanisms of self-regulation of population abundance in *Oncorhynchus nerka*. *Mitt. int. Verein. theor. angew. Limnol.* **13**, 66–87.

KALELA O. (1954) Über den Revierbesitz bei Vögeln und Saugetieren als populations-ökologischer Faktor. *Suomal. eläin- ja kasvit. Seur. van. Julk.* **16**, no. 2.

KALELA O. (1957) Regulation of reproduction rate in subarctic populations of the vole *Clethrionomys rufocanus* (Sund.). *Suomal. Tiedeakat. Toim.* A, **4**, 34.

KALELA O. (1962) On the fluctuations in the numbers of arctic and boreal small rodents as a problem of production biology. *Suomal. Tiedeakat. Toim.* A, **4**, 66.

KALLEBERG H. (1958) Observations in a stream tank of territoriality and competition in juvenile salmon and trout (*Salmo salar* L. and *S. trutta* L.). *Rep. Inst. Freshwat. Res. Drottningholm*, **39**, 55–98.

KEENLEYSIDE M.H.A. & YAMAMOTO F.T. (1962) Territorial behaviour of juvenile Atlantic salmon (*Salmo salar* L.). *Behaviour*, **19**, 139–69.

KING J.A. (1955) Social behaviour, social organization, and population dynamics of a black-tailed prairie dog town in the Black Hills of South Dakota. *Contr. Lab. vertebr. Biol. Univ. Mich.* **67**, 1–123.

KLEIN D.R. (1970) Food selection by North American deer and their response to over-utilization of preferred plant species. *Animal Populations in relation to their Food Resources* (Ed. by A. Watson), pp. 25–46. Oxford & Edinburgh.

KLUYVER H.N. (1951) The population ecology of the great tit, *Parus major* L. *Ardea*, **39**, 1–135.

214 ADAM WATSON AND ROBERT MOSS

KLUYVER H.N. & TINBERGEN L. (1953) Territory and the regulation of density in titmice. *Arch. néerl. Zool.* **10**, 265–89.

KREBS C.J. (1964) The lemming cycle at Baker Lake, Northwest Territories, during 1959–62. *Tech. Pap. Arct. Inst. N. Am.* **15**.

KREBS C.J. (1966) Demographic changes in fluctuating populations of *Microtus californicus*. *Ecol. Monogr.* **36**, 239–73.

KREBS C.J. & DELONG K.T. (1955) A *Microtus* population with supplemental food. *J. Mammal.* **46**, 566–73.

LACK D. (1954) *The Natural Regulation of Animal Numbers.* Oxford.

LACK D. (1964) A long-term study of the great tit. *J. Anim. Ecol.* **33** (Jubilee Suppl.), 159–73.

LACK D. (1966) *Population Studies of Birds.* Oxford.

LACK D. (1969) Population changes in the land birds of a small island. *J. Anim. Ecol.* **38**, 211–18.

LAGERSPETZ K. (1964) Studies on the aggressive behaviour of mice. *Suomal. Tiedeakat. Toim.* B, **131**, 1–131.

LARKIN P.A. (1956) Interspecific competition and population control in freshwater fish. *J. Fish. Res. Bd Can.* **13**, 327–42.

LAWS R.M. (1956) The elephant seal (*Mirounga leonina* Linn.). II. General, special and reproductive behaviour. *Scient. Rep. Falkld Isl. Depend. Surv.* **13**.

LE CREN E.D. (1965) Some factors regulating the size of populations of freshwater fish. *Mitt. int. Verein. theor. angew. Limnol.* **13**, 88–105.

LEYHAUSEN P. (1965) The communal organization of solitary mammals. *Symp. zool. Soc. Lond.* **14**, 249–63.

LICHT L.E. (1967) Growth inhibition in crowded tadpoles: intraspecific and interspecific effects. *Ecology*, **48**, 736–45.

LINDROTH A. (1965) The Baltic salmon stock. *Mitt. int. Verein. theor. angew. Limnol.* **13**, 163–92.

LOCKIE J.D. (1955) The breeding habits and food of short-eared owls after a vole plague. *Bird Study*, **2**, 53–69.

LOCKIE J.D. (1956) Winter fighting in feeding flocks of rooks, jackdaws and carrion crows. *Bird Study*, **3**, 180–90.

LOCKIE J.D. (1966) Territory in small carnivores. *Symp. zool. Soc. Lond.* **18**, 143–65.

LOCKLEY R.M. (1961) Social structure and stress in the rabbit warren. *J. Anim. Ecol.* **30**, 385–423.

LOWE V.P.W. (1969) Population dynamics of the red deer (*Cervus elaphus* L.) on Rhum. *J. Anim. Ecol.* **38**, 425–57.

MARSHALL A.J. (1954) *Bower-birds.* Oxford.

MCBRIDE G. (1966) Society evolution. *Proc. ecol. Soc. Aust.* **1**, 1–13.

MCCABE T.T. & BLANCHARD B.D. (1950) *Three Species of Peromyscus.* Santa Barbara, California.

MCCRIMMON H.R. (1954) Stream studies on planted Atlantic salmon. *J. Fish. Res. Bd Can.* **11**, 362–403.

MCLAREN I.A. (1962) Population dynamics and exploitation of seals in the eastern Canadian Arctic. *The Exploitation of Natural Animal Populations* (Ed. by E. D. Le Cren & M. W. Holdgate), pp. 168–83. Oxford.

MESLOW E.C. & KEITH L.B. (1968) Demographic parameters of a snowshoe hare population. *J. Wildl. Mgmt*, **32**, 812–34.

MOFFAT C.B. (1903) The spring rivalry of birds: some views on the limit to multiplication. *Ir. Nat.* **12**, 152–66.

MOSS R. (1969) A comparison of red grouse (*Lagopus l. scoticus*) stocks with the production and nutritive value of heather (*Calluna vulgaris*). *J. Anim. Ecol.* **38**, 103–12.

MOYER K.E. (in press) A preliminary physiological model of aggressive behaviour. *Fighting and Defeat* (Ed. by J. P. Scott and B. E. Eleftheriou), in press. Chicago.

MULLEN D.A. (1968) Reproduction in brown lemmings (*Lemmus trimucronatus*) and its relevance to their cycle of abundance. *Univ. Calif. Publs Zool.* **85**.

MUNDAY K.A. (1961) Aspects of stress phenomena. *Symp. Soc. exp. Biol.* **15**, 168–89.

MURTON R.K. (1968) Some predator-prey relationships in bird damage and population control. *The Problem of Birds as Pests* (Ed. by R. K. Murton & E. N. Wright), pp. 157–69. London & New York.

MURTON R.K., ISAACSON A.J. & WESTWOOD N.J. (1966) The relationships between wood-pigeons and their clover food supply and the mechanism of population control. *J. appl. Ecol.* **3**, 55–96.

MYERS K. (1964) Influence of density on fecundity, growth rates, and mortality in the wild rabbit. *C.S.I.R.O. Wildl. Res.* **9**, 134–7.

MYERS K. (1966) The effects of density on sociality and health in mammals. *Proc. ecol. Soc. Aust.* **1**, 40–64.

MYERS K., HALE C.S., MYKYTOWYCZ R. & HUGHES R.L. (1970) The effects of density on sociality and health in the rabbit. *The Use of Space by Animals and Men* (Ed. by A. H. Esser), in press.

MYERS K. & POOLE W.E. (1961) A study of the biology of the wild rabbit, *Oryctolagus cuniculus* (L.), in confined populations. II. The effects of season and population increase on behaviour. *C.S.I.R.O. Wildl. Res.* **6**, 1–41.

MYERS K. & POOLE W.E. (1963) A study of the biology of the wild rabbit, *Oryctolagus cuniculus* (L.), in confined populations. V. Population dynamics. *C.S.I.R.O. Wildl. Res.* **8**, 166–203.

MYERS K. & SCHNEIDER E.C. (1964) Observations on reproduction, mortality, and behaviour in a small, free-living population of wild rabbits. *C.S.I.R.O. Wildl. Res.* **9**, 138–43.

MYKYTOWYCZ R. (1960) Social behaviour of an experimental colony of wild rabbits, *Oryctolagus cuniculus* (L.). III. Second breeding season. *C.S.I.R.O. Wildl. Res.* **5**, 1–20.

MYKYTOWYCZ R. (1961) Social behaviour of an experimental colony of wild rabbits, *Oryctolagus cuniculus* (L.). IV. Conclusion: outbreak of myxomatosis, third breeding season, and starvation. *C.S.I.R.O. Wildl. Res.* **6**, 142–55.

NELSON J.B. (1966) Population dynamics of the gannet (*Sula bassana*) at the Bass Rock, with comparative information from other Sulidae. *J. Anim. Ecol.* **35**, 443–70.

NEWSOME A.E. (1969a) A population study of house-mice temporarily inhabiting a South Australian wheatfield. *J. Anim. Ecol.* **38**, 341–59.

NEWSOME A.E. (1969b) A population study of house-mice permanently inhabiting a reed-bed in South Australia. *J. Anim. Ecol.* **38**, 361–77.

NEWTON I. (1970) Irruptions of crossbills in northern Europe. *Animal Populations in relation to their Food Resources* (Ed. by A. Watson), pp. 337–57. Oxford & Edinburgh.

ORIANS G.H. (1961) The ecology of blackbird (*Agelaius*) social systems. *Ecol. Monogr.* **31**, 285–312.

PATTERSON I.J. (1970) Food-fighting in rooks. *Animal Populations in relation to their Food Resources* (Ed. by A. Watson), pp. 249–52. Oxford & Edinburgh.

PERRINS C.M. (1965) Population fluctuations and clutch-size in the great tit, *Parus major. J. Anim. Ecol.* **34**, 601–47.

PETRUSEWICZ K. (1957) Investigation of experimentally induced population growth. *Ekol. pol.* A, **5**, 281–309.

PETRUSEWICZ,K. (1963) Population growth induced by disturbance in the ecological structure of the population. *Ekol. pol.* A, **11**, 87–125.

PETRUSEWICZ K. (1966) Dynamics, organization and ecological structure of population. *Ekol. pol.* A, **14**, 413–36.

PETRUSEWICZ K. & ANDRZEJEWSKI R. (1962) Natural history of a free-living population of house mice (*Mus musculus* Linnaeus) with particular reference to groupings within the population. *Ekol. pol.* A, **10**, 85–122.

PINOWSKI J. (1965) Overcrowding as one of the causes of dispersal of young tree sparrows. *Bird Study*, **12**, 27–33.

PITELKA F.A. (1957) Some aspects of population structure in the short-term cycle of the brown lemming in northern Alaska. *Cold Spring Harb. Symp. quant. Biol.* **22**, 237–51.

PITELKA F.A., TOMICH P.Q. & TREICHEL G.W. (1955) Ecological relations of jaegers and owls as lemming predators near Barrow, Alaska. *Ecol. Monogr.* **25**, 85–117.

RATCLIFFE D.A. (1962) Breeding density in the peregrine *Falco peregrinus* and raven *Corvus corax. Ibis*, **104**, 13–39.

RICHARDS O.W. & SOUTHWOOD T.R.E. (1968) The abundance of insects: introduction. *Insect Abundance* (Ed. by T. R. E. Southwood), pp. 1–7. Oxford & Edinburgh.

RICHDALE L.E. (1957) *A Population Study of Penguins.* Oxford.

RICKER W.E. (1954a) Stock and recruitment. *J. Fish Res. Bd Can.* **11**, 559–623.

RICKER W.E. (1954b) Effects of compensatory mortality upon population abundance. *J. Wildl. Mgmt*, **18**, 45–51.

ROBERTSON F.W. & SANG J.H. (1944) The ecological determinants of population growth in a *Drosophila* culture. I. Fecundity of adult flies. *Proc. R. Soc.* B, **132**, 258–77.

ROSE S.M. & ROSE F.C. (1965) The control of growth and reproduction in freshwater organisms by specific products. *Mitt. int. Verein. theor. angew. Limnol.* **13**, 21–35.

ROWE F.P., TAYLOR E.J. & CHUDLEY A.H.J. (1964) The effect of crowding on the reproduction of the house-mouse (*Mus musculus* L.) living in corn-ricks. *J. Anim. Ecol.* **33**, 477–83.

SADLEIR R.M.F.S. (1965) The relationship between agonistic behaviour and population changes in the deermouse *Peromyscus maniculatus* (Wagner). *J. Anim. Ecol.* **34**, 331–52.

SAUNDERS J.W. & SMITH M.W. (1962) Physical alteration of stream habitat to improve trout production. *Trans. Am. Fish. Soc.* **91**, 185–8.

SCHAANNING H.T.L. (1907) Østfinmarkens fuglefauna. *Bergens Mus. Årb.* **8**, 1–98.

SCHOENER T.W. (1968) Sizes of feeding territories among birds. *Ecology*, **49**, 123–41.

SCHULTZ A.M. (1966) The nutrient-recovery hypothesis for arctic microtine cycles. *Grazing in Terrestrial and Marine Environments* (Ed. by D. J. Crisp), pp. 57–68. Oxford.

SEMEONOFF R. & ROBERTSON F.W. (1968) A biochemical and ecological study of plasma esterase polymorphism in natural populations of the field vole, *Microtus agrestis* L. *Bioch. Gen.* **1**, 205–27.

SHOEMAKER H.H. (1939) Social hierarchy in flocks of the canary. *Auk*, **56**, 381–406.

SILLIMAN R.P. (1968) Interaction of food level and exploitation in experimental fish populations. *Fishery Bull. Fish Wildl. Serv. U.S.* **66**, 425–39.

SILLIMAN R.P. & GUTSELL J.S. (1958) Experimental exploitation of fish populations. *Fishery Bull. Fish Wildl. Serv. U.S.* **58**, 214–52.

SMITH S.M. (1967) Seasonal changes in the survival of the black-capped chickadee. *Condor*, **69**, 344–59.

SMYTH M. (1968) The effects of the removal of individuals from a population of bank voles *Clethrionomys glareolus*. *J. Anim. Ecol.* **37**, 167–83.

SNOW D.W. (1958) *A Study of Blackbirds*. London.

SNYDER R.L. (1961) Evolution and integration of mechanisms that regulate population growth. *Proc. natn. Acad. Sci. U.S.A.* **47**, 449–55.

SNYDER W.D. (1967) Experimental habitat improvement for scaled quail. *Tech. Publs Colo. Game Fish Pks Dep.* **19**.

SOLOMON M.E. (1964) Analysis of processes involved in the natural control of insects. *Adv. ecol. Res.* **2**, 1–58.

SOMMER R. Spatial parameters in naturalistic social research. *The Use of Space by Animals and Men* (Ed. by A. H. Esser), in press.

SOUTHERN H.N. & LOWE V.P.W. (1968) The pattern of distribution of prey and predation in tawny owl territories. *J. Anim. Ecol.* **37**, 75–97.

SOUTHWICK C.H. (1955a) Regulatory mechanisms of house mouse populations: social behaviour affecting litter survival. *Ecology*, **36**, 627–34.

SOUTHWICK C.H. (1955b) The population dynamics of confined house mice supplied with unlimited food. *Ecology*, **36**, 212–25.

SOUTHWICK C.H. (1958) Population characteristics of house mice living in English corn ricks: density relationships. *Proc. zool. Soc. Lond.* **131**, 163–75.

SOUTHWOOD T.R.E. & CROSS D.J. (1969) The ecology of the partridge. III. Breeding success and the abundance of insects in natural habitats. *J. Anim. Ecol.* **38**, 497–509.

STENGER J. (1958) Food habits and available food of ovenbirds in relation to territory size. *Auk*, **75**, 335–46.

STEWART R.E. & ALDRICH J.W. (1951) Removal and repopulation of breeding birds in a spruce-fir forest community. *Auk*, **68**, 471–82.

STRECKER R.L. (1954) Regulating mechanisms in house mouse populations: the effect of limited food-supply on an unconfined population. *Ecology*, **35**, 249–53.

STRECKER R.L. & EMLEN J.T. (1953) Regulating mechanisms in house-mouse populations: the effect of limited food supply on a confined population. *Ecology*, **34**, 375–85.

SVÄRDSON G. (1957) The 'invasion' type of bird migration. *Br. Birds*, **50**, 314–43.

TERMAN C.R. (1962) Spatial and homing consequences of the introduction of aliens into semi-natural populations of prairie deermice. *Ecology*, **43**, 216–23.

TERMAN C.R. (1963) The influence of differential early social experience upon distribution within populations of prairie deermice. *Anim. Behav.* **11**, 246–62.

THOMPSON D.Q. (1955) The role of food and cover in population fluctuations of the brown lemming at Point Barrow, Alaska. *Trans. N. Am. Wildl. Conf.* **20**, 166–74.

TINBERGEN N. (1957) The functions of territory. *Bird Study*, **4**, 14–27.

TINBERGEN N. (1968) Territory in the three-spined stickleback *Gasterosteus aculeatus* L. *Anim. Beh.* **16**, 398–9. (Review.)

TOMPA F.S. (1964) Factors determining the numbers of song sparrows *Melospiza melodia* (Wilson), on Mandarte Island, B.C., Canada. *Acta zool. fenn.* **109**, 1–68.

VAN DEN ASSEM J. (1967) Territory in the three-spined stickleback *Gasterosteus aculeatus* L. *Behaviour*, suppl. 16.

WATSON A. (1957a) The behaviour, breeding, and food-ecology of the snowy owl *Nyctea scandiaca*. *Ibis*, **99**, 419–62.

WATSON A. (1957b) The breeding success of golden eagles in the north-east Highlands. *Scott. Nat.* **69**, 153–69.

WATSON A. (1964) Aggression and population regulation in red grouse. *Nature, Lond.* **202**, 506–7.

WATSON A. (1965) A population study of ptarmigan (*Lagopus mutus*) in Scotland. *J. Anim. Ecol.* **34**, 135–72.

WATSON A. (1967) Social status and population regulation in the red grouse (*Lagopus lagopus scoticus*). *Proc. R. Soc. Popul. Study Grp*, **2**, 22–30.

WATSON A. & JENKINS D. (1968) Experiments on population control by territorial behaviour in red grouse. *J. Anim. Ecol.* **37**, 595–614.

WATSON A. & MOSS R. (1970) Spacing as affected by territorial behavior, habitat and nutrition in red grouse. *The Use of Space by Animals and Men* (Ed. by A. H. Esser), in press.

WILSON F. (1968) Insect abundance: prospect. *Insect Abundance* (Ed. by T. R. E. Southwood), pp. 143–58. Oxford and Edinburgh.

WOOD-GUSH D.G.M. (1965) The social organization of domestic bird communities. *Symp. zool. Soc. Lond.* **14**, 219–31.

WYNNE-EDWARDS V.C. (1962) *Animal Dispersion in relation to Social Behaviour.* Edinburgh & London.

DISCUSSION

D. LACK: You have omitted what is to me the fundamental question. The red grouse has evolved a highly elaborate system of behaviour centred around its territory. What is the survival value of this behaviour? In other words, why (in the evolutionary sense) is it only the territory holders which breed?

A. WATSON: I agree that this is a fundamental question. The trouble is that it is impossible at present to do much better than guess about the selective value of territorial behaviour in a long-term evolutionary sense; we just don't have the factual data, in red grouse or any other species. There are of course several possible functions for territorial behaviour, which might all have selective value in different species or even in the same species. Territory may be an 'address' where the male and female, having been separated for any reason, can quickly find each other again. It may also allow mating without disturbance from other males. In the red grouse it provides an exclusive food supply in the critical few weeks before egg laying, and it also limits the size of the breeding population. It is possible that only one of these has important selective advantages, or that all of them, plus others, have selective value. The fact that one is shown to be important will not mean that the others are unimportant or irrelevant; the appropriate null hypotheses will need to be set up and tested for each function with a variety of species.

If we ask why territorial behaviour has evolved to provide such large territories in animals such as the red grouse, and such an excess failing to get territories, we are again unable to do more than guess. Professor Wynne-Edwards gave one possible explanation in his 1962 book, involving competition for a convention such as territory instead of for the food itself, apparent 'altruistic behaviour' on the part of the non-territorial animals in remaining non-territorial instead of trying to compete, provision against damaging both the food and the population through over-exploitation of the food resource, and group selection. Here again, we have no factual data in the grouse one way or the other.

In 1968 we began an attempt to explore some of these problems in the red grouse, but the difficulties in the field are formidable. The aim is to mark all the young reared on a fairly large area of moor, subsequently tag as many as possible for individual recognition later, and then find which get territories and which fail. If this is successful, we may be able to follow families from one generation to the next, and to find if grouse which take larger territories are more or less successful in contributing to future generations than grouse with smaller territories. At present, the number of young reared is often used in the literature for speculative discussions about natural selection and family size, but it has yet to be shown that this is valid. Really it is the number of these young which subsequently *survive to breed*, and the number of their grandchildren that get territories, and so on, that is important.

J. GREENWOOD: You said that the chicks produced in good years turned out to be less aggressive and to have smaller territories than average. Does 'good' in this context mean good in terms of the previous autumn or in terms of the breeding season itself? Secondly, were the parent's territories larger or smaller than average?

A. WATSON: 'Good' is a loose term, meaning the number of fully grown young per brood, and is determined by spring nutrition for the parents. The territories of the parents in good years may be larger or smaller than average; in other words, large broods may be reared at high as well as at low population densities in successive years.

T. B. REYNOLDSON: We have heard a great deal about the quality of food at this symposium, but what about the attributes of the quality of space in relation to population size in vertebrates?

A. WATSON: Quality of space is certainly important in relation to population size. There is good evidence with rodents and fish in confined conditions, and with rodents, fish, birds and mammals in the wild, that the physical structure of the habitat has a direct effect on spacing and thus on population density. In general, extra physical structures that are added will increase density—for instance, if more large stones are added to a stream, more trout are enabled to take territories (Kalleberg 1958), irrespective of changes in food. Territories of red grouse are smaller where hillocky ground or broken peat banks restrict the visibility or provide better cover.

INTRODUCTION TO LABORATORY
STUDIES

The effects of nutrition on the social behaviour and use of space by animals are very difficult to study in the wild (see Watson & Moss 1970, this symposium). Research on this with laboratory rodents, domestic animals and man is therefore likely to provide important leads for the field ecologist. Studies using domestic rats and mice, by psychologists and others, on the effects of early mother–infant relations (such as late or early weaning, or large versus small litters) on later social behaviour, aggression and exploratory activity, are likely to be relevant with rodents and other animals in the wild. Indeed they are beginning to be studied by ecologists using wild species (e.g. K. R. Barbehenn 1961, *Trans. N.Y. Acad. Sci.* **23**, 443–6; J. B. Calhoun 1963, *The Ecology and Sociology of the Norway Rat*, Bethesda, Maryland). The effects of nutrition and early social experience on problem-solving ability, exploratory behaviour and general activity are also another field where laboratory research by psychologists and nutritionists is likely to be relevant to the field ecologists' complex problems of social behaviour in relation to populations and nutrition. For example, exploratory behaviour may well be involved in determining the size of an animal's home range (see *Symp. zool. Soc. Lond.* No. 18), and the amount of general activity has been shown to be associated with degree of aggression by Calhoun (1963) and others. It was for these reasons that Dr Cowley, who is not an ecologist, was asked by the organizing committee to review laboratory work in this field.

Adam Watson

FOOD INTAKE AND THE MODIFICATION
OF BEHAVIOUR

BY J. J. COWLEY

Department of Psychology, The Queen's University, Belfast

INTRODUCTION

This paper reviews some of the literature on the effects on behaviour of altering the plane of nutrition of several species of vertebrates in the laboratory. Particular emphasis is given to studies that describe the close relationship between the nutritional state of the mother and that of her offspring. As an aspect of this, experiments are summarized which show the effects of delayed weaning on the exploratory behaviour of rats and mice.

THE EXPERIMENTAL MODIFICATION
OF FOOD INTAKE

Much of the literature on the effects of the experimental modification of food intake on behaviour is reviewed in a recent M.I.T. publication (Scrimshaw & Gordon 1968). A review by Brožek & Vaes (1961) is available and the same author has discussed, elsewhere, the Soviet contribution (Brožek 1962). Interest in the problem has been accentuated by an awareness of the high proportion of the human population which has access to limited food supplies, and our own interest (Cowley & Griesel 1966) in feeding laboratory rats a low protein diet was undertaken with this type of problem in mind. Rats fed a diet of low protein composition from weaning (21 days of age) until they were mature (90 days) showed little sign of retardation in their problem-solving ability, though the work of Barnes, Moore, Reid & Pond (1967) at Cornell has shown that pigs which have recovered from severe protein under-nutrition show retarded learning capacity.

Whether retardation in learning occurs no doubt depends on the amount of protein available and the age when the restriction is imposed. In our work on rats we found it necessary to take into account not only the nutritional history of the generation we were working with but also that of previous generations, and it was only on testing the *offspring* of the parent generation which had been retained on the low protein diet, that we were able to observe marked deviations in problem-solving from our control rats (Cowley & Griesel 1966). Indeed, in order to produce viable young it was necessary to keep the protein composition of our diet at the relatively high figure of 14%. A reduction in the protein content to a figure much below this would militate against the rearing of successive generations.

Feeding a low protein diet to dogs (Platt 1962) can affect the growth, development and behaviour of their offspring; changes in gait, the presence of athetoid movements, restlessness and excitability have been described and are more acute when congenitally undernourished animals are retained on the low protein diet. We do not know whether the learning capacity of these animals was affected but it would be surprising if it were not, particularly as they are described as running slowly, rapidly losing interest and less ready to 'follow' (Stewart & Platt 1968).

Thus, the plane of nutrition of the mother may be reflected in the behaviour of the offspring, and in the case of the laboratory rat, the mother's behaviour towards her young may itself be modified as a consequence of her nutritional state. Rats on a diet of a low protein composition built 'better' nests, suckled their young more frequently than control mothers, and were more reluctant to leave the home cage when tested on a cage emergence test (Cowley & Griesel 1964). It is not easy to see any clear pattern in the mothers' behaviour, for though there appeared to be a greater reluctance to leave the young, we also observed that they took longer to return to the home cage when placed on a retrieving platform in front of it.

A number of studies have shown that, by increasing or decreasing the number of young in the litter which the mother is suckling, a modification in growth can be effected. Lát, Widdowson & McCance (1960) reported that rats nursed in large litters showed less exploratory behaviour than rats reared in small litters. The modification of food intake by the control of litter size has been combined with more traditional methods. Thus, Barnes, Cunnold, Zimmermann, Simmons, Macleod & Krook (1966) reared rats in large litters (14–16) and then weaned them on to a low protein diet (3–4%) for 8 weeks after weaning. Their performance in a learning situation was compared with rats that had been reared in large litters but subsequently fed a normal diet, and also with rats that had been reared in litters of eight and subsequently fed a normal diet from weaning. These rats did significantly better in the task (a discrimination test involving a Y-shaped water maze) than the rats from the large litters and the rats fed the low protein diet. The testing was done some 5–9 months after the commencement of rehabilitation and did not show any consistent differences between the rats from the large litters, normal rats weaned on to the low protein diet, and the control rats.

Another simple method of modifying food intake is to wean at an early age. The method has been used extensively by a group of workers at the Czechoslovak Academy of Sciences and they have reported that rats weaned at 15 days did not develop conditioned reflexes as easily as those weaned at 30 days of age. Further, male rats show degenerative changes in the testes and impaired fertility. The feeding of a hypercalorie high fat diet to rats from the time they were weaned at 15 days prevented the impairment of spermatozoon production and of learning ability (Nováková, Faltin, Flandera, Hahn & Koldovský 1962).

The weight and size of newborn rats and mice varies and the rate at which they grow depends to a large extent on the amount of milk that the mother can provide. Other factors may, however, also be involved. Lát (1956) has shown that adult rats with a low level of 'excitability' grow more slowly under standard conditions of nutrition and environment, and their ultimate adult weight is less than that of excitable ones. The excitable animals showed a preference for large amounts of carbohydrates while the less excitable animals preferred protein (casein) and fat (margarine).

In other experiments, rats of approximately the same levels of excitability were fed diets of different composition. Lát reports that a diet with a high protein composition lowered excitability while carbohydrates increased the level of excitability. Excitability was assessed in terms of a number of behavioural criteria including the speed of acquisition of conditioned reflexes and the intensity of exploratory activity.

The food intake of excitable rats, when fed a standard diet over a 4-week period, was greater and a linear relationship between level of excitability and calories ingested per unit of body surface is reported. Animals with a low level of excitability grow slowly and their ultimate weights are low, while more excitable animals grow rapidly and for a shorter period of time though their ultimate weights are heavier. On the other hand extremely excitable animals, as judged by their exploratory reactions, do not reach the high body weights of the less excitable animals. The findings of Lát and his colleagues, if substantiated, may reaffirm an association between basal metabolic rate and activity. Rundquist & Bellis (1933) reported, for

example, that the basal metabolic rate of active animals was higher than
that of inactive animals.

MATERNAL NUTRITION AND
ACTIVITY OF THE OFFSPRING

Some of our own studies have concentrated on the activity of young
animals (Cowley & Griesel 1963). I would like to comment briefly on the
relationship between the movements made by infant animals, their early
development, and the wheel running (i.e. a measure of activity) of these
same animals when mature.

The offspring of mothers that had been fed a diet of low protein composition were photographed daily for 15 days from birth and the amount
of movement they showed in a small 'open field' or area was calculated.
The same was done for a control group of infant rats whose mothers
were on a normal laboratory diet.

There was some evidence to suggest that early motor movement was
correlated with such features of development as the opening of the eyes
and the first response to a sharp sound, more closely in the offspring of the
low protein sample than in the control sample. Similarly, birth weight in
the offspring of the low protein sample was better for predicting activity
during the 15 days of development ($r = 0.70$) than in the control sample
($r = 0.33$).

Infant activity was not correlated significantly with subsequent wheel
activity in the offspring of the rats from the low protein sample, and the
relationship between infant activity and subsequent wheel activity, in
the control rats, was low ($r = 0.4$, $P < 0.05$).

The results showed that the range of infant activity was greater for the
offspring of the low-protein mothers than it was for the control rats.
The low relationship between wheel activity and early infant activity may
indicate, as others have suggested in relation to adult rats, that the two are
quite separate aspects of spontaneous activity (Eayrs 1954).

WEANING AND EXPLORATORY
BEHAVIOUR

Reference has already been made to the effects of early weaning on rats
and I would like to report briefly on experiments in which baby rats and

mice were kept with their mothers well beyond the normal time for weaning. In both rat and mouse, males weaned at 21 days tended to be somewhat more active in exploratory behaviour when tested at 90 days of age in an Open Field Test than males weaned at 35 or 42 days. Rat mothers will continue to suckle their young well beyond 21 days if they are not removed. Weaning at 21 days produces a small drop in the weight curve but this does not persist and the weaned animals catch up with their litter mates within a day or two. Nutrition may play a part in producing the difference in exploratory behaviour after early weaning. However, mice removed from their mothers at 21 days were more active than those 'weaned' at 21 days but kept with a non-lactating foster mother till 35 days. Hence other variables are probably also involved.

ACKNOWLEDGEMENTS

I am indebted to Miss Kay Sim for her care of the animals.

SUMMARY

1. This paper reviews laboratory work on the effect of nutrition on behaviour in several vertebrate species.
2. Reducing the amount of food that infant rats receive either by increasing the size of the litter or by weaning them at an early age may affect their behaviour when mature. The changes range from a reduction in exploratory behaviour to a delay in the formation of conditioned reflexes.
3. Modifying the diet of the mother may have a profound effect on the growth, development and behaviour of the young, and these effects, which are present in a variety of species, are without doubt primarily due to the change in food intake. Altering the maternal diet may also affect the behaviour of the mother in relation to her offspring and this may, in turn, be of some consequence in influencing their behaviour.

REFERENCES

BARNES R.H., CUNNOLD S.R., ZIMMERMANN R.R., SIMMONS H., MACLEOD R.B. & KROOK L. (1966) Influence of nutritional deprivations in early life on learning behaviour of rats as measured in a water maze. *J. Nutr.* **89**, 399–410.
BARNES R.H., MOORE A.U., REID I.M. & POND W.G. (1967) Learning behaviour following nutritional deprivations in early life. *J. Am. Diet. Ass.* **51**, 34.

BROŽEK J. & VAES G. (1961) Experimental investigations on the effects of dietary deficiencies on animal and human behaviour. *Vitams Horm.* **19**, 43.

BROŽEK J. (1962) Soviet studies on nutrition and higher nervous activity. *Ann. N.Y. Acad. Sci.* **93**, 665.

COWLEY J.J. & GRIESEL R.D. (1963) Infant rat activity and its relationship to some aspects of development. *Psychologia Africana,* **10**, 117-22.

COWLEY J.J. & GRIESEL R.D. (1964) Low protein diet and emotionality in the albino rat. *J. gener. Psychol.* **104**, 89-98.

COWLEY J.J. & GRIESEL R.D. (1966) The effect on growth and behaviour of rehabilitating first and second generation low protein rats. *Anim. Behav.* **14**, 506-17.

EAYRS T.T. (1954) Spontaneous activity in the rat. *Br. J. Anim. Behav.* **2**, 25-30.

LÁT J. (1956) The relationship of the individual differences in the regulation of food intake, growth and excitability of the Central Nervous System. *Physiologia bohemoslov.* **5**, 38-42.

LÁT J., WIDDOWSON E.M. & McCANCE R.A (1960) Some effects of accelerating growth. 111. Behaviour and nervous activity. *Proc. R. Soc.* B. **153**, 347-56.

NOVÁKOVÁ V., FALTIN J., FLANDERA V., HAHN P. & KOLDOVSKÝ O. (1962) Effect of early and late weaning on learning in adult rats. *Nature, Lond.* **193**, 280.

PLATT B.S. (1962) Proteins in nutrition. *Proc. R. Soc.* B. **156**, 337-44.

RUNDQUIST E.A. & BELLIS C.J. (1933) Respiratory metabolism of active and inactive rats. *Am. J. Physiol.* **106**, 670-5.

SCRIMSHAW N.S. & GORDON J.E. (Ed.) (1968) *Malnutrition, Learning and Behavior.* Cambridge, Massachusetts.

STEWART R.J.C. & PLATT B.S. (1968) Nervous system damage in experimental protein-calorie deficiency. In *Malnutrition, Learning and Behavior* (Ed. by N. S. Scrimshaw and J. E. Gordon), pp. 168-80. Cambridge, Mass.

AGGREGATION BEHAVIOUR IN RELATION TO FOOD UTILIZATION BY APHIDS

By M. J. Way and M. Cammell

Department of Zoology, Imperial College, London

INTRODUCTION

Aphids are so-called sucking insects which feed on the contents of phloem or phloem-type cells—hence they are limited to parts of the plants where they can reach these cells with their stylets. Moreover, they can inbibe only soluble nutrients such as sugars and amino-acids. Consequently, unless they can alter the physiology of the plant, they thrive only on the young actively growing parts of plants which are physiological 'sinks' for soluble nutrients: on stems and petioles which are actively translocating such nutrients, or on senescing leaves where stored foods are being made soluble for translocation elsewhere (Kennedy & Stroyan 1959; Mittler 1953). These and other conditions discussed in more detail by Dixon (1970) create a situation where food, as indicated by abundant food plants, may be seemingly plentiful but is nevertheless severely limiting especially in mid season after active plant growth has ceased but before senescence has started.

The potentially enormous multiplication rate of many aphid species (Way & Banks 1967) based on parthenogenesis, viviparity and telescoping of generations is no doubt associated with the need to exploit the ephemerally suitable plants or parts of plants and this is reflected in the often violent seasonal changes in abundance (Way 1967). Adverse weather, natural enemies and seasonal scarcity of food plant are generally considered to be the major factors limiting aphid populations and yet the self-obliterating possibilities of their explosive powers of multiplication suggest that homeostatic mechanisms are as essential or even more essential in such animals, which are faced with continually changing quantity and quality of food, than in those which exist in more stable conditions (Wynne-Edwards 1962).

229

The pre-requisite for homeostasis, namely group life, exists in interesting variety according to the aphid species. Thus 'active' gregariousness was first demonstrated in the densely aggregating black bean aphid (*Aphis fabae* Scop.) by Ibbotson & Kennedy (1951), where tactile, and possibly chemical mechanisms were responsible for maintaining the aggregation. The aphid we have chosen for special study, the cabbage aphid (*Brevicoryne brassicae* L.) is among those which aggregate most densely. Solitary cabbage aphid larvae are restless and individuals put separately into the same cage quickly form groups. For example, when about fifteen first instar larvae were put into each of six different leaf cages each covering 3·0 sq cm of leaf, all had formed into thirteen groups within 3 days and into seven groups within 8 days, i.e. they were all in a single aggregate in five of the cages and in two aggregates in the sixth. Other aphids aggregate much less densely, but tactile mechanisms also determine the spaced-out gregariousness of the sycamore aphid (*Drepanosiphon platanoides* (Schrank), Kennedy & Crawley 1967) while, as Lees (1967a) has pointed out, the nut aphid (*Myzocallis coryli* Goeze) spaces out even more widely on the leaf but the mathematical precision of the spacing again demonstrates gregariousness. The peach-potato aphid (*Myzus persicae* Sulz.) also spaces out widely, in contrast to the cabbage aphid with which it often occurs simultaneously on the same brassica host plant (Bonnemaison 1951).

EFFECT OF AGGREGATION BEHAVIOUR ON ESTABLISHMENT OF A CABBAGE APHID POPULATION

An aggregate is normally established by a group of up to about twenty progeny of an immigrant alate adult. Unlike many aphids, cabbage aphid alatae will sometimes establish colonies on mature as well as on young leaves, so fully grown mature or almost mature leaves were chosen for the following experiments, to provide conditions where the plant food was not intrinsically highly favourable and where beneficial effects of aggregation would therefore be pronounced.

First, the development of three groups each of twenty 1st instar larvae was compared with that of a similar number kept solitarily in leaf cages on mature leaves of potted brussels sprouts plants (*Brassica oleracea gemmifera*). The grouped aphids grew larger (mean length of adult fore tibia = 0·530 mm compared with 0·511 mm for aphids reared alone; difference significant at $P < 0.01$). Furthermore, their first newly born larvae were larger

TABLE I

The effects of numbers of aphids per aggregate on their size when adult and on initial rates of reproduction

Initial no. of 1st instar larvae in the aggregate	5	10	20	40
Mean length of adult fore tibia (mm)	0·491 ±0·006	0·496 ±0·011	0·528 ±0·005	0·499 ±0·008
No. of progeny/adult when experiment ceased	1·7	2·7	5·2	—†

† Data unobtainable because leaf areas collapsed.

(mean length of hind tibia = 0·241 mm compared with 0·226 mm in one set and 0·259 mm compared with 0·237 mm in another set; differences significant at $P<0·01$).

An indication of optimum density was obtained by comparing the adult apterae developing from initial populations of five, ten, twenty and forty 1st instar larvae per 3·0 sq cm of leaf enclosed by each cage. Progeny were not removed because this unduly disturbed the adults in the crowded cages. The experiment was therefore stopped soon after reproduction started because leaf areas on which the largest aggregates were enclosed began to collapse. The measurements of the adults did not differ significantly except at $P<0·1$ (Table 1) but the results as a whole indicate that the aphids grew largest and perhaps began to reproduce earlier in aggregates of twenty than in smaller aggregates. An aggregate of forty was detrimental in the confined conditions. This initial beneficial effect of aggregation has also been demonstrated in the black bean aphid (Way & Banks 1967) and in the pea aphid (*Acyrthosiphon pisum* Harris) by Murdie (1969). It could be

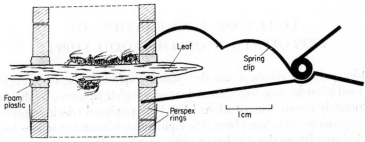

FIG. 1. Sectional diagram of a cage used to enclose a single cabbage aphid on the under side and a cabbage aphid aggregate on the upper side of a leaf.

due to mutual stimulating effects on other individuals in the groups or to group feeding improving the quality or flow of food from the plant. The effect on food supply was distinguished from that of mutual stimuli by setting up solitary 1st instar cabbage aphid apterae on the under side of leaves in leaf cages with or without an aggregate of 30–100 aphids directly opposite on the upper side of the leaf (Fig. 1). Pairs of leaf cages were set up on each leaf on either side of the midrib, one cage with and one without

TABLE 2

Sizes of apterous adults (length of fore tibia—mm) and their first progeny (length of 1st instar hind tibia—mm) when reared alone on the underside of leaves colonized or not colonized by aggregates on the upper surface

Leaf no.	Adult size		1st instar larval size	
	With aggregate	Without aggregate	With aggregate	Without aggregate
1	0·607	0·489	0·256	0·232
2	0·513	0·501	0·257	0·222
3	0·490	0·428	0·245	No progeny
4	0·621	0·522	—	—
5	0·585	0·545	0·246	0·219
6	0·602	0·581	0·233	0·240

an established aggregate. The adults reared opposite aphid aggregates were all larger than those reared on the same leaf without aggregates (Table 2) (differences significant at $P<0·05$) and the first born larvae were also larger except on leaf no. 6. This indicates that group feeding improved the food available to a solitary aphid and thus the beneficial effect of aggregation was not merely through direct mutual stimuli.

EFFECT OF AGGREGATION ON EXPLOITATION OF THE FOOD SUPPLY

Most species of aphids, unlike the sycamore aphid (Dixon 1970) exploit a food plant by successive generations of wingless (apterous) adults, but ultimately mostly winged adults (alatae) are produced which perpetuate the species on other host plants. The rapid multiplication rate and the large aphid populations that develop on plants are no doubt an insurance against the high probability of death during subsequent migration and colonization by alatae—especially as the alatae are poorly adapted to direct their flight and have limited powers of finding their host plants, which they first have to taste before the plant can be assessed as a host species and also as one which is nutritionally suitable. The highly discriminating behaviour of alatae of some species to food quality (Kennedy & Stroyan 1959)—frequently they will not colonize plants or parts of plants on which the apterae can still develop satisfactorily—adds to the hazards during plant colonization. This need to produce large numbers of alatae indicates that success in adaptation of an aphid population to the colonized plant can be judged in terms of the ability of the population to convert as much as possible of the

TABLE 3

Numbers of alate adult cabbage aphids produced on small cabbage plants with either one or all four leaves colonized

	Mean no. of alatae per plant	Mean area of colonized leaf (leaves) (sq cm)	Mean no. of alatae produced per sq cm of colonized leaf
One leaf colonized	3089	128	23·5
All leaves colonized	3323	493	6·8

plant food into viable adult alatae, subject to restrictions required to ensure that the plants are not so over-exploited that the food supply of future generations is jeopardized.

Dense aggregation, as by the cabbage aphid, tends to restrict populations to particular leaves on a plant, so the significance of such behaviour on the productivity of alatae was studied by comparing populations restricted to a single leaf of a plant with ones established on all the leaves. This was done using small potted brussels sprouts or cabbage plants kept in a constant environment room at a temperature of 23°C during the 16 h 'day' and 20°C during the 8 h 'night'. To maintain consistent amounts of available plant food throughout the experiment the plants were decapitated and the oldest leaves removed, leaving four young-mature leaves which were allowed to expand fully before the experiment was started. Cabbage aphid populations were established as in the field by alatae which were put on the leaves and allowed to reproduce. In one set of experiments eight cabbage or brussels sprouts plants caged singly were each colonized by sixteen larvae, four plants with all sixteen on one leaf and four plants with four larvae on each of the four leaves. The uncolonized leaves of the one-leaf colonized plants were kept free of colonizing aphids throughout the experiment which was continued until the aphid populations died out. Productivity was assessed from adult alatae which were collected daily, counted and measured (Tables 3 & 4). Similar numbers of alatae were produced per plant (difference not significant at $P<0.05$) but the one-leaf colonized plants produced many more per unit area of colonized leaf (Table 3)

TABLE 4

Mean weights of alate adult cabbage aphids produced at intervals on small cabbage plants with either one or all four leaves colonized

	Mean weights (mg)				Approximate mean wet wt biomass of total alatae per plant (mg)
	First alatae (22–24 days)	Time of peak production (31–32 days)	Time of declining production (38–39 days)	Last alatae (45–48 days)	
One leaf colonized	0·260±0·007	0·224±0·012†	0·196±0·007†	0·153±0·009	620
All leaves colonized	0·262±0·005	0·178±0·009†	0·134±0·008†	0·138±0·006	578

† Differences significant at $P<0.05$.

and except initially, when they were similar sized, the alatae were larger (Table 4) implying better nutrition. Consequently, the estimated wet weight biomass of the total alatae produced was greater (Table 4), differences not significant at $P<0.05$, even though this represented fewer aphids (Table 3). Moreover the uninfested leaves of the one-leaf colonized plants were healthy at the end of the experiment whereas the four-leaf colonized plants were virtually dead.

EFFECT OF AGGREGATION ON TRANSLOCATION OF FOOD BY THE PLANT

An explanation of the relatively better initial development of aphids in aggregates (Tables 1 & 2) and of what makes it possible for a cabbage aphid population to be as productive on a part as on the whole of a plant

TABLE 5

Effect of a cabbage aphid-colonized leaf on the translocation of ^{14}C labelled sucrose from another leaf of the same plant

Condition of treated leaves and plants		Translocation rate (mg labelled sucrose/hour)			Assimilation rate of labelled leaf (counts/minute/sq cm of leaf/h)
		Out of labelled leaf	Into aphid-colonized leaf	Into meristem and young leaves	
Young mature leaves of young plant	Aphid-colonized	3·35	0·81	0·19	10424
	Uncolonized	1·16	0·01	0·26	11440
Old mature leaves of older plant	Aphid-colonized	3·57	1·46	not recorded	2681
	Uncolonized	1·52	0·19	not recorded	448

(Tables 3 & 4) was examined in preliminary experiments undertaken with Dr D. Habeshaw. Populations of about 500 cabbage aphids per plant were established on one leaf of a set of intact young brussels sprouts plants grown in pots. On the following day an uncolonized leaf of each plant, slightly older and positioned opposite the infested one, was treated with $^{14}CO_2$. The distribution of ^{14}C-labelled sucrose in different parts of the plant was measured 3 hours after $^{14}CO_2$ treatment for 1 hour (Table 5). The aphid population had more than doubled the rate of movement of labelled sucrose from the labelled leaf, decreased the amount going to the growing leaves and greatly increased the amount ($\times$ 70–80) going to the aphid infested leaf; the aphid free 'controls' received very little. Furthermore, the aphids had seemingly increased the assimilation rate of the old-mature leaves in the experiment where relatively old plants were used.

The aphids appear to be acting as 'sinks' competing for nutrients with the natural 'sinks', namely, the growing parts of the plants and the storage organs. Unfortunately, as pointed out by Watson (1968), too little is known by plant physiologists about the capacity or strength of plant parts as sinks for nutrients. It seems, however, that the supply of photosynthetic materials may often exceed sink capacity; in these circumstances sugar in excess of sink capacity accumulates in the leaves and decreases their photosynthesis. Sink capacity can therefore indirectly influence photosynthesis and assimilation rates by a feed-back mechanism. This helps explain why the assimilation rate of mature leaves is often depressed, as in one of our experiments where it seemingly increased in response to the drain from the sink created by the feeding cabbage-aphid aggregate.

In some circumstances even solitary aphids can attract labelled assimilate to a colonized leaf (Canny & Askham 1967) and, no doubt, aphids which are dispersed as distinct from aggregated, benefit from their sink effect. However, it seems likely that aggregation strengthens the capacity of aphids to compete with natural sinks and this may explain why densely aggregating aphid species, such as the cabbage aphid and the black bean aphid can succeed on mature leaves whereas aphids like the peach-potato aphid, which do not aggregate densely, are successful only on young leaves which are natural sinks and on senescing ones which are making stored nutrients soluble (cf. Kennedy, Ibbotson & Booth 1950). Densely aggregating species such as the cabbage aphid, the black bean aphid and the mealy plum aphid (*Hyalopterus pruni* (Geoffroy)), nevertheless flourish on the natural sinks provided by very young leaves. Thus the mealy plum aphid forms extremely dense clusters completely covering both sides of very young plum leaves which become useless as food as soon as the leaf is separated from the stem. Similarly the cabbage aphid develops very poorly on leaf discs detached from immature leaves of the plant (Hughes 1963).

The aggregates of the elder aphid (*Aphis sambuci* L.) form dense collars around young growing shoots from which they are presumably extracting the soluble nutrients passing through the phloem to the meristem and growing leaves. In many plants the translocated nutrients consist primarily of sucrose, which to judge from the amount excreted unchanged in the aphids' honey dew, is invariably present in excess. In contrast the amino-acid supply is often critical as shown by Dixon (1969). Amino acid is translocated to young leaves in the stem phloem but that which is synthe-sized in the young leaves (Joy 1967) is seemingly unavailable to stem-feeding aphids. However, aggregating aphids such as the elder aphid on stems and the cabbage aphid on mature leaves may, as do young leaves, induce local amino-acid synthesis from other nutrients drawn to the aphid sink. This suggestion is supported by work on silver fir woolly aphids (*Adelges* spp.). These feed on peripheral bark parenchyma where they are usually confined sparsely to healing bark cracks and lenticels—the only places where amino-acids can be detected in the peripheral bark. Some-times, however, the sparse populations grow into large aggregates which in contrast stimulate development of much amino-acids in the bark around the aggregates. Aphids in these aggregates grow larger and appear to be much more fecund than in the sparse populations (Eichhorn 1968; Kloft 1955, 1960).

THE CONTROL OF DEVELOPMENT
OF THE APHID POPULATION

Results so far indicate that aggregation improves the food available to the newly established aggregate and also to the developing population which can divert food to the colonized leaf from other parts of the plant. Never-theless, we know that, although aggregation initially enhances multi-plication (Way 1968; Way & Banks 1967), the rate is quickly slowed as numbers in the aggregate increase above a relatively small critical level. Aggregation is therefore restricting the rate of increase. For example, the strong tendency to remain aggregated leads to highly contagious distri-bution of the black bean aphid on field beans and on the spindle tree *Euonymus europaeus* L. (Way 1968); severely overcrowded populations frequently develop on a very few plants in a field bean crop or on a few twigs of a spindle tree, leaving the rest virtually uncolonized though equally

17

acceptable nutritionally as shown when aphids are established artificially on them. Multiplication of the cabbage aphid is similarly restricted not only by very limited dispersion of apterae to neighbouring plants even from an overcrowded colonized plant but also in terms of the exploitation of a particular plant, at any rate when the plant is large, e.g. a large brussels sprouts plant. In these circumstances the population by restricting itself to one leaf produces many fewer alatae than does a more dispersed population. Unlike the small plant (Tables 3, 4) the uncolonized parts of the large plant are producing nutrients greatly in excess of the requirements of the aphid population on any one leaf.

The mode of development of a cabbage aphid population on one leaf of a small potted brussels sprouts plant was determined in a controlled environment room at 20–23 °C. The densely aggregated aphids cannot be counted or the instars determined without destroying the population so the detailed day to day changes were recorded indirectly from the cast exuviae which were collected and counted daily. The proportions of the different instars were determined in all, or in a sample of at least 200, exuviae from each day's collection (experiments showed that aggregation did not significantly affect the rate of development of the larval aphid). Knowing the durations of the different instars it was possible to calculate the number of aphids present on each day of the life of the population (Fig. 2a). Newly formed alate adults were also collected, counted and measured daily (Fig. 2b) and the area occupied by the aggregate was estimated.

It is remarkable that numbers on the plant altered so smoothly. They increased rapidly once reproduction had started on day 9 and then suddenly halted and remained steady between about days 25–33. Up to this time the aggregate had been developing on only one side of the leaf midrib (Fig. 3) but afterwards it extended to the other side. This caused a sudden increase to a peak population of over 5000 aphids on day 36; there was then an equally sudden decrease which slowed to a steady decline until day 99. The sensitivity of the population, and incidentally of the method, is indicated by the changes between days 47–50. These followed inadequate watering of the plant on day 46 which affected development and slowed moulting on day 47; this, however, was compensated for on day 50.

The numbers of adult alatae that were formed varied with the numbers of aphids on the plant but were most closely related to the density of the aphid population 8 days previously, i.e. about the time when the stimulus to produce wings is triggered in the developing larva. Thus the largest proportion of individuals were determined as alatae, rather than as apterae, when the population was small but dense (between days 13–17). The

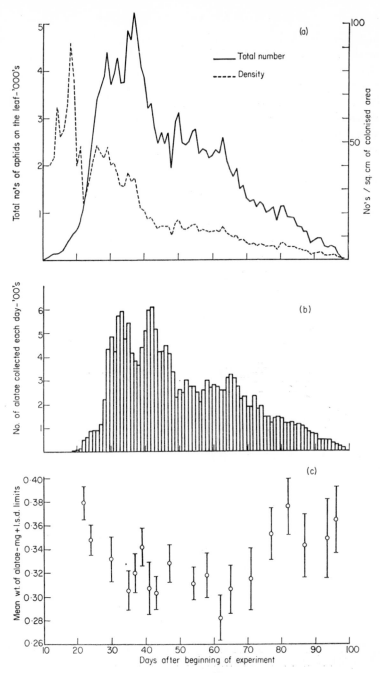

Fig. 2. Daily records of the development of a cabbage aphid population on one leaf of a plant kept at 20–23°C in a controlled environment room. (a) numbers and density of aphids on the leaf, (b) numbers of alatae produced, (c) mean weights of alatae and least significant difference limits at $P < 0.05$.

first alatae, like the last, were relatively large (Fig. 2c). Large alatae were also produced on about day 39. This was seemingly because the sample included first-formed large alatae from the aggregate developing on the previously uncolonized side of the leaf. The size range of individuals collected on any one day was always large (Fig. 2c) indicating that aphids developing simultaneously in the same aggregate were being nourished very differently. Such variation also occurs in field conditions but is not so marked as in the laboratory. The laboratory conditions also differed

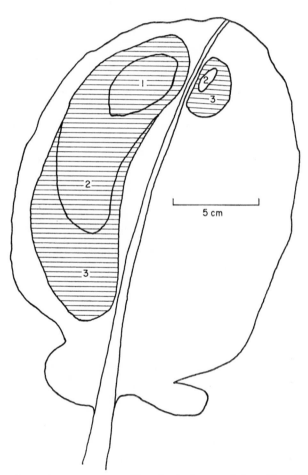

FIG. 3. Approximate areas covered by a developing population of the cabbage aphid. 1—after 18 days; 2—after 23 days; 3—after 33 days.

because aggregates do not grow so large or remain dense for as long as in the field where natural enemies also affect population development (e.g. Hughes 1963). Nevertheless the same intraspecific mechanisms are involved even when the delayed density dependent action of natural enemies severely curtails the life of the population (Way & Banks 1968).

NATALITY AND ALATE PRODUCTION
AS FACTORS REGULATING
POPULATION DEVELOPMENT

The self-restriction of the population in an aggregate of the cabbage aphid and black bean aphid is based primarily on varying natality and on varying production of emigrant alate adults (Hughes 1963; Way 1968). In the field the population in an aggregate maintains a surprisingly constant proportion of apterous adults, about 7% of the total number of aphids, irrespective of size of aggregate from about 200 aphids occupying about 0·5 sq cm to about 4000 aphids occupying about 15 sq cm of leaf (Way 1968). Natality is therefore regulated by the changing reproductive rate of the adult apterae and not by changing the proportion of the reproducing adults. That there are large variations in reproductive rate in relation to size of aggregate is indicated by the proportion of 1st and 2nd instar larvae in field populations of cabbage aphids which changed steadily from an average of about 70% in developing 200-aphid aggregates to about 10% in aggregates of about 4000 aphids (Way 1968). Present indications are that natality is controlled by a combination of competition for food and space. Thus solitary cabbage aphids feeding on the other side of the leaf from overcrowded aggregates are less fecund than in the absence of such competition. This indicates that nutrition is affected, but jostling from developing larvae also seems to slow reproduction even when nutrition appears to be adequate.

Factors governing alate production have been studied in detail in the laboratory and show that this is not merely an escape mechanism from overpopulation as is often thought nor is it a simple response to one particular set of conditions. Food is important both directly and indirectly.

Alate production in most aphids is unlike that of the sycamore aphid because it is facultative and, in species so far studied, tactile stimuli from other aphids acting on the parent or on the young larva are important in causing the switch from aptera to alate production (Bonnemaison 1951,

1967; Lees 1966). Hence, aggregation behaviour which ensures a characteristic form of dispersion for each species provides a population structure within which the proportion of individuals that become alate can be controlled according to the changing condition of the population.

Species differences are apparent from evidence that the peach-potato aphid produced an increasing proportion of alatae on chrysanthemums as the population increased to 0·21–0.27 per sq cm of leaf at which density it remained stable, with loss through departure of alatae counteracting natality (Wyatt 1965). A density of 0·27 aphid per sq cm is much less than that at which alate production would begin to affect population increase in the cabbage aphid and it also seems to be below that at which the peach aphid can stabilize on the radish (Miyashita 1954) which is a better food plant then chrysanthemum.

In the laboratory population of the cabbage aphid described above, the first alatae were produced early in development of the aggregate. Their large size suggests that they did not suffer nutritionally but that alate production was triggered by interference caused by self-induced competition for space in the very dense young aggregate. Later, food quality deteriorated as indicated by decreased size of alatae and, in these circumstances restlessness of the undernourished larvae was probably responsible for maintaining the frequency of alate-producing encounters despite decreased aphid density (Fig. 2a). This indirect effect of nutrition may be very powerful, as in the adult vetch aphid (*Megoura viciae* Buck.) and in the pea aphid which when starved produced alate progeny, providing they were starved in groups of two or more (Lees 1967b; Sutherland 1967). Furthermore the quality of food is important as shown by much greater production of alatae in populations of the vetch aphid, pea aphid, and *Aphis craccivora*, on unfavourable mature larves than on favourable young growth (Johnson & Birks 1960; Lees 1966; Sutherland 1967). Lees (1966) in particular showed that pairs of adult vetch aphids caged similarly on young and older leaves were visibly more restless on the latter and produced more alate progeny.

Although nutrition acts primarily through its influence on alate-producing encounters between aphids, inadequate nutrition can also directly induce the production of some alatae in the pea aphid and in *A. craccivora* (Johnson 1966; Sutherland 1967). Thus, indirectly and sometimes directly, nutrition controls a sensitive feed-back mechanism adjusting size of the population by altering the loss through alate production in relation to quality and quantity of the food supply (Wyatt 1965; Way 1968). The sensitivity is such that a presumptive alate larva of *A. craccivora* from a mother which was starved and crowded will nevertheless revert to an aptera if it is nourished favourably from soon after birth (Johnson 1966). Furthermore, even single chemical changes in the amino-acid content of the diet can change the proportion of alatae formed by a population of the peach-potato aphid (Mittler & Dadd 1966).

VARIATION IN QUALITY OF ALATE APHIDS

Alate production has the dual effect of helping regulate the multiplication rate of the local population while providing the essential means of maintaining the regional population. Alatae, however, vary greatly in quality as indicated by size, and Way (1968) concluded that, whereas the ability to decrease adult size enabled the maximum number to become adult on a dwindling food supply, the less well nourished small aphids were less fecund and were perhaps less able to find new host plants than the larger ones. Recent work with the bean aphid (Shaw 1968) indicates, however, a complex situation. Some of the large alatae produced early in the development of a population never depart, i.e. the self-induced crowding initiates wing development but not the 'drive' to migrate which seemingly requires the added stimulus of deteriorating nutrition. Two other kinds of alatae were distinguished, the 'fliers' which reproduce before they depart and probably colonize nearby hosts, and the 'migrants', with the largest wings and wing muscles, which depart before reproducing and seem to be the long-distance migrants. Almost all alatae produced at the time of peak numbers on the plant were 'migrants' whereas relatively more 'fliers' were produced before and afterwards. The changing conditions, notably changing nutrition and also population size or density, therefore determine the kind of alate that is produced and hence the pattern of dispersion of the aphid on new host plants.

It is significant that formation of seemingly long-distance migrants, which are subject to the greatest hazards, predominates when the bean aphid populations are largest, i.e. when abundance of migrants can compensate for the added risks in colonizing distant host plants.

As yet, the quantitative contribution of self-regulatory mechanisms to population regulation of aphids has not been examined in sufficient detail but field work on widely different kinds of aphids (Dixon 1970; Eichhorn

1968; Hughes 1963; Sanders & Knight 1968; Way & Banks 1968; Wyatt 1965) has already demonstrated its importance.

ACKNOWLEDGEMENTS

We gratefully thank Dr D. Habeshaw for the determinations of ^{14}C-labelled sucrose and also Mrs Carole Pinnock and Miss Cheryl Davies for valuable assistance.

CONCLUSIONS AND SUMMARY

1. Our results show that aggregation behaviour and the parasitic way of life of aphids feeding on the phloem of the plant have made possible the evolution of delicate mechanisms for exploiting the food supply and for adjusting their numbers to it. Thus aphids can act as 'sinks' diverting nutrients from distant parts of the plant, especially the dense aggregates of some species which can flourish on otherwise nutritionally unsuitable mature leaves. Initially, aggregation enhances the nutrition and hence the multiplication rate of the population. Subsequently, as the aggregate grows beyond a critical size, population growth slows or may be halted as a result of decreased natality and increased production of emigrant alatae. In these circumstances aphids with spaced-out populations like the peach-potato aphid may maintain stable low-density populations which do comparatively little direct harm to the host plant. Species which aggregate densely like the cabbage and bean aphids exploit the host plant piecemeal and this strong tendency to remain densely aggregated also helps to minimize over-exploitation.

2. Food quality and quantity control alate production mainly through their influence on restlessness and hence on alate-producing encounters between aphids in the aggregate. This provides a particularly sensitive mechanism helping to regulate numbers in the local population as well as ensuring optimum production of the alatae needed to colonize new host plants. Food supply also seems to influence the proportions of short- and long-distance migrants produced by a plant population and hence may control the pattern of dispersion of the regional population.

3. Aphids therefore have well-developed homeostatic mechanisms which can enable a population to adjust quickly to the highly variable and rapidly changing quality and quantity of their food supply in such a way

that over-exploitation is minimized and yet enough emigrants are produced to ensure perpetuation of the species on nearby and distant host plants.

REFERENCES

BONNEMAISON L. (1951) Contribution à l'étude des facteurs provoquant l'apparition des formes ailées et sexuées chez les Aphidinae. *Annls Épiphyt.* **2**, 1–380.

BONNEMAISON L. (1967) L'effet de groupe chez les animaux. *Colloques int. Cent. natn. Rech. scient.* **173**, 213–36.

CANNY M.J. & ASKHAM M.J. (1967) Physiological inferences from the evidence of translocated tracer: a caution. *Ann. Bot.* **31**, 409–16.

DIXON A.F.G. (1970) Quality and availability of food for a sycamore aphid population. *Animal Populations in relation to their Food Resources* (Ed. by A. Watson), pp. 271–87. Oxford.

EICHHORN O. (1968) Problems of the population dynamics of silver fir woolly aphids, genus *Adelges* (= *Dreyfusia*), Adelgidae. *Z. angew. Ent.* **61**, 157–214.

HUGHES R.D. (1963) Population dynamics of the cabbage aphid, *Brevicoryne brassicae* L. *J. Anim. Ecol.* **32**, 393–424.

IBBOTSON A. & KENNEDY J.S. (1951) Aggregation in *Aphis fabae* Scop. I. Aggregation on plants. *Ann. appl. Biol.* **38**, 65–78.

JOHNSON B. (1966) Wing polymorphism in aphids. III. The influence of the host plant. *Entomologia exp. appl.* **9**, 213–22.

JOHNSON B. & BIRKS P.R. (1960) Studies on wing polymorphism in aphids. I. The developmental process involved in the production of the different forms. *Entomologia exp. appl.* **3**, 327–39.

JOY K.W. (1967) Carbon and nitrogen sources for protein synthesis and growth of sugar-beet leaves. *J. exp. Bot.* **18**, 140–50.

KENNEDY J.S. & CRAWLEY L. (1967) Spaced-out gregariousness in sycamore aphids, *Drepanosiphum platanoides* (Schrank) (Hemiptera, Callaphididae). *J. Anim. Ecol.* **36**, 147–70.

KENNEDY J.S., IBBOTSON A. & BOOTH C.O. (1950) The distribution of aphid infestation in relation to leaf age. I. *Myzus persicae* (Sulz.) and *Aphis fabae* Scop. on spindle trees and sugar beet plants. *Ann. appl. Biol.* **37**, 651–79.

KENNEDY J.S. & STROYAN H.L.G. (1959) Biology of aphids. *A. Rev. Ent.* **4**, 139–60.

KLOFT W. (1955) Untersuchungen an der Rinde von Weibtannen (*Abies pectinata*) bei Befall durch *Dreyfusia* (*Adelges*) *piceae* Ratz. *Z. angew. Ent.* **37**, 340–8.

KLOFT W. (1960) Wechselwirkungen zwischen pflanzensaugenden Insekten und den von ihnen besogenen Pflanzengeweben. *Z. angew. Ent.* **46**, 42–70.

LEES A.D. (1966) The control of polymorphism in aphids. *Adv. Insect Physiol.* **3**, 207–277.

LEES A.D. (1967a) Cited by Bonnemaison (1967) p. 235.

LEES A.D. (1967b) The production of the apterous and alate forms in the aphid *Megoura viciae* Buckton, with special reference to the rôle of crowding. *J. Insect Physiol.* **13**, 289–318.

MITTLER T.E. (1953) Amino-acids in phloem sap and their excretion by aphids. *Nature, Lond.* **172**, 207.

Mittler T.E. & Dadd R.H. (1966) Food and wing determination in *Myzus persicae* (Homoptera: Aphidae). *Ann. ent. Soc. Am.* **59**, 1162–6.

Miyashita K. (1954) Cited by Shiga M. (1967) Ecological studies on the green peach aphid *Myzus persicae* (Sulzer) and the cabbage aphid *Brevicoryne brassicae* (Linnaeus) in Japan, with special reference to biological control. *Mushi*, **41**, 75–89.

Murdie G. (1970) Some causes of size variations in the pea aphid *Acyrthosiphon pisum* Harris (Hemiptera: Aphididae). *Trans. R. ent. Soc. Lond.* (in press).

Sanders C.J. & Knight F.B. (1968) Natural regulation of the aphid *Pterocomma populifoliae* on bigtooth aspen in northern Lower Michigan. *Ecology*, **49**, 234–44.

Shaw M.J.P. (1968) Polymorphism in relation to migration by alate alienicolae of *Aphis fabae* Scopoli. Ph.D. thesis, Univ. of London.

Sutherland O.R.W. (1967) Role of host plant in production of winged forms by a green strain of pea aphid *Acyrthosiphon pisum* Harris. *Nature, Lond.* **216**, 387–8.

Watson D.J. (1968) A prospect of crop physiology. *Ann. appl. Biol.* **62**, 1–9.

Way M.J. (1967) The nature and causes of annual fluctuations in numbers of *Aphis fabae* Scop. on field beans (*Vicia faba*). *Ann. appl. Biol.* **59**, 175–88.

Way M.J. (1968) Intra-specific mechanisms with special reference to aphid populations. *Insect Abundance* (Ed. by T. R. E. Southwood), pp. 18–36. London.

Way M.J. & Banks C.J. (1967) Intra-specific mechanisms in relation to the natural regulation of numbers of *Aphis fabae* Scop. *Ann. appl. Biol.* **59**, 189–205.

Way M.J. & Banks C.J. (1968) Population studies on the active stages of the black bean aphid, *Aphis fabae* Scop. on its winter host *Euonymus europaeus* L. *Ann. appl. Biol.* **62**, 177–97.

Wyatt I.J. (1965) The distribution of *Myzus persicae* (Sulz.) on year-round chrysanthemums. *Ann. appl. Biol.* **56**, 439–59.

Wynne-Edwards V.C. (1962) *Animal Dispersion in relation to Social Behaviour.* Edinburgh & London.

DISCUSSION

A. Macfadyen: Am I right in assuming that your figures show that a senescent plant, which is under attack by aphids, will actually photosynthesise more than if it had not been attacked?

M. J. Way: This requires confirmation. The assimilation rate of mature leaves is often depressed by the accumulation of nutrients not required by natural plant sinks. Our results indicate that the aphid sink may draw upon these accumulated nutrients, thereby removing the check to photosynthesis.

R. L. Kitching: As an aphid colony acts as a sink for food material from all parts of the plant, and not merely from the leaf on which it is situated, do you think that two or more colonies on the same plant will compete, and if so, what effect will this have on the numbers of aphids in the colonies?

M. J. WAY: We have demonstrated that competition can occur between two cabbage aphid (*Brevicoryne brassicae*) aggregates on the same leaf, but have not yet developed the work to show whether the population of one leaf can affect the development of a population on another.

D. C. SEEL: You compared the growth of aphids on the undersides of leaves when aphids were present and absent on the uppersides. Is this a reasonable experiment? Do your aphids normally utilize both sides of cabbage leaves? My impression of aphids on nettle leaves, particularly old ones, is that they usually occupy the undersides. Was it necessary to take the side of the leaf into account?

M. J. WAY: Cabbage aphids do occur naturally on the uppersides of leaves, though less commonly than on the underside. In our experiments the test aphid was always kept on the underside—the aphids on the upper surface were used solely to provide a group-feeding stimulus.

R. D. HUGHES: In Australia, cabbage aphids may occur equally on both sides of leaves, but any heavy rains wash them off the upper surface.

FOOD FIGHTING IN ROOKS

By I. J. Patterson

This paper reported incomplete work in progress and is being published here in summary only; it will be published in full when the study is complete.

The paper by Watson & Moss in this section has emphasized the importance of behavioural mechanisms in the regulation of animal populations to their food supply. This study examines one potential mechanism and assesses its importance.

Rooks (*Corvus frugilegus*) interact frequently over food, usually by one bird supplanting another from a food item. If this food fighting has an important regulatory effect, then the frequency of interaction should be related to the food supply, and should be highest when food is scarce.

Preliminary data on food supply strongly suggested that in Aberdeen-shire food was scarce in summer through absence of grain, emergence of leatherjackets, and aestivation of earthworms. Other evidence is consistent with this; the mean weight of shot rooks was lowest in summer, rooks were most easily caught in summer in baited traps, mortality was highest in summer and rooks at that time spent least time in rookeries. Observations by C. J. Feare (pers. commun.) show that food is not scarce in winter even during deep snow cover.

The rate of interaction in feeding flocks was highest in winter and lowest in summer when food was scarce. The low summer rate was probably caused by the very low flock density in summer. Similarly the high winter rates were probably caused by high flock density especially when this was associated with snow cover, when the rooks tended to concentrate on localized food resources (grain stacks and dung).

The frequency of food fighting thus did not have the expected relationship with variations in food supply, being affected much more by flock density and being highest in winter when food was not scarce and lowest during summer food shortage. This suggests that this behaviour is unlikely to be an important regulatory mechanism relating population size to food supply, though this conclusion must be tentative until the completion of current work on seasonal changes in food supply and the effect of food fighting on the feeding rate of individuals.

DISCUSSION

V. C. WYNNE-EDWARDS: If rooks are short of food in June–July, why don't they breed in the autumn?

I. J. PATTERSON: Food quality may be important. Grain is abundant in winter, but this is probably inadequate for the rearing of young.

I. NEWTON: You showed that most fighting occurs in late winter, when the birds are in, or coming into, breeding condition, and least fighting occurs in summer, after breeding. Have you considered the possibility that fighting, although occurring on the feeding grounds, might be related in part to the reproductive state of the birds?

I. J. PATTERSON: The peak of fighting was in February. It is almost certain that some of the interactions observed were over mates or individual space, but this fighting was indistinguishable from food fighting. However, in many cases the winning bird took over food or a feeding place left by the loser.

K. PAVIOUR-SMITH: You have shown that grain, as food, is absent in summer, but not that invertebrate food is short. I think that this needs to be measured in the actual place and year where and when the rooks are being studied.

I. J. PATTERSON: I must first emphasize again that none of the food supply data are conclusive by themselves. The data on earthworm abundance shows that both the number and weight of this important food fall in July to about one-quarter of the spring and autumn levels. I agree that we need more information on the intake of other foods.

D. C. SEEL: It is not strictly true that rooks are in poor condition only in mid-summer. Lockie's (1956, *Bird Study*, 3, 180–90) data show that the female has a low weight also in April. Could you detect any differences between males and females in the amount of food fighting in April?

I. J. PATTERSON: The low weights of females in April are reached during incubation, when they are being fed on the nest by the males. They are, therefore, not involved in food fighting. The sexes cannot be distinguished in field flocks.

D. LACK: I have two comments: (1) Food fighting is surely related to the type of food, since your results show that it occurs when rooks are feeding on grain, but not when they are feeding on insects in grassland.

It will be of value only if food is gained, and this will not apply to the dispersed feeding on insects in summer. Even if food is scarce in summer it does not mean that food fighting is not related to food shortage. (2) In winter, stacks and dung heaps (on which most food fighting is observed) are utilized primarily during snow. However, these resources are there all the time, suggesting that food is short during snow.

I. J. PATTERSON: Farmers deliberately site stacks near their farms so that they can disturb the birds. Stacks do not, therefore, constitute a preferred feeding situation.

D. LACK: This, I think, supports my view that the birds are short of food at this time. There must be disadvantages to feeding on the localized food resources (stacks and dung), or such feeding would not be limited to periods of snow. I think your observations are consistent with the view that food fighting is related to food shortage, because the birds have to be closely aggregated at such times.

I. J. PATTERSON: Even when snow forces rooks to feed on these localized resources, they are still able to spend time in the rookeries.

D. R. KLEIN: Is it not possible that food fighting has evolved in the past as a population regulatory mechanism in relation to food supply, but that man has altered the seasonal food cycle of rooks through his agricultural activities? Food fighting may then be a regulatory mechanism which is relatively rigid and, therefore, a carry over from the different ecological conditions that existed in past times.

I. J. PATTERSON: Man has certainly altered the seasonal pattern of food supply, particularly in providing a large grain supply during the winter. In the past there might well have been winter food shortage when food fighting could have been an important regulatory mechanism.

J. C. COULSON: Do you consider loss in crude weight in rooks or any other bird to be a good measure of food shortage, particularly since lipid deposition is also related to gonad state and endocrine secretion?

I. J. PATTERSON: The question of weight loss is complex, but I think that the low summer weights are suggestive when taken in conjunction with the other data.

J. C. COULSON: *Tipula paludosa* is the only common leatherjacket on agricultural land. During July in much of Scotland these are in final instar or pupal stages and readily available to rooks, mainly emerging as adults in August.

I. J. PATTERSON: A study of leatherjackets carried out by Dunnet (1955) showed that numbers in Aberdeenshire dropped sharply in early July.

P. BLAZKA: Does the percentage of young birds in early summer influence your mean weight data at that time as well as your trapping results? Young birds, being less experienced, may wander into traps more readily.

I. J. PATTERSON: The weights discussed were those of shot adult rooks only.

J. M. HINTON: If space, such as at nest sites, limits the breeding population, could competition for feeding space during snow serve to establish a hierarchy?

I. J. PATTERSON: I am not sure that nest sites are limiting. Fighting may have a stress effect causing emigration, or alternatively a hierarchy could be established.

A. WATSON: One piece of extra circumstantial evidence is that large flocks of rooks appear every summer and feed on mountain grasslands up to 750 m in this part of Scotland, often 10–15 km from breeding colonies and up to 10 km from the nearest arable land. This occurs from mid-June to the end of August, coinciding with the postulated period of food shortage. No flocks are seen on the hills at other times of year.

PART III · POPULATION PROCESSES IN RELATION TO THE QUANTITY, QUALITY AND AVAILABILITY OF THE FOOD RESOURCES

The local abundance of animals and the timing of various physiological events are frequently related to food resources which become available at particular places and seasons. In Part III, growth, breeding, density, survival and movement are considered in relation to variations in food quantity, quality and availability.

The first two papers discuss the effects of food quality and quantity upon reproduction and other population processes. In the Australian bushfly (Hughes & Walker) many population processes, including larval and pupal survival and adult reproduction, are affected by the quality of the dung upon which the larvae feed. When the density of the larvae becomes high, competition for food may occur, and larval mortality increases. In sycamore aphids (Dixon) reproductive rate again varies according to the quality of the food available. In summer, the food quality of phloem sap from the mature leaves is low; and at high densities under these conditions, interaction between adults results in reproductive diapause. This is not the result of competition for quantity of food, but is a direct effect of poor quality of food causing more interaction in relation to group size, movement, and suitable space. At other times of year when food quality is good, diapause does not result from high numbers, but dispersal does.

The next two papers discuss variations in density in relation to behaviour and food. The density of feral sheep within one island on St Kilda (Gwynne & Boyd) varies greatly between years, due to heavy mortality in springs with severe weather. Seasonal and annual changes in the quality of the food available, and its amount, are also involved. Differences between areas in the territory size and breeding density of the insectivorous dunlin (Holmes) on arctic and subarctic tundra in Alaska are related to the abundance and availability of the bird's high-quality food. This is a clear quantitative demonstration of something that is often assumed but is seldom measured. Possible mechanisms to account for this are described.

The effects of food quality and quantity on territorial and dispersive behaviour, with consequent changes in breeding density, are considered next.

The paper on red grouse (Miller, Watson & Jenkins) shows that in a situation where the quantity of food is in excess, an improvement in food quality improves breeding success. The density of the breeding population also increases, but as a result of changes in territorial behaviour associated with the improved breeding, and apparently not as a direct response to the food. By contrast, in some migratory finches such as the crossbills of Scandinavia, eruptive dispersal and a consequent decrease in breeding density occur directly when the amount of food available from the cone seeds of coniferous trees becomes insufficient to support high populations (Newton).

Many situations are more complex than simple dependence of a feeder upon its food resources. The relation between hyaenas (Kruuk) and their prey is to some extent reciprocal. The hyaenas do not limit the populations of their main prey of ungulates but do affect the population structure and turn over of their main prey—the wildebeeste. The population of hyaenas is limited by density and availability of their prey. Numbers of young plaice (Steele, McIntyre, Edwards & Trevallion), however, are limited by mortality factors other than food supply. These predatory fish eat siphons of the bivalve *Tellina* in preference to other foods, but change to alternative prey species when the density of *Tellina* drops below a certain threshold. This has the effect of conserving this food supply until it is again abundant. The energetics of this marine food chain are explored, and possible mechanisms in predator-prey relationships are discussed.

The final paper (Varley) describes energy flow in a woodland ecosystem, at Wytham, Oxford. The results of many workers' researches on the population studies of different animal species are compared with work on production and physiology of the trees. From this synthesis, he gives examples of food limitation due to variation in both quality and quantity.

R. Moss and A. Watson

THE ROLE OF FOOD IN THE
POPULATION DYNAMICS OF THE
AUSTRALIAN BUSHFLY

By R. D. Hughes and Josephine Walker

Division of Entomology, C.S.I.R.O.,
P.O. Box 109, Canberra, Australia

The bushfly (*Musca vetustissima* Walk.) is widely distributed throughout the drier areas of Australia, but contracts away from the colder southern-third of the continent in winter. The very active and long-lived adult flies are highly dispersive and as a result their local distribution seems largely independent of physiographic and vegetational features of their environment. Low temperatures ($<15°C$) limit their activity and survival but apart from this their prolonged existence in an area depends only on minimal supplies of moisture and carbohydrates which can probably be supplied by vegetation.

The flies repeatedly approach large animals settling temporarily on or near them. Occasionally the flies obtain food and moisture directly from the animals, but in general the behaviour seems to be developed for maintaining contact with the sources of their major requirement—freshly dropped faeces. As a result flies are often locally concentrated near groups of large animals, e.g. herds of cattle.

The association of the bushfly with faeces is well defined. The adults are strongly attracted to, and feed on them. The protein so obtained enables immature female flies to start the development of their eggs (Tyndale-Biscoe & Hughes 1969). The nourishment of the first batch of about thirty eggs may require one or more additional feeds during their development. Gravid females respond to freshly exposed dung as an oviposition site, but further feeding on dung appears to be a normal preliminary feature of oviposition behaviour, as the development of the next batch of eggs is not immediately begun without it (Tyndale-Biscoe & Hughes 1969).

The dung is usually suitable for oviposition only for a short period after it has been dropped because of crust formation and slow drying, so the larval population that results tends to be uniform in development. The larvae feed on the fluids and suspended solids of the faeces, passing rapidly through three instars before becoming fully fed. They then migrate from the dung and pupate below the surface of drier soil around the periphery. Eclosion from the pupa and emergence from the soil take place after a relatively long pupal instar.

While man, dingo, emu and various herbivorous marsupials must have been the principal sources of faeces in the past, introduced animals such as cattle, horses and sheep now provide most of the available dung. Beef cattle and dairy herds currently provide most breeding sites for bushfly. Studies have therefore been concentrated on the ecology of bushfly in relation to the dung pads of cattle.

BUSHFLIES AND COW DUNG

The major physical factors influencing bushfly larvae in the dung pads of cattle are temperature and moisture. Over its normal ranges, temperature merely affects the rate of development, but extremes of moisture content reduce survival.

At present few natural enemies are known, but one species of nematode (Hughes & Nicholas 1969) appears to have a substantial impact on the bushfly populations.

When the data for eleven laboratory populations were examined there was a strong suggestion of numerical regulation within the dung pad. In these experiments different numbers of eggs, ranging from 1406 to 4254 were seeded on to equal volumes (2 litres) of dung. In Fig. 1, the points show how the percentage survival from egg to adult varied with the initial number of eggs placed on the dung. The number of adults emerging ranged from 754 to 1196, averaging 915 per pad. If the mortality of the initial number of individuals (eggs) had been adjusted to give 915 adults in each case, the points of Fig. 1 would lie on the superimposed (calculated) curve.

The geographical distribution of cattle tends to be stable. Furthermore, dung pads are produced regularly at a rate of about twelve per animal per day (Hancock 1953), so that only under extremely good or bad pasture conditions is there much variation in the number of discrete fresh dung pads occurring in any area. The only quantitative effects on population processes normally seen are therefore those of differing larval densities within the dung pads.

In contrast the quality of the dung varies directly with seasonal changes in the pasture vegetation. The variation of dung quality shows in many features of the dung, e.g. consistency, moisture content, fibre content, nitrogen content, etc. Observation shows that changes in these features tend to be interrelated and so almost any one can be used as a general index of quality. Considering cattle dung for its food value, it seemed reasonable

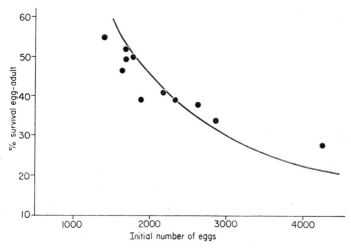

FIG. 1. The relationship between survival of individuals during development and the number of eggs inoculated into the dung pad. The superimposed curve shows the survivals needed to stabilize the populations to 915 adults per dung pad (see text).

to regard the percentage of nitrogen as an index of food quality, although its actual meaning as an ecological parameter is only now being investigated.

A graph of the annual variation in nitrogen content of dung collected from a grass/clover pasture on the tablelands of S.E. Australia (Fig. 2a, generalized from Heath 1966) shows spring and autumn maxima and a pronounced summer minimum. For comparison, a typical population curve for adult bushflies from the same area is given (Fig. 2b, after Norris 1966). Bearing in mind the time-lag of a few weeks between the population events involving dung and the appearance of the resulting flies, there seems to be a clear association between population increase and the normal spring period of high dung quality, and between population decline and the low quality of summer dung. The autumn rise in dung quality

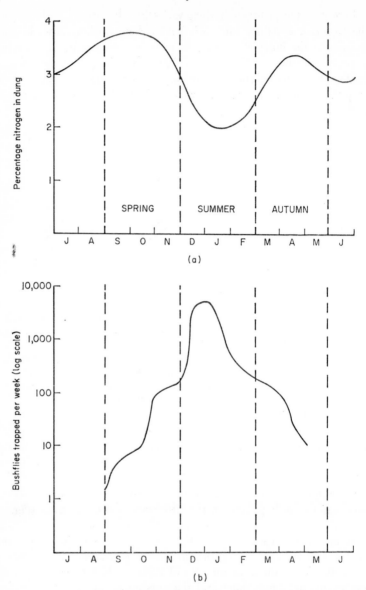

Fig. 2a. The annual cycle of nitrogen content of dung taken from a grass/clover pasture on the tablelands of S.E. Australia. After Heath (1966).
Fig. 2b. The annual cycle of numbers of bushflies trapped near Canberra, on the tablelands of S.E. Australia. After Norris (1966).

cannot be exploited because the onset of cooler weather limits fly activity and survival. For the same reason, no flies utilize the dung of good quality available during the winter. It is not until the spring that immigrant flies from warmer regions reoccupy the niche.

The time-lag in the association of quality change and numerical change suggested that the effect of population increase during quality decline should be investigated further.

EFFECTS OF FOOD QUALITY AND QUANTITY ON SURVIVAL AND REPRODUCTION

A crude 3 × 3 factorial design of treatments was used. Analytical facilities for estimating total nitrogen were not available initially so use was made of three types of dung from cattle feeding in midsummer on differing pastures: good—irrigated legume pasture; fair—dry natural grass pasture; poor—overgrazed dry pasture. (Crude estimates made later of percentage nitrogen in these types suggested that it probably ranged from just above 3% to 1·5%). Three population densities in the ratios of 4:2:1 were used—newly-hatched larvae being inoculated directly into 1 litre volumes of dung. Three sets of each of the nine regimes were set up and the numbers of each subsequent developmental stage of the fly were recorded for the twenty-seven cultures, as were the numbers of emerging females and their average egg complement after being fed on slices of fresh liver, which is considered to be an optimum source of protein.

By summing, first over replicates and dung qualities, and then over replicates and population densities, average values of nine observations each were derived, showing the effects of variation of population density and of food quality (Table 1a, b). These results show that, whereas only larval mortality is seriously affected by the numbers of larvae used, food quality shows marked effects on larval mortality, pupal mortality, and the fecundity of the females.

Density-related larval mortality would thus seem to be the mechanism causing population regulation within dung pads (see Fig. 1). This mortality could be the result of intraspecific competition for food, oxygen, or moisture. Cannibalism is unlikely because the larva does not use its mouth parts in imbibing food; but purely physical effects of tunnelling by larvae, causing more than normal levels of water loss, chemical change and solidification of the dung are possible alternative causes of the mortality.

It is not yet known how dung quality causes larval and pupal mortality, but the reduction of fecundity with food quality during the larval stage was associated with a 19% reduction in the average size (head width) of the emerging flies, thus:

Good 2·06 mm   Fair 1·75 mm   Poor 1·67 mm

It was linked also with a 30% prolongation of the period needed for eggs to mature after feeding on liver. It seems possible therefore, that larval food of poor quality reduces body size and food reserves in the adult

TABLE 1a

Effects of varying food quantity (larval density) on bushfly life-tables

	Numbers resulting from different initial densities			Percentage reductions in densities between different stages		
	× 4	× 2	× 1	× 4	× 2	× 1
Larvae	520	260	130	69	65	54
Pupae	160	92	60	39	28	41
Adults	98	66	35	51	50	49
Females	48	33	18	8	8	0
Eggs/female	23	23	25			
Total eggs	1104	759	450			
Rate of increase eggs per larva	2·12	2·92	3·46			

TABLE 1b

Effects of varying food quality on bushfly life-tables

	Changes in numbers with different food qualities			Percentage reductions in numbers between different stages		
	Good food	Fair food	Poor food	Good food	Fair food	Poor food
Larvae	300	300	300	49	72	75
Pupae	154	83	75	25	43	51
Adults	116	47	37	53	50	48
Females	54	23	19	0	36	42
Eggs/female	33	21	18			
Total eggs	1782	483	342			
Rate of increase eggs per larva	5·94	1·61	1·14			

females, causing a reduction in the number of ovarioles per ovary and more prolonged oogenesis.

The 'potential progeny' of females as shown in Table 1 was ascertained after the flies had been fed on liver. Therefore if they had fed from sub-optimal protein sources, such as dungs of poorer quality, a further reduction of egg numbers would result. Dissections of gravid females caught in the field from spring to early summer were used to compare the egg numbers produced in the field with those produced by laboratory-reared flies of the same size (head width) fed on liver. During three monthly periods, October to December, the average fecundity of flies caught in the field, calculated as a percentage of that of liver-fed equivalent flies, fell from 85% to 73%, and then to 65%.

In further experiments the numbers of eggs produced by females allowed to feed during one short period on seasonally-variable cow dungs of known nitrogen content were again compared with the egg numbers produced by equivalent females fed on liver. The cultures of flies were started with batches of 400 larvae set up in 1 litre of the cow dung. The larval densities in this series were comparable with the average of those in the factorial experiment. In a concurrent series, using the same dungs, flies derived from cultures started with 800 larvae in 1 litre volumes were used to show the effect of doubling the population density.

The results of these experiments are shown in Fig. 3, in which the numbers of eggs produced after feeding on dung, are expressed as a percentage of the number usually produced after feeding on liver. The generally linear relationship of egg numbers to dung quality is apparent, and there tends to be a reduction in the egg numbers produced by flies reared at the higher density.

Taking these results in conjunction with those of Table 1, it is evident that the differences in the quality of dung used could have resulted in enormous differences in the potential rate of increase of the bushfly, expressed here in an oversimplified way by the number of eggs produced per larva present initially. Under the best conditions provided, the increase would be five to sixfold. At the other extreme, the small flies emerging could be prevented from developing any eggs, so that severe population decrease or even local extinction would have occurred.

It can thus be seen that in the observed field situation of rapid population increase concurrent with a decline of food quality, a subsequent rapid decrease in fly numbers is inevitable.

Furthermore, field observations on the proportion of dung pads used for egg laying suggest that their acceptability as oviposition sites is almost certainly affected by their quality. In the laboratory newly-gravid females generally do not lay eggs on dungs of poorer quality although older flies may eventually do so. This would be an additional effect of declining dung quality on the local reproductive performance of the bushfly.

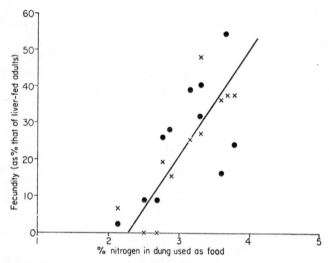

FIG. 3. The relation between fecundity and the percentage nitrogen in the dung used as adult food, at two levels of larval density: ● ● 400/pad and × × 800/pad. The fitted line is the regression ($P < 0.001$) shown by the high density series.

EFFECTS OF FOOD QUALITY AND QUANTITY ON FLIGHT ACTIVITY

The huge and rapid decline in bushfly numbers from midsummer onwards (see Fig. 2b), so long before conditions become unfavourable for the adult flies, suggests that the species must possess a highly adapted mechanism for survival over winter in the colder parts of Australia. No physiologically differentiated phase is known, but there is circumstantial evidence that long-distance movements by adult bushflies occur. As some wind patterns, in association with rain-bearing frontal systems, could assist the flies to reach areas with high-quality pastures (cf. locusts), close attention has been paid to factors likely to affect the flight activity concerned in the long-distance spreading of bushfly populations.

In the laboratory, the effects of variation of environmental conditions

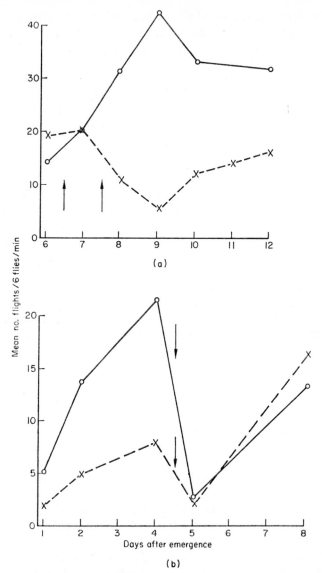

FIG. 4a. The flight activity indices of two groups of similar flies, one of which X—X was allowed access to dung on the days indicated.

FIG. 4b. The flight activity indices of flies reared as larvae either on poor quality dung O—O, or on good quality dung ×—×. Both sets of flies were given access to liver on the day indicated.

could be demonstrated on an index of flight activity obtained by observing the number of flights made per minute by a group of six flies in a cage of standard pattern, 1 minute after a sudden change in light intensity. It was found that if flies were fed only on sugar and water, a rapid increase in the number of flights by young individuals was followed by a steady high level of activity as they grew older. But when flies of any age obtained a substantial feed of protein, this resulted in a marked decline of activity for a period of several days. An example of this response after week-old flies had been allowed access for 2 days to dung containing 2·5% nitrogen, is shown in Fig. 4a.

A comparison has also been made between the indices of flight activity of flies reared as *larvae* on dung of either high or low quality. Figure 4b shows the very different indices (which are each the average of eight counts) recorded each day over a period of 4 days. The flies from both treatments were then allowed to feed on liver, when their activity fell to the same low level.

The effect of three levels of larval density in the dung pad on the flight activity of the emerging flies is being examined in the same way. After seven sets of replicated observations using only better-quality dungs, the results show that the indices of flight activity in flies reared at densities of 200 and 800 larvae per litre pad of dung are 28% and 15% higher than those reared at a density of 400 per pad.

DISCUSSION

Figure 5 has been constructed to give a synopsis of the influences that food is known to exert on the life system of bushfly.

The interplay of the effects of seasonal changes of dung quality, with those of population densities and, perhaps, of natural enemies (such as the nematode parasite), appears to be able to account for the bushfly population curve of Fig. 2b.

Although the quantity of dung lying around is almost constant in any (large) area, each mass is available as food only for a short period after defaecation. Changes in flight activity that reinforce the contact between the flies and groups of animals providing dung thus clearly affect food availability. Furthermore, if the species in fact depends on long-distance movement to avoid the effects of periods of unfavourable conditions, its widely dispersive behaviour, coupled with mechanisms increasing or decreasing flight activity in association with seasonal cycles of food quality and with population density, would have great survival value.

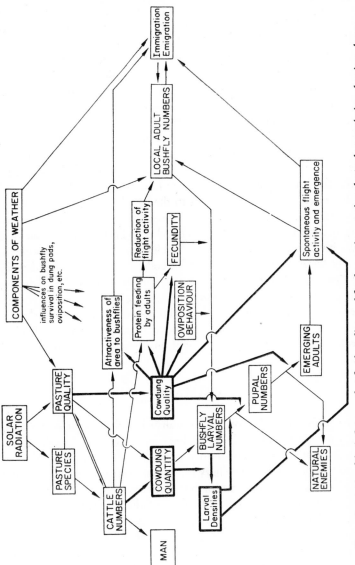

FIG. 5. Synopsis of the influences that food quality, and food quantity—acting through population density—have on the life system of the bushfly.

Unlike many other animals, the larval population of the bushfly lives within its food source. Desiccation and crust formation change the physical environment as the larvae develop, and the size of the faecal mass and its other characteristics strongly influence the rate at which physical conditions change. It is therefore difficult to separate the direct from the induced effects of food quality and quantity. An interesting example of this is the slight change of sex ratio with food quality shown in Table 1. Further experiments showed this to be an effect of the more prolonged moist conditions in the dungs of better quality. They had an adverse effect on the survival of the more rapidly developing female larvae.

Whether food quality and quantity can actually be separated for study, raises an interesting philosophical question for general discussion. In the results with bushflies there is some evidence for a separation of food effects. Whereas food quality and quantity seem to have similar effects on larval mortalities (see Table 1), subsequent to the larval stage the effects of the quality differences seem to persist more strongly. Presumably the more lasting effects of the differences in food quality lie in the nature of the food reserves involved. Recent work on the biochemistry of metamorphosis of the fly *Lucilia cuprina* by Birt and his colleagues (Birt & Christian, in press; Crompton & Birt 1967; D'Costa & Birt 1966) showed that, during the pupal stage, fat metabolism was almost the only source of energy utilized. The slight changes in the carbohydrates, and particularly the stability of the reserves of nitrogen compounds present, suggest a mechanism to account for the relative persistence of quality effects beyond the larval stage.

SUMMARY

1. The adult population of the Australian bushfly (*Musca vetustissima*) is highly mobile and dispersive. The flies occur all over the continent but disappear from the southern-third during the colder months. Flies are attracted to, and stay in the vicinity of, groups of large mammals. Freshly dropped faeces constitute a major food source and the oviposition sites for the adult females, as well as the food of the larvae. Currently cattle dung provides the bulk of favourable food in Australia.

2. In each locality there is deposition of a more or less regular number of dung pads throughout the year, but there are marked seasonal variations in the food residues they contain. In S.E. Australia these cyclic changes in food quality are associated during the warm months with numerical changes in the adult fly populations which show a rising and falling pattern

lagging a little behind that of food quality. Changes with the season in the fecundity and the fertility of the field-caught flies were also noted.

3. This situation was analysed in the insectary by constructing life-tables for the bushfly for nine combinations of food quality and quantity (i.e. larval density). The effects of larval density seem limited to larval survival and are reflected in some regulation of fly numbers emerging from dung pads. In contrast food quality affects both larval and pupal survival, and also the sex ratio, fecundity and fertility of the emerging adults. Indices of rates of increase derived from each life-table show sixfold variation, and, where low quality dung is the only available adult food, may fall to near zero.

4. Besides the effects on survival and reproduction, food quality affects the flight activity of the adults. Experiments suggest that such effects would result in flies remaining longer in areas with food of high quality, and dispersing from areas where food quality was low.

REFERENCES

BIRT L.M. & CHRISTIAN B. (1969) Changes in nitrogenous compounds during the metamorphosis of the blowfly, *Lucilia cuprina*. *J. Insect. Physiol.* **15**, 711–9.

CROMPTON M. & BIRT L.M. (1967) Changes in the amounts of carbohydrates, phosphagen, and related compounds during the metamorphosis of the blowfly, *Lucilia cuprina*. *J. Insect Physiol.* **13**, 1575–92.

D'COSTA M.A. & BIRT L.M. (1966) Changes in the lipid content during metamorphosis of the blowfly, *Lucilia*. *J. Insect Physiol.* **12**, 1377–94.

HANCOCK J. (1953) Grazing behaviour of cattle. *Anim. Breed. Abstr.* **21**, 1–13.

HEATH D.D. (1966) Studies on the ecology and epidemiology of gastrointestinal nematodes. M.Sc. thesis, Univ. of New England.

HUGHES R.D. & NICHOLAS W.L. (1969) *Heterotylenchus* sp. parasitising the Australian bushfly: additional information on the origin of the parasite of the facefly. *J. econ. Ent.* **62**, 520–1.

NORRIS K.R. (1966) Notes on the ecology of the bushfly *Musca vetustissima* Walk. (Diptera, Muscidae), in the Canberra district. *Aust. J. Zool.* **14**, 1139–56.

TYNDALE-BISCO, MARINA & HUGHES R.D. (1969) Changes in the female reproductive system useful as age indicators for the bushfly (*Musca vetustissima* Walk.). *Bull. ent. Res.* **59**, 129–41.

DISCUSSION

P. BLAZKA: What are the larvae actually eating from the faeces?

R. D. HUGHES: They take in fluids and small particles of dung, but I do not know what is used.

M. J. WAY: What is the nature of the competition between larvae in the dung pad in view of your evidence that similar numbers of adults develop from a fixed amount of dung, irrespective of the number of eggs laid in the pat?

R. D. HUGHES: I am not sure. It could be physical: the more larvae there are, the quicker the dung dries, and the less suitable it becomes. It is probably not cannibalism.

T. H. COAKER: Is it possible that the ovipositing flies recognize suitable pads for egg laying by their odour?

R. D. HUGHES: Possibly they do, but flies continually visit pads, whether they are about to lay or not, with no obvious difference in their behaviour. However, a fly lays only after feeding, so laying may also be associated with food quality.

J. PHILLIPSON: (a) You have shown that the quality of the dung pads, in terms of percentage nitrogen, varies seasonally. Have you any measurements of variation in quality within a shorter period, say 1 day? If such variation does occur, then it would suggest, on the basis of earlier papers, that the cows as a group are relatively non-selective feeders even though particular individuals may be selective. (b) How long is the dung available to flies for oviposition purposes?

R. D. HUGHES: (a) I do not know of any relevant measurements. There is not much variation in the quality of the dung from similar animals in any one paddock, but dung from different types of animals (such as cows and calves) does differ. (b) It depends on the weather. In Australia it is usually sunny, so dung remains available for only a few hours until it dries and forms a hard crust, but if the dung is dropped at night, it remains available at least until the next morning. In wet weather, it remains available for about 36 hours.

P. R. EVANS: What other animals are present in the cow dung community, and do these affect population processes in the bushfly?

R. D. HUGHES: Few other species are present. Most local mammals produce pellet dung, and the local beetles cannot cope with pats. Sometimes cow pats remain in fields for up to 3 years.

D. C. SEEL: You said that the flies require fresh dung, which you attribute in the female to a need for materials for maturation of the eggs. Why should male flies need to visit dung at this time?

R. D. HUGHES: I do not know the answer; possibly the males need a protein meal as well.

N. WALOFF: Have you investigated the carbohydrate content of the good and bad flyers?

R. D. Hughes: Not yet, but we shall be looking into it.

M. J. Coe: I believe that many dipterous larvae produce a great deal of metabolic heat when encrusted in a good insulator, such as dung. Have you noticed a relationship between the number of larvae, volume of dung, and the temperature generated within the faecal matter?

R. D. Hughes: We have not looked into this but I doubt whether they cause a noticeable temperature rise. The laboratory animals were reared at a constant 30°C, but under field conditions, temperatures in the dung may vary diurnally by 30°C, so that heat from the sun is likely to greatly exceed any from the animals themselves.

QUALITY AND AVAILABILITY OF
FOOD FOR A SYCAMORE APHID
POPULATION

By A. F. G. Dixon

Zoology Department, Glasgow University

INTRODUCTION

Most aphids feed on phloem sap by inserting their stylets into the phloem elements of plants. Phloem sap contains high concentrations of sugars but little amino-nitrogen (Mittler 1958; Ziegler 1956). The quantity and quality of the amino-nitrogen present in the phloem sap of plants changes with the progress of growth and maturation of the leaves and shoots. When shrubs and trees are actively growing or senescing, as in the spring and autumn, the phloem sap contains relatively high concentrations of amino-nitrogen composed of many amino-acids; in the summer when growth has ceased the sap is poor in amino-nitrogen and it then contains relatively few amino acids (Lindemann 1948; Mittler 1958; Ziegler 1956). As the concentration of amino-nitrogen in the phloem sap falls, the aphids become smaller and their birth rate drops (Mittler 1958; Lindemann 1948). This has also been confirmed by rearing aphids on synthetic diets (Dadd & Mittler 1965). Thus the limiting factor in the food of aphids is the level of amino-nitrogen.

The sycamore aphid, *Drepanosiphum platanoides* (Schr.) spends most of its life on the leaves of its host plant which supply the aphid not only with food but an area on which to live. Thus the availability of leaves suitable as living space also influences the quantity of food that is available to the aphid population.

The aim of the present paper is to illustrate the influence of the quality and availability of food on sycamore aphid populations.

271

THE EFFECT OF THE QUALITY OF PHLOEM SAP ON REPRODUCTION AND BODY SIZE IN THE SYCAMORE APHID

(a) REPRODUCTION

The reproductive rate of the sycamore aphid changes markedly during the season, and is strongly correlated with the level of amino-nitrogen in the leaves on which the aphids are feeding (Fig. 1, $r = 0.91$, $P < 0.01$). Reproduction is at a high rate when the leaves are actively growing and senescing, and at a low rate, or ceases altogether, when the leaves are mature. The

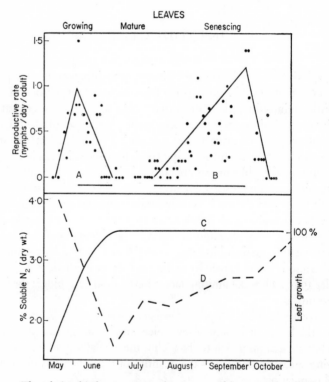

FIG. 1. The relationship between reproductive rate of the sycamore aphid and the progress of growth and maturation of sycamore leaves (C = leaf growth, D = percentage of soluble nitrogen present in the leaves).

increase in the reproductive rate at the beginning of the year occurs because the adults are at first mostly teneral or immature and then come into reproduction and finally reach their maximum reproductive rate; in the autumn the decline results from the drying out of the leaves which then no longer proffer a satisfactory food source.

Both the nutritive status of the host plant and the population density of the sycamore aphid affect the reproductive rate of the aphid during the summer. As previously described (Dixon 1963), the low level of amino-nitrogen in the phloem sap of the mature leaves can be compensated for by the aphid feeding at a higher rate and with greater efficiency than when it is feeding on young or senescent leaves. Presumably under field conditions the aphid can achieve this only when the population density is very low and individuals do not disturb one another, as disturbance reduces the time available for feeding. Evidence to support this comes from year to year variations in the proportion of nymphs in the population during the summer, and in fact the proportion of nymphs in the population is negatively correlated with the population density of the aphid (Dixon 1966). Further, the reproductive rate of the aphid on eight trees during the summer in 1968 was negatively correlated with the population density of aphids on those trees ($r = 0.93$, $P < 0.001$). Aphids similar to the non-reproducing adults present in the field during most summers can also be induced by partial starvation of aphids (Dixon 1963). Therefore under normal field conditions in summer the population density is sufficiently high to give rise to considerable mutual interaction and resultant frequent interruption of feeding which would result in the aphid being unable to feed at the high rate necessary to compensate for poor phloem sap.

Therefore with improving quality of the phloem sap there is a corresponding improvement in the reproductive rate of the sycamore aphid. There is in addition, when the nutritive quality of the leaves is poor, a density dependent reduction in the reproductive rate of the aphid, which can result in the complete cessation of reproduction at the higher densities.

(b) BODY SIZE

As well as changes in the reproductive rate there are also changes in the size of the aphid with the season (Dixon 1966). In spring and autumn the adults are two to four times heavier than they are in the summer. Under field conditions it is difficult to separate the role of nutrition and interaction between aphids in determining body size. Rearing aphids in isolation from the first instar to maturity on plants of different physiological ages results in adults of different sizes, comparable with those observed in the field when the leaves are growing, mature or senescing, according to the season (Fig. 2). Aphids reared in crowds on plants of similar physiological age are always smaller than those reared in isolation (Fig. 2). Thus the effect of crowding during the development of an aphid, as well as the nutritive quality of the host plant, influences the size of the adult aphid.

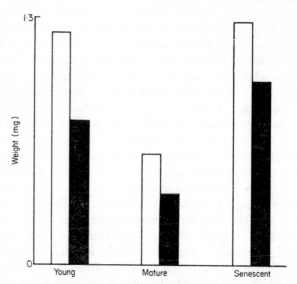

FIG. 2. Weight of adult aphids reared in crowds and in isolation, on young, mature and senescent sycamore leaves. □ reared in isolation, ■ reared in crowds.

The quality of these adults varies according to their size. After an aphid becomes adult there is a delay of several days before the onset of reproduction. The duration of this delay is correlated with the size of the aphid (Fig. 3). Large aphids come into reproduction sooner than do smaller ones ($r = -0.835$, $P < 0.001$). The larger aphids also give birth to more nymphs during the first 10 days of their reproductive life than do the smaller aphids ($r = 0.902$, $P < 0.001$), and the larger adults produce larger nymphs ($r = 0.63$, $P < 0.01$). The larger aphids are also less likely to die before reproducing than are the smaller ones.

Thus the nutritional quality of the host plant can have a marked effect on the aphids feeding on it. Good nutrition results in large high-quality aphids with a greater reproductive potential.

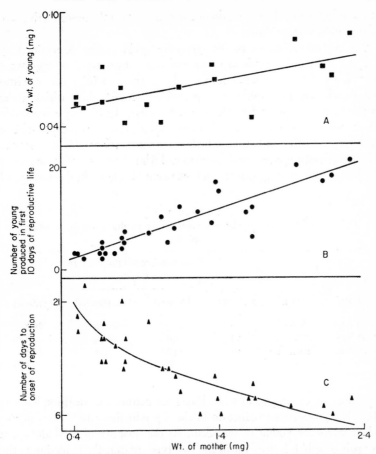

FIG. 3. The relationship between the weight of the adult aphid and (A) the average weight of its young, (B) the number of young produced in the first 10 days of reproductive life, and (C) the number of days from adult moult to onset of reproduction.

THE EFFECT OF THE QUALITY OF PHLOEM SAP ON THE INTERACTION BETWEEN APHIDS

Previously I suggested that interaction between aphids is responsible for the reproductive depression during the summer (Dixon 1963, 1966). In the spring and autumn there are often very high population densities of the aphid which, however, do not appear to affect the reproductive rate of the aphid at these times.

The reproductive rate of the sycamore aphid can be determined by keeping individual aphids in small leaf cages. Two types of cage were used, both of the same area but one had a partition dividing the cage into two equal parts. Two aphids were placed in each cage, the partition in the divided cages serving to keep the two aphids apart. In this way it is possible to test whether the interaction between aphids results in a depression in their reproductive rate, and whether nutrition affects this interaction. This experiment was repeated three times during the course of a year, when the leaves were young, mature and senescent (Table 1). In the absence of

TABLE 1

The effect of the interaction between aphids on reproduction in the sycamore aphid

Date	Leaf	No. of replicates	Total number of nymphs born Cages		Significance
			Divided	Undivided	
23–31 May	Young	10	223	272	N.S.
6–18 July	Mature	15	46	12	$P < 0.001$
1–15 Aug.	Senescing	13	218	235	N.S.

a partition in the cage, and when leaves are mature and nutrition is poor, there is a pronounced reduction in the reproductive rate. This is not so when leaves are young or senescent. As the quantity of food and space per pair of aphids is the same in the two types of cage the reduction in the reproductive rate of the aphids in undivided cages on mature leaves must result from the interaction between the aphids. If this is correct, then increasing the size of the group of aphids while keeping the area of leaf available to each aphid constant should also result in a reduction of the reproductive rate of individual aphids in the larger groups.

This was tested using groups of two and six aphids. The area of leaf available for each aphid was the same in both treatments. Increasing the size of the group from two to six aphids has a highly significant effect in depressing the reproductive rate (Table 2). This effect is still apparent, though less marked, after the leaves have begun to senesce. Under field conditions in the summer, aphids space themselves out over a leaf at a relatively uniform distance from one another. Thus the area available to

TABLE 2

The effect of the number of aphids in a group on reproductive rate in the sycamore aphid

| | | Cage | | Average no. of |
Leaf	No. of aphids	Area (cm²)	No. of replicates	nymphs produced per adult per day
Mature	2	3·8	12	0·8
	6	11·4	9	0·07
Senescing	2	3·8	14	0·8
	6	11·4	9	0·31

aphids in the field can be determined and it is 43% less than that available to each aphid in either type of cage.

Thus interaction between aphids is influenced by the nutritive quality of their food, and the number of aphids in the group.

Interaction between aphids possibly results when aphids meet and the chances of meeting would be increased if the aphids moved frequently. The nutritive quality of the plant has a marked effect on the incidence of movement. An aphid allowed to settle on a mature leaf is five times more likely to move than an aphid allowed to settle on young or senescent leaves (Table 3). Movement is also more marked at higher temperatures. An increase of 8°C approximately doubles the incidence of movement.

In early spring and late autumn, individuals of the sycamore aphid are often seen together in compact groups because both the high nutritive status of the host and low temperature reduce the restlessness of both adults and nymphs. In summer when the nutritive quality of the host is poor and temperatures are higher the aphids move frequently from leaf to leaf,

TABLE 3

The effect of the age of a leaf on an aphid moving off a leaf within a 24-hour period

| | No. of observations Aphid | | Proportion of aphids that moved |
Leaf	Remained	Moved	(%)
Young	68	4	6
Mature	32	16	33
Senescent	75	5	6

which results in more encounters with other aphids; thus when population density is high, mutual disturbance leads to the complete cessation of reproduction. It is also possible that poor nutrition heightens the response of aphids to the presence of other aphids. Partial starvation of aphids results in the production of small adults which reproduce slowly (Dixon 1963) and non-reproductive and slowly reproducing aphids are smaller than the highly reproductive ones. It is possible therefore that the mutual disturbance when aphids meet could limit the time available for feeding and result in partial starvation, especially when the nutritive quality of the host is poor.

THE ROLE OF NUTRITION IN THE FLUCTUATIONS IN THE NUMBERS OF THE SYCAMORE APHID

Fluctuations in the numbers of the sycamore aphid follow a similar pattern each year (Fig. 4). In the spring the aphids that are present on the leaves at bud burst suffer a heavy mortality and relatively few reach the adult stage. These adults have a high reproductive rate resulting in a rapid increase in the number of aphids as the year progresses until there is a peak in June or July. There is then a big decline in numbers followed by another increase in the autumn. The peak number of aphids in the summer and autumn varies from year to year.

During the course of the year the reproductive rate of the sycamore aphid is correlated with the nutritive quality of its host plant. Although the pattern of increase and decline in reproductive activity through the season is the same from year to year, the level of reproductive activity at a particular time of year can vary markedly from year to year. The extent to which aphids reproduce during the summer reproductive depression depends partly on the population density of the aphid. However, in spring and autumn when the nutritive quality of the host plant is good and there is less interaction between the aphids there are still unexplained variations in the level of reproductive activity.

In the spring the reproductive rate rapidly increases to a maximum and then gradually falls off with the approach of summer (Fig. 1). The maximum reproductive rate attained and the rate at which it falls off with the approach of summer vary from year to year. An index of the reproductive rate can be determined by calculating the area of the triangle above the line A in

Fig. 1. Similarly, an index of the reproductive rate in the autumn can be obtained by calculating the area of the triangle above line B in Fig. 1.

In addition to the reproductive rate realized at a particular time in the year, the peak numbers could also depend on the numbers of adult aphids present in the population at the start of the increase in numbers. This latter possibility was examined (Table 4) for 4 years on two trees, and for 2 years on another two trees.

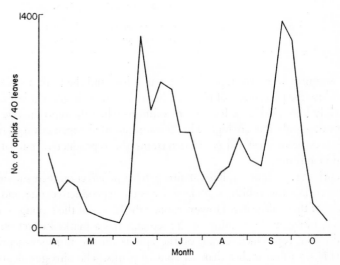

FIG. 4. Changes in numbers of aphids on a sycamore tree through a year.

The summer peak numbers are strongly correlated with the peak numbers of adult aphids present in the same spring. They are not significantly correlated with the spring reproductive rate, but this lack of correlation may be due to the small number of observations. The reproductive index for the spring is correlated with tree growth for that year. In years of above average tree growth the reproductive index is also high ($r = 0.6276$, $P<0.05$).

In the autumn the peak numbers are significantly correlated with both the numbers of adults present at the start of the increase in numbers and also with the reproductive rate of the aphid. The difference in reproductive rate from year to year cannot be attributed to temperature differences as the average autumn temperature over the period of the observation has remained constant.

TABLE 4

Partial correlation coefficients relating summer and autumn peak numbers of aphids (1), the number of adult aphids present in the population at the start of the increase in numbers (2), and the index of the reproductive rate over the period of the increase in numbers (3)

Peak nos.	No. of observations	Partial correlation coefficients	Significance
Summer	12	$r1,2.3 = 0.7874$	$P < 0.01$
		$r1,3.2 = 0.2941$	N.S.
Autumn	12	$r1,2.3 = 0.6294$	$P < 0.05$
		$r1,3.2 = 0.6723$	$P < 0.05$

In autumns when the reproductive rate is low and the peak population is also low the percentage of nitrogen in the leaves of the trees at leaf fall is relatively high (Table 5). It is low in autumns when the reproductive rate and peak populations are high. Thus the amount of nitrogen salvaged from the leaves before leaf fall could determine the reproductive rate of the aphid in autumn.

Depriving sycamore saplings of nitrogen has no effect on their suitability in the spring for aphids, as judged by their reproductive rate and the weight of their offspring. However, once the leaves of the sapling start to senesce, nitrogen deprivation of the saplings has a marked effect on the aphids. The reproductive rate of the aphids on the nitrogen-deprived plants is 1.7 times higher than on control plants. The nitrogen-deprived plants salvaged 1.6 times more nitrogen, via the phloem sap, from their

TABLE 5

The relation between the index of the reproductive rate of the aphids in the autumn and the percentage of nitrogen in the leaves at leaf fall for two trees

Year	Index of reproduction	% nitrogen in leaves at leaf fall
Tree 1		
1966	44.3	2.7
1967	35.8	3.4
1968	36.2	3.9
Tree 2		
1966	49.5	2.4
1967	43.2	2.9
1968	33.1	3.1

leaves before leaf fall, than did the control plants. The translocation of this nitrogen back into the stem of the plant therefore results in a highly nutritive food becoming available to the aphids.

There is evidence that shoot growth of many woody plants depends on stored food rather than the products of current photosynthesis and mineral uptake (Kozlowski 1963; Meyer & Tukey 1965). The growth from the terminal bud of branches of sycamore from which all other buds are removed is 20% greater than from branches which are not debudded. Thus when competition for the available food reserves is lowered by debudding, the growth from the terminal bud is greatly increased, or alternatively, the increased growth is due to the absence of growth-inhibiting hormones from the lateral shoots.

The number of aphids present on the growing leaves in the spring is related to the number present the previous autumn. Analysis of the effect of number of aphids in the spring and the previous autumn, on tree growth, reveals that it is only the number in the spring that affects growth. It is possible that the aphids in the spring are competing with one another and with the growing leaves and shoots for available nutrients. This results in below-average tree growth in years of high aphid numbers in the spring, and the aphids reproduce at a lower rate in such years (p. 279).

AVAILABILITY OF FOOD AND
SPACE TO THE SYCAMORE APHID

Not all the leaves of a sycamore are equally suitable for the sycamore aphid and this is associated with a leaf's position within the canopy of the tree and its resultant micro-environment. With the approach of summer most of the aphids descend from the upper parts of the tree canopy and aggregate on the leaves of the lower canopy (Dixon 1969). This happens at a time when the intensity of solar radiation is increasing. Temperatures on the leaves in the upper canopy on sunny days at this time of year can be as much as 10°C higher than on leaves at the bottom of the tree. Thus for short periods of time on sunny days the temperature in the upper canopy may approach or even exceed the lethal temperature for the sycamore aphid (Jackson, unpublished). Aggregation on the leaves of the lower canopy during the summer enables the aphid to avoid the lethal temperatures it would otherwise experience. Recolonization of the upper canopy in the autumn, when reproduction recommences, appears to be related to population density (Dixon 1969), but other factors such as weather and differences in the nutritive quality between the leaves of the upper and lower canopy may also play a part.

Even within the lower canopy the leaves are not all equally suitable for colonization by aphids. Some leaves are consistently well populated and others consistently not so all through the year. This is associated with a leaf's position within the canopy and its resultant micro-environment (Dixon &

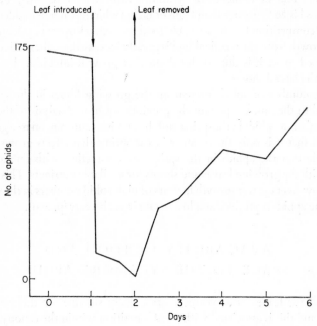

FIG. 5. The effect on the number of aphids on a leaf of inserting another leaf immediately below it.

McKay, in preparation). Less suitable leaves differ from suitable leaves by usually being close to other leaves immediately below them. The extent to which the undersurface of a leaf is brushed or touches other leaves, as a result of wind, is a factor which determines the suitability of a leaf for aphid colonization. Placing a leaf 3 inches below and parallel to the lamina of a leaf which has consistently borne large numbers of aphids over several days results in a sudden and dramatic decline in the numbers of aphids present on that leaf. This response to the presence of a leaf immediately below is extremely rapid (Fig. 5). Aphids begin to move and leave the leaf as soon as the two leaves touch because of wind. In still air in the laboratory the presence of another leaf close below does not repel the aphids or affect the suitability of a leaf. It is also possible that certain leaves are more exposed to wind than others and, either because of air currents alone, or the resulting leaf movements, are less suitable for colonization.

Solar radiation also affects the suitability of leaves. Shaded leaves are more suitable than unshaded ones.

Therefore sycamore trees as a habitat for aphids are far from uniform because of the effects of solar radiation and wind; a few leaves consistently provide optimum conditions and are similar to the outbreak centres described in locust biology. Thus the sycamore aphid, despite an apparent abundance of space, may often fully exploit the whole area available to it.

DISCUSSION

Fluctuations in the numbers of the sycamore aphid both within a year and also between years depend to a great extent upon the nutritive quality of the host plant. The seasonal changes in the nutritive quality of the host plant depend on the growth and senescence of leaves. It is not known what causes the differences in nutritive quality in the autumn between years. However, over a large area all sycamore trees are nutritively good or bad in the autumn. This could be the result of weather factors. Whatever determines these changes in tree physiology, the resulting differences from year to year in the nutritive quality of the host has a profound effect on the fluctuations in the numbers of the sycamore aphid.

As yet there is no evidence to support the idea that parasites or predators are capable of regulating sycamore aphid populations. Indefinite increase in numbers of the aphid appears to be checked by interaction between individuals in the population. The degree of mutual interaction between aphids is related to population density. Mutual interaction increases with population density, and results in a reduction of the reproductive rate and, or, an increase in the incidence of dispersal (Dixon 1969). However, these self-regulatory processes are not independent of the nutritive quality of the host plant. When the nutritive quality of the host is poor the aphids move about more frequently, and consequently the degree of mutual interaction between individuals is much greater. Thus the number of individuals per unit area alone is not a satisfactory index of mutual interaction. An aphid's tendency to move about on a leaf is determined by the numbers of aphids per unit area, the nutritive quality of the host plant, and temperature. High nutritive quality and low temperature reduce the restlessness of the aphids and in so doing allow the accumulation of large numbers of aphids per unit area. Movement in the field is not only affected by population density, nutritive quality and temperature; the effect of wind on the leaves, exposure to solar radiation and possibly other components of the leaf's micro-environment, are all very important. At any one time the leaves of a sycamore tree are dissimilar in their suitability for aphid colonization. The ortho-kinetic response of the aphid to the various components of its environment results in it aggregating on the leaves that are most suitable, and in rapidly leaving those that are not suitable. Thus the adaptive significance of movement, an important factor in mutual interaction, is to enable aphids to colonize the more suitable leaves. The avoiding response observed when two aphids meet is similar to the response shown by this aphid to the presence of insect predators (Dixon 1958). The adaptive significance of the avoiding response is that it is an anti-predator response. Whether the degree of disturbance of an aphid resulting from the meeting of two aphids is the same when an insect predator meets an aphid is unknown. It is possible, however, that as in the vetch aphid, *Megoura viciae* Buckt., where tactile stimulation induces the appearance of alate forms, the stimulus is not species specific and can be supplied by other species of insects (Lees 1967).

The fact that most aphids aggregate on certain parts of their host plant leaving much of the plant unexploited has led to the hypothesis that competition for food or space in such situations is self induced (Kennedy & Crawley 1967; Way & Banks 1967). This assumes that the rest of the plant is equally suitable to the aphids. This is not so for the sycamore aphid where very little of the leaf area is consistently suitable for colonization. Therefore despite an apparent abundance of food or living space the sycamore aphid often suffers from a relative shortage of these resources. Aggregation of the sycamore aphid on the more suitable leaves leads to an increase in mutual interaction between aphids with the resultant negative effects on population growth. Huffaker (1957) and Voûte (1957) also believe that only special parts of plants are suitable for plant-feeding insects and as a consequence material resources are often not plentiful, as is often thought.

SUMMARY

1. Young and senescent sycamore leaves are rich in amino-nitrogen which leads to the development of large adult aphids having a higher reproductive rate and producing heavier nymphs than do those aphids

feeding on mature leaves. During the summer, when the leaves are mature, the aphids suffer a shortage of amino-nitrogen in their food.

2. During the summer months, in years of high population density, mutual interaction between adults results in reproductive diapause. At other times of the year when the leaves supply a rich food, mutual interaction between adults does not influence their reproductive rate but does result in dispersal.

3. The size of the autumn peak in abundance of the sycamore aphid is correlated with the average reproductive rate of the aphid in that autumn. In certain years trees translocate large quantities of nutrients out of the leaves back into the tree. When this occurs the phloem sap is a very rich food and the aphids reproduce at a high rate. Such autumns are followed by below-average growth in sycamore trees.

4. Not all the leaves of a sycamore tree are equally suitable as feeding sites for the sycamore aphid because of the leaf's position within the canopy and its resultant micro-environment. Sycamore trees as a habitat for aphids are far from uniform; a few leaves consistently provide optimum conditions; some are uninhabitable. Thus the sycamore aphid, despite apparent abundance of space, often fully exploits the whole area available to it.

REFERENCES

DADD R.H. & MITTLER T.E. (1965) Studies on the artificial feeding of the aphid *Myzus persicae* (Sulzer)—III. Some major nutritional requirements. *J. Insect Physiol.* **11**, 717–43.

DIXON A.F.G. (1958) The escape responses shown by certain aphids to the presence of the coccinellid *Adalia decempunctata* (L.). *Trans. R. ent. Soc. Lond.* **110**, 319–34.

DIXON A.F.G. (1963) Reproductive activity of the sycamore aphid, *Drepanosiphum platanoides* (Schr.) (Hemiptera, Aphididae). *J. Anim. Ecol.* **32**, 33–48.

DIXON A.F.G. (1966) The effect of population density and nutritive status of the host on summer reproductive activity of the sycamore aphid, *Drepanosiphum platanoides* (Schr.). *J. Anim. Ecol.* **35**, 105–12.

DIXON A.F.G. (1969) Population dynamics of the sycamore aphid, *Drepanosiphum platanoides* (Schr.) (Hemiptera, Aphididae): migratory and trivial flight activity. *J. Anim. Ecol.* **38**, 585–606.

HUFFAKER C.B. (1957) Fundamentals of biological control of weeds. *Hilgardia*, **27**, 101–57.

KENNEDY J.S. & CRAWLEY L. (1967) Spaced-out gregariousness in sycamore aphid *Drepanosiphum platanoides* (Schrank) (Hemiptera: Callaphididae). *J. Anim. Ecol.* **36**, 147–70.

KOZLOWSKI T.T. (1963) Growth characteristics of forest trees. *J. For.* **61**, 655–62.

20

LEES A.D. (1967) The production of the apterous and alate forms in the aphid *Megoura viciae* Buckton, with special reference to the role of crowding. *J. Insect Physiol.* **13**, 289–318.

LINDEMANN C. (1948) Beitrag zur Ernährungsphysiologie der Blattläuse. *Z. vergl. Physiol.* **31**, 112–33.

MEYER M.M. & TUKEY H.B. (1965) Nitrogen, phosphorus and potassium plant reserves and the spring growth of Taxus and Forsythia. *Proc. Am. Soc. hort. Sci.* **87**, 537–44.

MITTLER T. (1958) Studies on the feeding and nutrition of *Tuberolachnus salignus* (Gmelin) (Homoptera, Aphididae). II. The nitrogen and sugar composition of ingested phloem sap and excreted honeydew. *J. exp. Biol.* **35**, 74–84.

VOÛTE A.D. (1957) Regulierung der Bevölkerungsdichte von schädlichen Insekten auf geringer Höhe durch die Nährpflanze (*Myelophilus piniperda* L., *Retinia buoliona* Schff., *Diprion sertifer* Geoffr.). *Z. angew. Ent.* **41**, 172–8.

WAY M.J. & BANKS C.J. (1967) Intra-specific mechanisms in relation to the natural regulation of numbers of *Aphis fabae* Scop. *Ann. appl. Biol.* **59**, 189–205.

ZIEGLER H. (1956) Untersuchungen über die Leitung und sekretion der assimilate. *Planta*, **47**, 447–500.

DISCUSSION

T. HUXLEY: Does the leaf-spot fungus have any influence on the numbers of aphids?

A. F. G. DIXON: Not as far as we can tell. Of eight trees sampled, four have leaf spot, four do not (because they are within the zone of industrial pollution), and the numbers of aphids on all trees appear not to be significantly different.

M. J. WAY: (a) Do the adults reproduce only on the leaves where they themselves aggregate?

(b) Is there any evidence that, because they aggregate, the aphids utilize only some of the nutritionally suitable leaves available to them?

A. F. G. DIXON: In most years, aphids fill the available space to the full.

J. M. CHERRETT: Did you find any evidence of a relationship between the availability of amino-nitrogen, and the feeding rate of the aphids, as perhaps indicated by honeydew production?

A. F. G. DIXON: To a large extent aphids compensate for a decline in the nutritional content of their food by feeding more, but this can be achieved only if they are not disturbed.

J. C. COULSON: To what extent does temperature account for seasonal variation in the behaviour and biology of the sycamore aphid?

A. F. G. DIXON: Temperature has a marked effect, for when temperatures are low aphids move around very little and this affects their dispersion. In summer, when temperatures are high, aphids tend to be well spaced, but in spring and autumn, when temperatures are lower, aphids are closer together. In summer, nutrition is poor, in spring and autumn good, so it is not easy to separate the effects of temperature from those of nutrition.

RELATIONSHIPS BETWEEN
NUMBERS OF SOAY SHEEP AND
PASTURES AT ST. KILDA

By D. C. Gwynne

The West of Scotland Agricultural College, Glasgow

and J. Morton Boyd

Nature Conservancy, Edinburgh

INTRODUCTION

St. Kilda lies 73 km west of the Outer Hebrides and is of interest to biologists because of its indigenous fauna, vast seabird assemblies and maritime vegetation. In 1932 a flock of 107 Soay sheep from the neighbouring isle of Soay was introduced to Hirta (630 ha) from which domestic blackface sheep had been cleared in 1930, with stragglers shot in 1931. Since then numbers have increased to a high level and have fluctuated considerably, with insignificant interference by man. There has been no burning or other ground treatment. There are no predators except for great black-backed gulls (*Larus marinus*) and ravens (*Corvus corax*) taking small lambs. There are no competitors; the only other mammal is the field mouse (*Apodemus sylvaticus hirtensis*) and there are no reptiles. Herbivorous birds such as geese (*Anser* and *Branta* spp.) have an insignificant effect. The system is, therefore, unusually simple: one large herbivore on a fairly simple pasture.

The sheep were counted annually from 1955 with an earlier count in 1952 (Boyd 1953). In 1959 the study was intensified by a team drawn from the Nature Conservancy, Hill Farming Research Organisation, Wellcome Institute of Comparative Physiology, Royal Veterinary College London, Rowett Research Institute and the West of Scotland College of Agriculture. The vegetation was mapped by Poore & Robertson (1949) and McVean (1961). The summarized information in this progress report is drawn from the work of the Soay Sheep Team and acknowledgement is made particularly to P. A. Jewell, P. Grubb, C. Milner, D. J. Martin, R. G. Gunn and R. N. Campbell. The work will later be published in full as a monograph.

THE STUDY AREA

Village Glen is a steep amphitheatre rising from sea-level to the summit of Conachair at 419 m. The eastern part of Hirta is composed of granophyre, and the western of breccia (Cockburn 1935). The soils are nearly all podsolic, but less so under grassland than under heaths. The soils of the derelict cultivation on the seaward side of the village street have a 22 cm layer of brown loam above clay and iron pan; the slopes above the village have a thin layer of up to 10 cm of peat, above peaty grits and clays with iron cementation (McVean 1961).

The vegetation consists of mixed grassland and a *Molinia* grassland within the village area (defined by the perimeter wall). This is surrounded by moorland on the steep sides of the glen, with inliers of mixed grassland, dwarf Callunetum and *Nardus-Rhacomitrium* heaths. The mixed grass-lands are the main sheep pastures, and have been much modified by sheep grazing, manuring, salt spray and seepage.

SHEEP NUMBERS AND BEHAVIOUR

The sheep population in the study area has fluctuated greatly (Table 1). There was probably a rapid increase in grazing pressure from the original stock of 107 in 1932, and since 1955 the population has fluctuated around a mean of about 1100 with an amplitude of about 1000 and a mean period of about 4 years. There are ten groups of ewes in home ranges throughout the study area, eight within the village perimeter wall and two on the seaward slopes of Oiseval and Ruaival. Marked ewes seldom move from one home range group to another; ewes appear strongly hefted and the great majority spend their entire lives in the same home range, dying close to the place of birth.

Adult rams graze independently of ewes in six or more groups, though ram lambs and yearling rams mostly run with ewe groups until about 18 months old. Ram groups and ranges change greatly during the rut; in early October rams move to other parts of Village Glen and to the remainder of the island, while others arrive from outside the Glen. The return move-ment is probably complete by December. Home ranges of rams are usually superimposed upon those of ewes.

St. Kilda is very exposed with about 23 days of gales and 113 cm precipitation per annum. Shelter for sheep is provided by natural niches among the cliffs and also by the hundreds of *cleitean* or 'cleits', dry-stone cells built by the St. Kildans who lived on Hirta for centuries prior to the evacuation in 1930. Within the study area and its immediate environs there are 175 cleits, and more than twice as many throughout the entire Glen. The home ranges of both sexes have their complement of cleits,

TABLE I
Numbers of Soay sheep on Hirta

Year	Study area			Whole island Total sheep
	Rams	Ewes	Total sheep	
1932	—	—	—	1114
1952	—	—	—	1114
1955	—	—	—	710
1956	—	—	—	775
1957	—	—	—	971
1958	—	—	—	1099
1959	140	272	412	1344
1960	40	132	172	610
1961	121	190	311	910
1962	152	239	391	1056
1963	164	260	424	1589
1964	104	212	316	1006
1965	131	264	395	1469
1966	126	279	405	1598
1967	70	232	302	876
1968	91	231	322	1096

certain of these being used habitually by each group of sheep; rams particularly may use the same cleit from their first visit as a weaned lamb or yearling. Sheep go to the cleits when ailing, many dying there out of sight of carrion-eating birds. Bodies decompose there and the products are scattered by mice, invertebrates and seepage. Some cleits have large numbers of ram skeletons and others contain the remains of ewes and young rams.

An army camp was built within the study area in 1957. Sheep are not fed and edible refuse is put in the sea. The sheep in the village, living continuously in the sight of man, have become less wary than those in outlying parts of the island, their flight distance varying from a few metres unhurried walk to a gallop of over 100 m, depending on the abruptness of man's appearance. Therefore, although the pasturing of the sheep within the study area has been much affected by man's presence, their adaptability has resulted in few areas escaping heavy grazing, including those among the camp buildings.

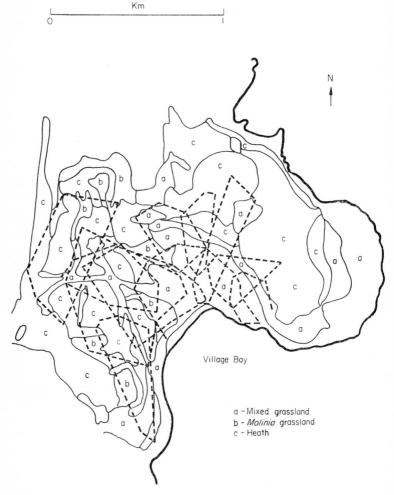

FIG. 1. Home ranges of ewe hefts (groups) during September–October 1965 (broken lines), superimposed upon the plant communities of Village Glen.

UTILIZATION OF PASTURE BY SHEEP

Grubb & Jewell (1966) have described the daily movements of sheep in the village area and their seasonal patterns of activity in relation to the general condition of the pastures. During the period of pasture production from April to September, the home ranges are centred on the *Holcus–Agrostis* grasslands within the village area; from September until early March they are larger, occupying wider areas of *Holcus–Agrostis* grassland and spreading over the heaths on the high hinterland.

TABLE 2

Proportion (%) of vegetation types occurring in the ranges of ewes in September–October 1965

| | | | Vegetation types* | | | |
Range	1	2	3	4	5	Others
1	27	32	17	12	12	
2	43	13		30	13	3
3	34	1		6	17	42
4	25	9		47	19	
5	24	7		50	9	11
6	28·5	15		39	17	1
7	41	24		35		
8	30	13		43		13
9	14	20		66		
Mean	30	15	2	36	9	7

* 1 *Calluna* heath, 2 *Calluna–Trichophorum–Molinia* heath, 3 *Nardus–Rhacomitrium* heath, 4 *Holcus–Agrostis* grassland, 5 *Molinia* grassland.

Figure 1 shows the vegetation map of Village Glen, with superimposed the home-ranges of nine of the ten groups of ewes in Village Glen in September–October 1965, when the sheep were in best condition.

Grubb calculated the size of ranges for individual sheep in different months, of which those showing the maximum and minimum ranges are shown here, in hectares:

| | Ewes | | Rams | |
Month	Bl 70	W 18	W 21	W 6
Dec.	3·1	19·6	3·9	16·0
Jan.	4·8	15·6	6·0	15·2
Mar.	2·1	6·4	4·0	1·9
May	3·4	11·6	3·2	1·6
Sept. Oct.	4·0	24·1	20·8	21·6

The ranges for September–October are probably close approximations to the maximum ranges of the sheep and Table 2 shows that about 36% of the entire area covered by sheep at the time was *Holcus–Agrostis* grassland, 30% was *Calluna* heath, and 15% was *Calluna–Trichophorum–Molinia* wet heath. Ranges 1 and 3 did not enter the village area and both have a higher proportion of heaths to grassland.

LONG TERM CHANGES IN THE PASTURES

Certain changes have been observed in the vegetation of Hirta since 1931, when the previous stock of domestic sheep had gone, and prior to the introduction of the foundation stock of 107 Soays. These changes can be assessed from the surveys of Petch (1933), Poore & Robertson (1949), McVean (1961) and the present work.

CHANGES IN HEATHS

In 1931 the moorland was dominated by heath grasses, and though *Calluna* was widespread it did not grow above turf-level and was apparent only on close inspection. Today *Calluna* is dominant over much of the moorland, in places almost to the exclusion of other species. Poore & Robertson (1949) suggest that *Calluna* attained dominance when grazing was at a minimum following the evacuation in 1930, but the present authors agree with McVean (1961) that the process has been a more continuous one, largely independent of the activities of the sheep. The Soays appear to graze the heath communities rather lightly, and this may be accentuated by the absence of management, which means that the sheep are not confined to the hill, and also that the population is held at a low level by severe winter mortality.

CHANGES IN GRASSLANDS

In 1931 a rank growth of *Holcus lanatus* dominated the village pastures, with abundant *Agrostis* spp. and *Anthoxanthum odoratum*. Today, *Agrostis tenuis* is dominant over most of the area, and three grassland types have been identified: species-rich short, species-rich tussocky, and species-poor tussocky. Growth starts early in the year in the *Agrostis* tussocks, which soon reach an unpalatable condition, and are little utilized by the sheep during the summer. This leads to the build-up of a heavy growth some 30–45 cm high which in late summer becomes increasingly prostrate under the effects of wind, rain and trampling by the sheep. This mat of moribund vegetation inhibits the growth of other species, and ensures the continued dominance of *Agrostis tenuis* by the growth of shoots from the protected rhizomes.

SEASONAL CHANGES IN THE PASTURES

Changes in the *Holcus–Agrostis* grasslands were studied from January 1965 to September 1965, by measuring fresh and dry weights, percentage organic matter, percentage digestibility and the performance of the main constituent species.

At fortnightly intervals during the period, the herb-poor tussocky grassland was sampled by cutting all the green material within fifty 15 cm square quadrats randomly distributed in Signals meadow and the adjoining fields. Tussocks were cut at the base of the green shoots and open turf at near ground level. Sampling was carried out in as dry weather as possible, and the material obtained was bulked, dried and weighed. The dry weights and *in vitro* digestibility figures in the table were obtained after milling and re-drying at the West of Scotland Agricultural College, Dept. of Chemistry, at Auchincruive, Ayr.

To assess the availability of the main grazing species, monthly analyses were made of the above samples. Complete samples in winter, and at other times 50 g sub-samples, were separated by hand into the following constituents:

(1) *Agrostis* spp. and *Anthoxanthum odoratum*.
(2) *Festuca* spp.—almost entirely *F. rubra*.
(3) *Holcus lanatus*.
(4) *Poa* spp.—mostly *P. pratensis* and *P. trivialis*,
(5) Miscellaneous spp.—including rushes, sedges, herbs and mosses.
(6) Remainder—including only dead or moribund material collected in the course of sampling.

Fresh and dry weights of these separated portions were expressed as percentages of total green material (classes (1) to (5) above), and the results are shown against time in Fig. 2.

PERFORMANCE OF MAIN SPECIES

Figure 2 shows an increasing preponderance of *Agrostis* spp. throughout the year, illustrating the build-up of these species, mentioned in the previous section's discussion of changes in the grasslands. *Holcus lanatus*, the co-dominant in this community, forms the greater part of the samples in late winter but thereafter declines; *Poa* species show a similar, though less marked trend. *Holcus* and *Poa* occur mainly between the tussocks of *Agrostis*

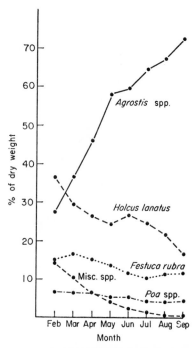

Fig. 2. Proportions of the major species in quadrat samples from the study area grasslands, February–September 1965.

tenuis and are therefore more heavily grazed and trampled, so accentuating the increase of *Agrostis* in the sward.

Preferential grazing of *Festuca rubra* depresses its performance compared with *Agrostis* and *Anthoxanthum*. The late-summer increase in *F. rubra* perhaps shows that it makes most of its growth later in the season, when plenty of grazing is available. The Miscellaneous category at this site consisted mainly of mosses. The fairly large proportion of these in the samples in February and March indicates how very short the sward is at this time, compared with late summer when mosses are very insignificant.

PHYSICAL AND CHEMICAL CHANGES

Because of variation in weather and possible discrepancy in cutting, running means of the fresh and dry weights have been given in Table 3 and the figures for moisture content are calculated from these.

TABLE 3
Quadrat samples from the Village grasslands in 1964–1965

	Fresh weight (g)	Dry weight (g)	Organic matter (%)	Digestibility (%)
27 Dec.	80	10·8	90	41
16 Jan.	95	12·9	89	31
6 Feb.	48	14·3	—	—
20 Feb.	52	22·7	87	39
8 Mar.	46·5	16·8	—	—
22 Mar.	46·5	15·3	90	45
8 Apr.	65	21·4	—	—
19 Apr.	100	36·0	90	52
2 May	133	39·6	—	—
15 May	175	50·7	90	67
9 June	367	88·0	—	—
20 June	605	102·3	91	71
9 July	885	180·8	90	70
9 Aug.	1025	237·8	—	—
23 Aug.	685	169·0	91	62
7 Sept.	545	154·6	—	—
23 Sept.	510	95·9	90	57

As a measure of standing crop, the data on fresh weight are affected by the grazing of the sheep. In view of the large excess of production in summer, the reduction due to grazing must be much greater in winter; in the absence of grazing the difference between summer and winter figures would be somewhat reduced. However, very large seasonal variations clearly do occur, from a minimum in February and March to a maximum in late July/early August.

The chemical composition and digestibility of a grass species is closely related to its stage of growth (Raymond 1959), young actively-growing herbage being much more digestible than older tissue. Minimum digestibility occurred in January and February, increasing well in advance of fresh weights and reaching a maximum in June (Table 3). Dry weight remained fairly constant from January until March, while fresh weight declined, indicating a marked drying out of the herbage. The data in Table 3 suggest a complex situation early in the year, with new growth in the pastures being offset by the continued die-back of the old herbage, giving a minimum standing crop of green material. After 22 March 1965 this moribund material was rapidly replaced by fresh growth, which formed the greater part of the samples from that date.

DISCUSSION

MORTALITY, PASTURE CONDITIONS AND WEATHER

In all years, and particularly in years of heavy mortality, sheep typically die in late winter and early spring. Both sexes are then under stress; the ewes are in the later stages of pregnancy and early lactation, and the rams are in poor condition following the demanding period of the rut in October and November. During this critical time from January until mid-April, the sheep may not be taking in enough digestible energy to meet their requirements for maintenance, movement and production (Milner, pers. commun.).

The most important factor governing food intake by sheep is the digestibility of the herbage (Blaxter 1962). Herbage digestibility in the study area is low throughout the year (Table 3), and in late winter may so limit the sheep's intake that it must use its body reserves to stay alive. The weather, too, is severe at this time, with frequent gales and showers of rain, sleet and snow. In such conditions the sheep are often reluctant to leave shelter, and will return readily to the cleits during the day. Hard weather thus affects the animals directly, by increasing their energy requirements and reducing their foraging time; and also indirectly, by delaying new growth and so prolonging the period of minimum standing crop.

A succession of cold, late springs might possibly reduce the population to a very low level, whereas a succession of mild winters and early springs might allow it to rise to a level above any reached so far. The question is whether poor nutrition is the sole cause of the heavy winter mortality, or if some aspect of the sheep's behaviour is also involved, through competition for food, perhaps aggravated by crowding at high numbers. No study of behaviour in relation to nutrition and numbers has yet been done. Many sheep do die when there is new growth available, and possibly they are unable to utilize this to the full because of behavioural interactions, especially in years of higher population densities. Alternatively, the sheep may not have the energy to forage for this scattered new growth. In any case, an excess of food potential is soon reached, in late April or early May (Table 3). Hence if competition for new grazing does occur in spring, it is probably short-lived.

SUMMARY

1.   A study has been carried out over ten years of the population of feral Soay sheep on the island of Hirta, St. Kilda, and in particular of those animals inhabiting the area of the old village of Hirta.

2.   The population consists of a number of distinct hefts, each grazing a particular home range. The part of the home range which is occupied varies throughout the year; the village sheep mainly graze the mixed Holcus–Agrostis pastures in the deserted meadows, using the surrounding heaths from September until early March.

3.   Chemical and botanical analysis of the grasslands has shown marked seasonal variations in the proportions of the major grass species, standing crop, and digestibility of the herbage. It is suggested that the extremely low digestibility of the herbage in late winter and early spring may reduce the sheep's intake at a time when energy requirements are greatest. Many die at this time, causing major fluctuations in population from year to year.

4.   The effects of animal numbers on the pastures and vice-versa are discussed.

REFERENCES

BLAXTER K.L. (1962) The Energy Metabolism of Ruminants. London.

BOYD J.M. (1953) The sheep population of Hirta, St. Kilda, 1952. Scott. Nat. 65, 25–9.

COCKBURN A.M. (1935) The geology of St. Kilda. Trans. R. Soc. Edinb. 58, 511–48.

GRUBB P. & JEWELL P.A. (1966) Social grouping and home range in feral Soay sheep. Symp. zool. Soc. Lond. 18, 179–210.

McVEAN D.N. (1961) Flora and vegetation of the islands of St. Kilda and North Rona in 1958. J. Ecol. 49, 39–54.

PETCH C.P. (1933) The vegetation of St. Kilda. *J. Ecol.* **21**, 92–100.
POORE M.E.D. & ROBERTSON V.C. (1949) The vegetation of St. Kilda in 1948. *J. Ecol.* **37**, 82-99.
RAYMOND W.F. (1959) The nutritive value of herbage. *Proc. Easter Sch. agric. Sci. Univ. Nott.*, **1959** (Ed. by J. D. Ivins), pp. 156–64. London.

DISCUSSION

S. L. SUTTON: Already some attention has been paid at this Symposium to the question of food quality as opposed to quantity. To what extent is the quality of the spring grass important? Is it very much richer than the winter food, and if so, is it perhaps too rich for a digestive system adapted to food of poor quality?

J. M. BOYD: In some springs many animals die. In 1959/60, the 'vet.' suggested enterotoxaemia caused by *Clostridium welchii*, which increases in the intestines with the flush of spring vegetation. Many of the animals were emaciated, and parasites were very numerous in the weakest. In some years with low numbers, there is probably no epidemic, and enterotoxaemia may occur only in very warm springs, when there is an unusually rapid flush of vegetation.

H. V. THOMPSON: How important is shelter for the sheep during the few critical weeks of mid-winter?

J. M. BOYD: We think shelter is important in all months. Within their range, the sheep have a variety of shelter places which are used both day and night. The sheep tend to remain overnight on the high ground, move down quickly in the morning, and climb slowly during the day. There are shelters available at all levels, which the sheep use while cudding and during periods of rain, strong wind and even hot sunshine.

J. EADIE: (a) Have you any detailed information on reproductive performance of the sheep and how this changes as the total population increases?

(b) Have you any data on the body composition, in particular the fat content, at the peak of body condition?

(c) When do the sheep mate?

(d) Could you enlarge on what happens to the home ranges when the population rises? Are the existing ones more densely occupied, or are the extra sheep pushed out to new ones?

J. M. BOYD: (a) Lambing success varies annually between say 20–25% and 100–120%. Some details of barrenness are given in a thesis by P. Grubb.

(b) We have no information on body composition.

(c) The animals have a silent oestrus at the end of September, and come into positive oestrus in mid October–early November. The mating season varies slightly from year to year; and yearling ewes mate later than older ones and have their young later.

(d) Home ranges seem to vary over a 12-month period, but we have not followed them in detail. There are probably also long-term changes associated with changes in the vegetation: those shown were for 1964–65.

T. G. O'DONNELL: Have you any information on the extent to which the various vegetation types are used at different seasons?

J. M. BOYD: Utilization changes seasonally, and in the study area it differs from elsewhere on the island. We are beginning to examine the faeces, which in due course should tell us how each range is used. Broadly speaking, the animals feed from the heather areas mainly in late summer and winter and from the grassland mainly in late winter, spring and early summer.

D. R. KLEIN: It seems strange that the Soay sheep, which have been on Hirta for an extremely long period, have not adjusted their cycle of nutritive requirements to coincide with the nutritive cycle of the vegetation. In short, I do not agree with your statement that food requirements are highest in late winter. I would suspect, on the basis of comparison with other species, that the period of greatest nutritive requirement (in both quality and quantity) comes somewhat later than you have suggested, during the period when the females are lactating, the growth of the young is accelerating, and the adult males are beginning to recover. This period would be at least a few weeks later, when newly growing vegetation would be of extremely high quality and digestibility, and greater in quantity than in late winter.

J.M. BOYD: The digestibility changes in the pasture may not coincide with the demands of the population, which has nonetheless survived there for a thousand years or more. During this time, however, the sheep could not have changed genetically to adjust to conditions prevailing at St. Kilda. Lambing takes place towards the end of March. Some ewes in poor condition (say 8–9 kg) lose their lambs, and others (say 11–12 kg) raise them. I cannot explain why the reproductive behaviour does not coincide with the climate. During January–March, when digestibility is low, these animals cannot take in enough food. The tups don't die during the tupping season, but lose weight from then on. Further, the early grass might be deficient anyway, because some grass species begin growing much before others, and only later in the season do the sheep have a full range of grass species to feed on.

K. R. ASHBY: What is the origin of the Soay sheep? If, like some other domestic animals, it was originally derived from the Middle East, it might still be genetically adapted to a very different climatic regime to that in which it is now living.

J. M. BOYD: From the scanty information available, it seems that the Soay sheep shows affinities with the Moufflon and with the native sheep of Faeroe and Norway. Medieval British sheep were also similar, and some Bronze Age skeletons from Salisbury Plain also resemble those of the Soay sheep. I suspect it had its origin somewhere in the Mediterranean region, moved north with the Bronze Age and Celtic people, ultimately surviving only on Soay.

D. LACK: There is some misunderstanding here. The Soay sheep has not been on Hirta for a thousand years. If it had, there would have been ample time for genetic change. It was introduced in 1932, I believe, and this is probably too short a time for important genetic changes. It has been on Soay for a thousand years or so, and is doubtless well adapted to this island. But Hirta provides a rather different habitat, and in particular a rich lowland area near the village, which has much better grazing than anything on Soay. So if this difference is important, the Soay sheep might not be perfectly adapted to Hirta. Of course, the animal was probably on Hirta a long time back, but ordinary sheep were present later, without Soay sheep, and this state of affairs continued until 1931.

DIFFERENCES IN POPULATION DENSITY, TERRITORIALITY, AND FOOD SUPPLY OF DUNLIN ON ARCTIC AND SUBARCTIC TUNDRA

By Richard T. Holmes

Department of Biological Sciences, Dartmouth College
Hanover, New Hampshire, U.S.A.

INTRODUCTION

The factors regulating animal populations are not well understood (e.g. Lack 1954, 1966; Wynne-Edwards 1962). Two points, among others, which require further clarification are (1) the relationship between the density of a population and food abundance, and (2) the role of territorial behaviour in relating the size of a bird population to its food supply (see Watson & Moss 1970).

This paper compares the densities, territorial activities, and food supplies of the dunlin or red-backed sandpiper (*Calidris alpina*) at Barrow in high arctic Alaska and at Kolomak River, 10° latitude further south.

THE STUDY AREAS

BARROW, ALASKA

Dunlin populations were studied at Barrow (71°51′N, 156°39′W) in the summers 1959–1964 (see Holmes (1966a,b)). The tundra there consists mostly of a flat plain covered by grasses and sedges, dotted with numerous freshwater lakes and ponds, and crossed by occasional rivers and streams. There are local patches of relatively well-drained tundra (e.g. old beach ridges and polygon ground resulting from freeze-thaw action), but most of these rise less than 1 metre above the surrounding marshes.

The snow cover at Barrow is continuous from early September till late May. Snow melting in spring lasts 2 to 3 weeks, so that the tundra is

303

usually free of snow by about 20 June in most years. By early July, the tundra becomes fairly well-drained and the ponds frequently shrink in size. This process is reversed with the onset of generally cold rains in late July and early August which then re-saturate the tundra. Rains or snow can occur at any time in summer, and midsummer temperatures average between 2°C and 5°C.

In three summers, 1966–1968, breeding dunlin were studied in the Yukon Delta region of western Alaska. The study site was along the Kolomak River (61° 30′ N, 164° 50′ W), 30 km north-east of Hooper Bay and about 16 km inland from the Bering Sea.

The habitat for dunlin at the Kolomak and on the Delta as a whole is an extensive marsh, of which 50% or more is covered with water in the form of numerous lakes and ponds, but also in an interconnecting system of river channels, sloughs, and small streams which comprise part of the Yukon Delta's drainage system. The water levels in these rivers and channels, even many kilometres inland from the coast, have a tidal fluctuation of 2–4 m. During severe storms, which occur mainly in September and October, the brackish water in the tidal channels overflows, flooding many of the tundra ponds and lakes. However, for most of the summer and certainly for the period when the dunlin are breeding, the water level in the ponds remains relatively stable and does not fluctuate. The vegetation on the marsh consists primarily of sedges (*Carex* spp.), grasses (*Elymus* sp. *Poa* sp.), and a few scattered forbs. The ponds and lakes are surrounded by dense growths of *Carex*, and an aquatic herb, *Hippurus* sp., grows in shallow water.

The climate at the Kolomak is less severe than further north. The snow-free season lasts from late May or early June through mid to late September. Summer temperatures range between 7°C and 15°C. Rains occur frequently through the summer, but snow or freezing temperatures are rare after 1 June and before 1 September.

TERRITORIAL SYSTEMS AND DENSITIES OF DUNLIN

From their arrival on the tundra during the snow-melt period until the time of hatching, adult dunlin are found in specific areas, where they advertise their presence by aerial displays accompanied by song, and actively chase out intruding dunlin. Males are the more aggressive in territorial display and chasing, although females occasionally chase and give a weak version of the song (Holmes, MS). At Barrow, the breeding pairs are restricted to their territories during this period; all courtship, pair-bonding behaviour, nesting, and feeding occur there. At the Kolomak, most activity is confined to the territory, although some dunlin move off their territories to favoured feeding ponds late in the incubation period; this was discovered by following colour-ringed individuals. At these common feeding areas, birds which have only minutes before been display-ing next to each other and chasing one another, feed side by side with no sign of aggression. At both Barrow and the Kolomak, territorial defence terminates after hatching, and the adults and their broods wander freely on the marshy tundra.

For determining territory size and density, gridded census plots were established on the tundra; and each pair's positions, movements, and encounters with neighbours were recorded throughout the territorial period. At Barrow, a 40 ha (100 acre) plot on polygon tundra was censused for 4 consecutive years, 1960–1963; the results are reported in Holmes (1966a). For comparisons here, a sample of the census data (made on part of the census plot in one year, 1962) is given in Fig. 1e. At Barrow, the sizes of territories defended through the season were 5·5 to 7·5 ha, with an average of 6·5 ha/territory or approximately 6 pairs/40 ha. These sizes remained relatively constant during the 4 years of census work, and as has been documented by Holmes (1966a), this appears to be characteristic of dunlin densities elsewhere on the coastal tundra of northern Alaska.

At the Kolomak, a census plot of 15 ha (37·5 acres) was established in 1966 and the birds followed there for three seasons. The data on territory sizes on this plot are shown in Fig. 1a, b, and c, while the distribution of lakes and ponds on the plot is indicated in Fig. 1d. At the Kolomak, territories ranged in size from 0·25 to 2·25 ha, averaging 1·25 ha. The mean density for the 3 years was 1·3 ha/territory or about thirty pairs of dunlin/40 ha. The relative constancy of territory size between years at Barrow is also found at the Kolomak, although data over a longer series of years are still needed. It is evident that the densities of these Alaskan dunlin do not fluctuate widely from one year to the next as occurs in some tundra-nesting wader species such as *C. melanotos* (Pitelka 1959; Holmes 1966a).

Even with this five-fold difference in density, the intensity of territorial display, judged subjectively, did not differ noticeably in the two study

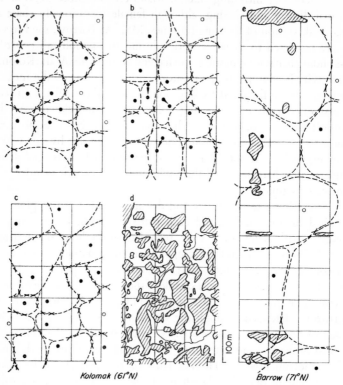

FIG. 1. Territories of dunlin at the Kolomak River, Alaska in (a) 1966, (b) 1967, (c) 1968; and (e) at Barrow, Alaska in 1962 (the latter adapted from Holmes, 1966a). Ponds and lakes are shown by cross-hatching in (d) for the Kolomak and in (e) for the Barrow plot. Paired lines represent actual territorial boundaries based on observed contact points between adjacent males; broken lines approximate territorial boundaries. Located nests are indicated by solid dots, while those nests not actually found but known to be present in the general vicinity are shown by open circles.

areas, at least until late in the incubation period. At that time, more active displaying did occur at the Kolomak, probably due to the increase in trespassing by birds leaving their own territories en route to feed on favoured communual feeding ponds (see p. 305). The territorial systems at the two localities are therefore essentially identical, the main difference being the sizes of areas defended and the fact that some feeding at the Kolomak occurs off the territory.

FOOD ABUNDANCE AND DUNLIN DIETS

METHODS

At both places, foods taken by dunlin were determined from analyses of the stomach contents of collected specimens. At Barrow, collections were made during five summers (1959–1963); the results are reported in Holmes (1966b). At the Kolomak, adult dunlin were collected at regular intervals from mid May to mid July in the 3 years of study there (1966–1968); their stomachs were preserved and examined later.

The availability and abundance of insect prey at Barrow were measured during three summers (1961–1963); the techniques and results are reported and discussed by Holmes (1966b). For estimates of insect abundance at the Kolomak, ten to seventeen samples, 20 × 20 cm in area and 5 cm deep, were taken weekly during the 1968 field season, at the margins (half in and half out of the water) of lakes and ponds on the census plot. Trial samples were taken in other regions of the marsh. Adult insects and other surface-active organisms were not sampled quantitatively at the Kolomak, since it was the larvae and not adults that were predominant in the diet (Table 1).

FOOD ABUNDANCE AND DIET AT BARROW

At Barrow, the most important insect prey for dunlin are the larvae and adults of two families of Diptera: Tipulidae (three species) and Chironomidae (many species, not yet taxonomically distinguished). Their abundance varies with the season, tipulid larvae being the most abundant in June, adult insects in early July, chironomid larvae in mid to late July, and then in August tipulid larvae on the tundra and chironomid larvae in lagoons near the coast or along the major river channels (Holmes 1966b). The diets of adult dunlin follow closely this seasonal succession in prey availability, and are associated even with the availability of particular species. Variation in availability of the insects is determined largely by the characteristics of their life-cycles and by variable and undependable weather (Holmes 1966b). Weather affects most critically the rate and timing of snow melting in the spring, the numbers of surface-active prey in early July, and the availability of chironomid larvae in late July.

DIET AT THE KOLOMAK

The foods available to and taken by dunlin at the Kolomak differ markedly from those at Barrow. Analyses of the stomachs of 131 adult dunlin collected at the Kolomak are shown in Table 1. The most important part of the diet throughout the breeding season was chironomid larvae, which in four of the five time periods, made up 73% or more of the diet. Only in mid June did the dunlin switch to another food item, in this case adult insects, most of which were chironomids. By late June, they had switched

TABLE 1

Proportion (%) of prey items in stomachs of adult *Calidris alpina* in three summers from 1966–1968, at Kolomak River in the Yukon–Kuskokwim Delta, Alaska

	18–31 May %	1–10 June %	11–20 June %	21–30 June %	1–12 July %
Larval Diptera					
Tipulidae	7·6	0·8	10·5	1·9	0·2
Muscidae	3·7	0·5	1·2	0·0	11·0
Chironomidae	73·4	97·1	10·5	85·3	75·1
Other larvae	2·0	0·1	5·7	5·5	10·2
Adult Diptera	3·2	0·1	50·0	2·5	1·5
Adult Coleoptera	5·8	1·4	18·6	4·8	2·0
Arachnida	4·3	0·0	3·5	0·0	0·0
'Seeds'	+	+	+	+	+
Total (%)	100	100	100	100	100
Total no. of items	563	873	86	361	862
No. of crops examined	23	15	26	24	43

back to chironomid larvae. The lack of other items in the diet reflects the scarcity of other prey organisms on the marsh and conversely the extreme abundance of chironomids. In the majority of stomachs, small seeds, mostly of the aquatic plant *Hippurus*, were found; these probably were not taken as food but came from eating caddis fly larvae (Trichoptera) whose cases in this locality are made from these seeds. This suggests that Trichoptera larvae, which are present in the ponds, may be more important in the diet than is indicated by Table 1.

By mid to late July, dunlin at the Kolomak begin to flock and shortly after most leave the marshes and move to the coast of the Bering Sea. At this time, they continue feeding on chironomid larvae which they obtain from ponds and lagoons near the coast, but they also begin taking some marine organisms (mostly small molluscs, and annelids) from the large intertidal flats that occur in this region (Holmes, MS).

FOOD ABUNDANCE AT THE KOLOMAK

The occurrence, distribution, and abundance of insect food items and dunlin diets at the Kolomak differ considerably from those at Barrow. At the high latitude, much of the food (tipulid larvae, many adult insects, beetles and spiders) is found in or on the sod of the grass and sedge-covered tundra. At the Kolomak, in contrast, there are essentially no sod-dwelling insect larvae, and relatively few surface-active organisms are found there

TABLE 2

Numbers of insect larvae/m² at pond-margins in marsh tundra at Kolomak River in Alaska, 1968

		Diptera Chironomidae			Muscidae	Trichoptera	Total
		< 5 mm	5–10 mm	> 10 mm			
	(N)						
30–31 May	10	690·0	900·0	477·5	0·0	0·0	2067·5
3–9 June	16	137·5	690·0	477·5	12·5	3·0	1320·5
13–16 June	17	50·0	585·0	270·0	3·0	5·0	913·0
19–23 June	17	195·0	307·5	127·5	26·0	1·5	657·0
26–30 June	17	247·0	200·0	152·0	14·5	0·0	613·5
3–6 July	17	90·0	107·5	47·7	58·2	4·3	307·7

during most of the summer. Instead, the insect larvae of the Kolomak marsh occur in the ponds and are available to dunlin only at the pond margins. The results of samples taken along the pond margins at the Kolomak are given in Table 2; these are expressed for convenience in numbers/ m², although it must be realized that in most cases, these larvae are available in a linear band along the pond margin. In some places on the marsh, however, there are ponds where several square metres of mud are exposed and therefore available as feeding areas for dunlin. These are the communal feeding sites where some birds go during temporary absences from their territories late in the incubation period (see p. 305).

From the data in Table 2, it is evident that chironomid larvae provide the most abundant food; they are particularly common in early summer, their numbers declining sharply through the season. The cause of this drop in numbers is not entirely clear, but it represents at least in part the hatch of adult chironomids, which in most years appear in large swarms over and near the ponds during the second and third week of June. The rather sharp decline in abundance of chironomids through the dunlin's breeding season may be one reason why dunlin at the Kolomak begin breeding at the earliest part of the summer and finish by early or mid July even though 2 months or more of favourable weather remain.

Other types of larvae, mostly tipulids (*Prionocera* spp.) and caddis flies, occurred in the samples, and were taken in small numbers by the dunlin. The caddis fly larvae were probably under-represented in the samples of food abundance, since they are found mostly in the deeper water of the ponds and thus wandering only occasionally to shallow water, and thus might have been easily missed by the samples taken at the water's edge.

COMPARISON OF FOOD ABUNDANCE AT
BARROW AND THE KOLOMAK

There are difficulties in comparing the abundance of insects at the two places. Sampling at Barrow was more intensive, and also included estimates of those surface-active organisms which were used as prey. At the Kolomak, larvae were sampled only at pond edges. However, since insect larvae comprise the most important parts of the food at both places, it is useful to compare their abundance alone. In Table 3, the data from the pond-edge samples at the Kolomak are summarized and contrasted with the abundance (average of three seasons' data from Holmes (1966b) and original figures) of insect larvae from pond edges and from the tundra sod at Barrow. The time scales in the table have been adjusted, so that food abundance can be compared at equivalent stages of the nesting cycles.

This comparison shows that the total number of larvae is highest at the Kolomak, especially early in the season. Even at the end of the breeding period, insect density is still higher there than at any part of the summer at Barrow. The many ponds and lakes at the Kolomak, where water levels remain relatively stable through the dunlin's breeding season, provide a long linear feeding zone within each territory, giving the dunlin easy access to the chironomid larvae. At Barrow, the larvae in the tundra sod are rather uniformly distributed, but the chironomid larvae at pond edges are available only for a short time in mid to late July when the pond levels are low enough to make them accessible. It is evident that these environmental factors have important effects on the availability of prey.

TABLE 3

Abundance of insect larvae (numbers/m²) at the Kolomak River, Alaska and at Barrow, Alaska

Kolomak, 1968		Phenology of the dunlin	Barrow, 1961–3 average	
30–31 May	2067·5	egg-laying	125·6	16–21 June
3–9 June	1320·5	incubation	150·6	25–29 June
13–16 June	913·0	hatching	162·2	3–7 July
19–23 June	657·5	growth of young	262·0	11–15 July
26–30 June	613·5		201·0	19–23 July
3–6 July	307·7		188·9	27–31 July

DISCUSSION

ENVIRONMENT AND FOOD SUPPLY

The results of this study show (1) that the *C. alpina* population in western Alaska at 61°N is about five times more dense than in northern Alaska at 71°N, and (2) that insect prey is more abundant at the lower latitude. Other differences between the two places must be considered. These include the general characteristics of the tundra habitat, the impact of weather, the nature of the insect populations themselves, and the occurrence of possible competitors. A summary of dunlin densities and the main environmental features at the two study areas is given in Table 4.

The high latitude environment at Barrow is severe. The snow-free season lasts only 2 to 2·5 months, during which the dunlin arrive, settle, pair, lay eggs, incubate, raise young, moult, and depart before winter sets in. Thus, the time schedule of events in the breeding cycle is critical, and in adapting to the rigours of this environment, the dunlin have compressed or modified certain activities normally associated with the breeding period (Holmes 1966a,c). Not only is the summer short at Barrow, but the weather is extremely harsh and unpredictable, which in turn has important effects on the availability of insect food. Moreover, at any particular part of the summer, only one or a few types of food is present, due primarily to the impoverished insect fauna in arctic regions. Thus, when adverse weather conditions occur during the breeding season there are few if any alternative foods available to a dunlin. When such catastrophes happen in midsummer, the result can be poorer survival of young, while the adult dunlin respond by leaving the breeding habitat for other regions, e.g. the coast, where food can be found (Holmes 1966b).

At the Kolomak, the longer summer season, the more moderate summer weather, and the fact that the availability of the major food item (chironomid

TABLE 4

Summary of dunlin densities and environmental differences at the two Alaskan study sites

	Barrow, Alaska 71°N	Kolomak River, Alaska 61°N
Pairs/40 ha	6	30
Snow-free period	June–August	Mid May–Sept.
Weather	Harsh and unpredictable	Relatively mild
Breeding habitat	Upland tundra, few fresh-water lakes and ponds	Marsh tundra, semi-tidal, many ponds
Food	Tipulid larvae, adult Diptera, chironomid larvae	Mainly chironomid larvae; only few adult Diptera
Food availability and dependability	Seasonal succession; sudden changes in availability due to weather	Little seasonal change

larvae) is not strongly affected by rains, combine to produce a more favourable environment and more importantly a dependable food source available for breeding dunlin. The importance of tidal flooding, which occurs at the Kolomak only once or at most a few times a year and then usually after the birds have left, is uncertain; it may contribute, in combination with the longer and warmer season, to a high productivity of the marsh vegetation and subsequently to the consumer elements of this community, i.e. the insect and bird populations. Unfortunately, adequate data on this point are not available.

The composition of the avifauna at the two study sites also differs, particularly in the numbers of sympatric *Calidris* sandpipers. At Barrow, there are four *Calidris* species—*alpina*, *melanotos*, *bairdii*, and *pusilla*—of which the first two co-occur over the widest area and overlap most in diet (Holmes & Pitelka 1968). At the Kolomak, no other *Calidris* sandpiper breeds in the marsh, although *C. mauri* nest very abundantly on nearby upland tundra and frequently feed in the marsh alongside the dunlin. At Barrow, occasional aggressive encounters do occur among the sympatric *Calidris* sandpipers, but these interspecific actions do not appear to have an important effect on breeding densities. For example, the numbers of *C. melanotos* at Barrow vary markedly from one year to the next (Pitelka 1959; Holmes 1966a), with no apparent effect on the stable density of *alpina*. Thus, although some questions still exist about the utilization of the food resource at Barrow by several sympatric congeneric species (Holmes & Pitelka 1968), I suggest that the presence or absence of congeneric species probably does not account for the density differences between these two populations of dunlin.

A similar association with a favourable environment and a relatively high breeding density (16–17 pairs/40 ha) occur in dunlin populations in southern Finland at about 61°N lat. (Soikkeli 1967). In this area, the breeding habitat is a coastal meadow and the climate is relatively mild. However, in these populations the adult dunlin feed mostly off their territories in nearby intertidal zones, and thus the role of food in affecting density there seems questionable.

POPULATION REGULATION AND TERRITORIALITY

The data in this paper show a clear relation between the density of breeding dunlin and the abundance, availability and dependability of their food supply. This implies that dunlin settle more densely where there is more food. If so, this raises the basic questions of where and when these populations are limited in size and what factors are principally involved in such limitation. Since critical information on population dynamics, especially on longevity and mortality, is lacking for the Alaskan dunlin, the ensuing discussion is confined to considering some indirect evidence and some speculations on the regulation process, particularly the importance of spacing through territorial behaviour.

(a) REGULATION OUTSIDE THE BREEDING SEASON?

First consider the possibility that these dunlin populations are regulated by factors outside the breeding season. Lack (1954, 1966, 1968) believes that the main controls on bird populations in general, are those density-dependent factors affecting post-juvenile and perhaps winter mortality. Dunlin are in the breeding areas only for 2 to 3 months, the remainder of the year being spent in migration or in winter quarters. Some recently discovered evidence (Holmes & MacLean, studies in progress) suggests that these two Alaskan dunlin populations winter in different areas; those from western Alaska migrate to the west coast of North America, while those from northern Alaska winter elsewhere, perhaps along the north temperate coast of eastern Asia. Thus, if post-breeding mortality is important, the difference in density in the two dunlin populations might hypothetically be accounted for by differential mortality in their separate migrating and wintering areas. The result would be that different numbers

of birds would be returning to each breeding area. However, as will be
seen below, there is evidence from Barrow that territorial individuals
exclude others, so that even at the high latitude, low-density site, some
control occurs within the breeding population.

One other explanation for the observed density differences, assuming
that limitation occurs outside the breeding season, is that birds moving
north in spring might settle more densely in those breeding areas first
encountered; as these sites fill up, the late-comers would then be forced
further north. The result would be a latitudinal gradient in density, with
fewer birds at the higher latitudes because fewer individuals arrived there.
This is unlikely for at least two reasons: (1) dunlin tend to return to the
same breeding area year after year (Heldt 1966; Soikkeli 1967; Holmes,
MS) so that year to year shifts in settling patterns for individuals probably
do not occur, and (2) the two Alaskan dunlin populations not only differ
in migratory routes and wintering areas but also in certain morphological
characteristics which are relatively uniform and distinct, indicating some
genetic homogeneity of each population (Holmes & MacLean, studies in
progress).

(b) REGULATION IN THE BREEDING SEASON?

Regulation of a bird population in the breeding season would operate
mainly through an effect on clutch size and/or on the numbers of individuals
actually breeding. In dunlin, the former is unlikely, since clutch size is
constant; thus, any regulation at this season would concern the number
of breeders, probably through some aspect of social behaviour such as
territoriality (Wynne-Edwards 1962). There has been much debate on the
role of territoriality in population control (see other papers in this symposium; Wynne-Edwards 1962; Lack 1966). The correlation between territory
size and food abundance, as shown here for dunlin, occurs in many other
bird species (see extensive review and analysis by Schoener (1968)), but
conclusive experimental evidence for territoriality actually limiting the
number of breeding individuals has only recently been demonstrated by
Watson & Jenkins (1968) for red grouse (*Lagopus lagopus*).

REMOVAL EXPERIMENTS AND A SURPLUS
POPULATION

The evidence which suggests that territoriality may play a role in adjusting
dunlin populations to their food supply is based on the results of experimental removal plots, on the existence of a 'floating' population in the

breeding area, and on a consideration of the function of territoriality in breeding dunlin populations.

In two seasons at Barrow, territorial vacancies were experimentally produced in a small area, and as new individuals settled, they were continuously removed by shooting. The results (Holmes 1966a) showed that some members of the population were being prevented from settling by already-established territorial individuals. Vacancies created before 20 June were filled almost immediately; after that, re-occupancy did not occur, probably because there would have been insufficient time for a bird to get a territory, attract a mate, produce young and moult before adverse weather set in (Holmes 1966a). Although no removal experiments were done at the Kolomak, a few flocks of dunlin, some containing as many as 100 individuals, were present there near the study area in the breeding period; these remained in groups, mostly along the river channels, and were not territorial, nor did they breed. The age and sex composition of these flocks was not determined, but since 1-year-old individuals can breed (Holmes 1966a), presumably these flocking individuals were not just immature birds and may have been potential breeders. Thus, there does appear to be a floating surplus of birds, certainly at Barrow and probably at the Kolomak, which are capable of obtaining a territory when vacancies occur. Or conversely and more importantly, there is a surplus population of dunlin which are prevented from breeding perhaps by their inability to gain a territory.

ADAPTIVE SIGNIFICANCE OF TERRITORY

Since the defence of a breeding territory in many birds results in a pattern of dispersion, it is worth considering what selective factors might favour the development of spacing and what the adaptive significance of territoriality may be. Various functions have been ascribed to bird territoriality (review by Hinde 1956). One that has received considerable support (e.g. see Lack 1966) is to provide an area in which pairing, courting, and nesting and related activities can take place with least disturbance. Soikkeli (1967) considers this to be the main feature of dunlin territories. However, it seems to me unlikely that for the two populations of Alaskan dunlin, which are all members of the same species and have essentially identical breeding behaviour, one would require 6 ha and the other 1 ha for courting and mating.

Hinde (1956), Lack (1966), and others have argued that wide spacing through territorial behaviour may evolve to lessen the effects of predation.

This has some support from the experiments on gulls by Tinbergen, Impekoven & Franck (1967). However, it does not apply to dunlin territories, since predation is heavier at the Kolomak where territories are smaller (Holmes, unpublished). A further possibility is that territorial spacing acts against the spread of disease, but there are so few data on this that it cannot be evaluated.

The one remaining function of territoriality is that it spaces a population in relation to some environmental resource, the most likely of which is food supply. The correlations between territory size and food abundance shown here for the Alaskan dunlin and for various other bird species by Schoener (1968) indicate that territorial behaviour does space birds out and that this is directly related to the abundance of food. A further point is that since most work on territoriality has been with passerine birds, the food value of territory has usually been thought of in terms of food for the young in the nest. This is not the case for dunlin, whose territorial defence stops before hatching and whose young are precocial and feed for themselves. The size of dunlin territories appears to have evolved in relation to food abundance for the adults in early summer, thus serving a dispersive function at that time. As Schoener (1968) points out, even though territorial behaviour may or may not have evolved specifically for population control, it can disperse a population in relation to food resources.

ACKNOWLEDGEMENTS

The summer field work was supported at Barrow by a grant from the Arctic Institute of North America to Dr F. A. Pitelka, and at the Kolomak by a Summer Faculty Fellowship from Tufts University (1966), a grant from the Arctic Institute of North America (1966), and the National Science Foundation grant GB-6175 (1967–1968). The assistance of C. Black, M. Dick, and especially of Dr C. Lensink, U.S. Fish and Wildlife Service, Bethel, Alaska, is acknowledged. I am grateful to Prof. Pitelka for his advice and criticism, especially in the early part of this work.

SUMMARY

1. Dunlin or red-backed sandpipers (*Calidris alpina*) breeding in Alaska defend territories five times bigger on arctic tundra (71°N lat.) than on subarctic (61°N lat.) tundra. At the high latitude, a low density (6 pairs/40 ha) is associated with a harsh rigorous environment and an insect food

supply which varies unpredictably in abundance and availability during the summer. At the subarctic site, where the environment is less stringent, the summer season longer, and the food source more abundant and dependable, dunlin maintain a higher (30 pair/40 ha) density.

2. The conclusions are that the density of breeding dunlin is related to the abundance and availability of their food supply and that the main function of territorial behaviour is to disperse the populations in relation to food.

REFERENCES

HELDT R. (1966) Zur Brutbiologie des Alpenstrandläufers, *Calidris alpina schinzii*. *Corax* 1 (17), 173–88.

HINDE R. (1956) The biological significance of the territories of birds. *Ibis*, **98**, 340–69.

HOLMES R.T. (1966a) Breeding ecology and annual cycle adaptations of the red-backed sandpiper (*Calidris alpina*) in northern Alaska. *Condor*, **68**, 3–46.

HOLMES R.T. (1966b) Feeding ecology of the red-backed sandpiper (*Calidris alpina*) in arctic Alaska. *Ecology*, **47**, 32–45.

HOLMES R.T. (1966c) Molt cycle of the red-backed sandpiper (*Calidris alpina*) in western North America. *Auk*, **83**, 517–33.

HOLMES R.T. & PITELKA F.A. (1968) Food overlap among co-existing sandpipers on northern Alaskan tundra. *Syst. Zool.* **17**, 305–18.

LACK D. (1954) *The Natural Regulation of Animal Numbers*. Oxford.

LACK D. (1966) *Population Studies of Birds*. Oxford.

LACK D. (1968) Bird migration and natural selection. *Oikos*, **19**, 1–9.

PITELKA F.A. (1959) Numbers, breeding schedule, and territoriality in pectoral sandpipers of northern Alaska. *Condor*, **61**, 233–64.

SCHOENER T.W. (1968) Sizes of feeding territories among birds. *Ecology*, **49**, 123–41.

SOIKKELI M. (1967) Breeding cycle and population dynamics in the dunlin (*Calidris alpina*). *Suomal. eläin- ja kasvit. seur. van. Julk*, **4** (1967) 158–98.

TINBERGEN N., IMPEKOVEN M. & FRANCK D. (1967) An experiment on spacing-out as a defence against predation. *Behaviour*, **28**, 307–21.

WATSON A. & JENKINS D. (1968) Experiments on population control by territorial behaviour in red grouse. *J. Anim. Ecol.* **37**, 595–614.

WATSON A. & MOSS R. (1970) Dominance, spacing behaviour and aggression in relation to population limitation in vertebrates. *Animal Populations in relation to their Food Resources* (Ed. by A. Watson), pp. 167–220. Oxford.

WYNNE-EDWARDS V.C. (1962) *Animal Dispersion in relation to Social Behaviour*. New York.

DISCUSSION

I. NEWTON: You discussed the average difference in territory size of birds nesting at the two localities but there was also considerable variation in the size of territories at any one locality. Were these differences related to the food situation too?

R.T. HOLMES: We do not yet have sufficient information on this point, but it is a possibility, due to the local variation in habitat and therefore in insect food.

R. MOSS: Do you feel that the differences in average territory size which you describe between the two areas have evolved over a number of generations, or do the birds assess the food supplies each year and adjust their territories accordingly?

R.T. HOLMES: During the several years of our study the average territory size has remained roughly the same. It is presumably therefore a product of natural selection. In other species, the size of area defended seems to depend on local conditions. For example, the territories of the pectoral sandpiper vary in size from year to year, perhaps in relation to food.

R.C. BIGALKE: From the answer given to a previous question (on whether the two populations have evolved different territory sizes), I presume that your two populations derive from two discrete wintering areas and remain quite distinct. Is this so?

R.T. HOLMES: There is a difference in bill length and plumage colour between the two populations, suggesting genetic distinctness. Some evidence now being acccumulated suggests that these two populations do winter in different areas.

D. LACK: The key question seems to be where did the newcomers come from that displaced the shot birds? (a) Would these have bred elsewhere if the previous owners had not been shot? (b) Do you see non-territorial non-breeding birds in your area?

R.T. HOLMES: (a) The season is so short that the incomers cannot have attempted to breed elsewhere first, but whether they would have bred elsewhere if they had not occupied the vacated territories, I cannot say. (b) At Barrow, there are a few non-breeding birds around, and at Kolomak flocks are present, but I have no information about the birds involved.

A. WATSON: Do you have any information on the survival of the young? There are two distinct problems here. One is whether, within any one of your areas, some territories contain better food supplies than other territories, and so produce larger broods or young that are more viable when fully grown. The other problem is to consider breeding success in relation to differences between your two areas. Are the smaller territories and higher population density at the southern area with more abundant food related to or maintained by a consistently better breeding success there?

R.T. HOLMES: Such information is very hard to come by, because of the difficulty of following individual broods. I do have figures for the number of young produced by the whole population, but these are fairly crude (see Holmes 1966a).

P. R. EVANS: Do any males hold territory but fail to get a mate? I ask this because, in the pied flycatcher populations which I have studied, males are territorial but some remain unmated; thus the density of breeding pairs is determined by the number of females which settle, and not by the number of males which hold territories. Does a similar situation occur in the dunlin?

R. T. HOLMES: This situation does not arise in the dunlin. Every male in my study area was paired.

M. J. COE: A similar situation to that described by Dr Holmes exists with *Nectarinia i. johnstoni* on Mt. Kenya. This bird occurs between 3500 and 4700 m and feeds on insects associated with the giant lobelias. Since at 4700 m these plants are sparse and at their altitudinal limit and the birds' territories are twice as large as they are at the lower limit, in this case also the available food appears to be associated with territory size.

GENERAL DISCUSSION

J. PHILLIPSON: I understand from the organizers that the aim of this part of the programme is to develop a discussion involving problems of principles which, though related to this afternoon's papers, will not necessarily be specific to any one contribution. As many of you will realize, I am a long way from my own territorial boundaries, but nevertheless trust that I shall not be considered irruptive—indeed whilst in Aberdeen, I have not been short of food and in no way feel aggressive.

In one sense I am fortunate in that this afternoon's session has comprised two invertebrate papers and two vertebrate ones—a somewhat different ratio to the overall pattern of this Symposium. It would seem to me that we should be pleased to hear from any one who has invertebrate comparisons with the vertebrate papers. Further, the ratio of herbivore : carnivor contributions in the Symposium as a whole seems right, as it lies in the region of 2·5:1, but we have so far heard mainly of herbivores involved in the grazing as opposed to the decomposer chain—contributions dealing with decomposers would be welcome for this discussion.

The inclusion of population processes in the title of this session implies rates, and hence a temporal basis. Yet we have seen how as both food quality and availability change with time, so do the animals' habits. This raises the problem of defining quality; for example, by man's assessment or by the animal which uses it? In the Australian bushfly and aphids, individual performance clearly alters with food quality. But what of animals which feed for trace elements, for example, woodlice for copper or other animals feeding for water or calcium ions? As Gwynne & Boyd have shown, climate can influence the time of vegetation growth, and availability appears to be the key here, rather than quality, but what is availability?

Further, how do we decide and measure important rate processes? Dr Hughes had a compartment diagram with key pathways, some marked thicker than others to indicate their relative importance. Can we quantify this sort of system further and produce a realistic systems-analysis model for manipulation by computer?

O. W. HEAL: It has often been suggested that litter-feeding animals ingest litter to obtain the attached microflora. It is also known that there is a rapid burst of microbial activity on faeces immediately after defaecation. It is reasonable, therefore, that a litter-feeding animal should practise coprophagy if it is actually feeding on microbes and microbial products.

A. DUNCAN: On the subject of decomposers, raised by Dr Phillipson, research has been done on the variability of assimilation efficiency in the detritus feeder *Asellus* by Levanidov and Prus. Levanidov's experiment on consumption was a long-term one and he obtained an assimilation efficiency of 50%; Prus did short-term experiments and obtained a much lower assimilation efficiency of about 10%.

P. BLAZKA: As a partial reply to one of Dr Phillipson's questions—how to measure the availability of protein, or whether an animal population is limited by protein—measurements of nitrogen excretion related to oxygen consumption, where possible, should answer these points (see Blazka 1966, Metabolism of natural and cultured populations of *Daphnia* related to secondary production. *Verh. int. Verein. theor. angew. Limnol.* **16**, 380–5.).

C. P. MATHEWS: With reference to Drs Phillipson's and Blazka's remarks about quality, in the field of fish biology there is disagreement about whether energy or nitrogen is limiting. Gerking has analysed populations and examined the nitrogen turnover and excretion, and concludes that nitrogen is limiting. Others, however, believe that energy is crucial, since nitrogen may be recycled but energy may not. In fact, I suspect that when a complete energy and nitrogen budget is drawn up, both energy and nitrogen may be limiting at once. In any case, it seems to me that an animal dying through deprivation of any food material is likely to die of incapacity to transmute energy; the 'ultimate' cause may be lack of nitrogen or some other essential component that puts stress on the animal, so that the 'proximate' may be quite different from the 'ultimate' cause of death.

M. B. USHER: I should like to ask what form of benefit coprophagy takes. Is it a form of extra-gut digestion by the gut flora?

K. L. BLAXTER: Coprophagy has, of course, been well studied. Eden, working in the thirties, showed a recirculation of copper, and work since then has shown the value of coprophagy and refection in meeting certain needs for vitamins of the B-complex. As a source of nitrogen, faeces from omnivores are not of very great value, since bacterial cell walls are hydrolysed only with difficulty.

RESPONSES OF RED GROUSE
POPULATIONS TO EXPERIMENTAL
IMPROVEMENT OF THEIR FOOD

By G. R. Miller, Adam Watson and
David Jenkins

The Nature Conservancy, Banchory, Scotland

INTRODUCTION

Miller, Jenkins & Watson (1966) showed by field surveys that (a) the average stock of red grouse (*Lagopus lagopus scoticus*) which bred on moors in spring was negatively correlated with the average age of the birds' main food plant, heather (*Calluna vulgaris*), and (b) their breeding success was positively correlated with the amount of heather growth in the summer of the previous year. Moreover Picozzi (1968) found that the average numbers of grouse shot on different moors were positively correlated with the amount of recently burned heather there.

These findings suggest that average grouse stocks can be kept high by regular burning of the heather. On the other hand, the production of young birds may be less easy to control, and may depend on the previous summer's weather influencing the growth and chemical composition of the heather. Thus weather might control year-to-year fluctuations in spring numbers since these are correlated with variations in breeding success in the intervening summer (Jenkins, Watson & Miller 1967).

This paper reports preliminary results from two experiments designed to test whether or not there is a causal connection between the age and performance of heather and the population processes of red grouse. Firstly, the effect of the age of heather on the density of a breeding population was assessed by burning to produce numerous small patches of young heather and by recording the numbers of breeding grouse. Secondly, the effect of heather growth and quality on breeding success was studied by fertilizing the heather and counting the number of young grouse reared subsequently. This paper is a preliminary report on experiments still in progress, and the data will later be published in full elsewhere.

BURNING

STUDY AREA AND PROCEDURE

Burning was done on a 49-ha block at 120–220 m on a moor at Kerloch in north-east Scotland. The 47-ha control area, where there was no burning, was 2·5 km away at 200–300 m. On both areas the soils were

TABLE I

Number and size of patches burnt during the burning experiment

Burning season	No. of patches burnt in autumn	No. of patches burnt in spring	Proportion of study area burnt (%)	Mean area of burnt patch (ha)	Mean perimeter of burnt patch (km)
1961–62	1	11	10	0·40	0·65
1962–63	1	24	8	0·15	0·35
1963–64	9	8	4	0·13	0·06
1964–65	19	10	9	0·15	0·29

peaty podsols, dominated by heather, with small amounts of *Erica tetralix*, *E. cinerea*, grasses and rushes. A few rabbits (*Oryctolagus cuniculus*), and brown and mountain hares (*Lepus europaeus* and *L. timidus*) grazed freely. No cattle or sheep grazed on the experimental area and only a few on the control area. In summer 1961 there were only six patches of young heather, amounting to 10% of the experimental area (Fig.1), which had all been burned 4–5 years before. New patches were burned each season (October–April) from autumn 1961 until spring 1965, when about 30% of the area had been burned to create a mosaic of young and old heather (Fig. 1). There was no burning in 1966 and only one patch of 0·12 ha was burned in spring 1967. The number, size and shape of the patches varied from year to year (Table 1). Many patches were burned alongside areas burned the year before, thus forming a larger patch of young heather. However, new patches were mostly less than 0·25 ha, with a perimeter of less than 0·5 km.

REGENERATION OF HEATHER

Regeneration was assessed by visual estimates of the proportion of ground covered by young heather. This was done in a sample of the patches burned

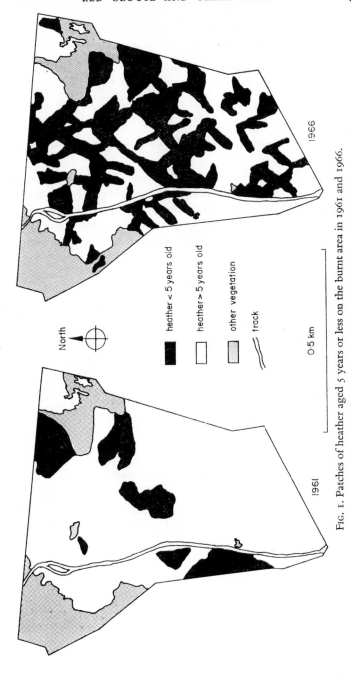

North

0·5 km

heather < 5 years old

heather > 5 years old

other vegetation

track

1966

1961

FIG. 1. Patches of heather aged 5 years or less on the burnt area in 1961 and 1966.

TABLE 2
Regeneration of heather after burning

Patches burnt in season	No. of patches sampled	Proportion (%) of ground covered by heather in spring 1965	summer 1967
1961–62	9	31	67
1962–63	12	19	61
1963–64	5	11	55
1964–65	29	0	47

each season, in March 1965 and in July 1967 (Table 2). Regrowth after burning was generally slow; even after six growing seasons, young heather occupied only 67% of the ground on patches burned in 1961–62. Faster regeneration on patches burned in 1964–65 was probably due to the predominance of autumn burning in this season (Miller & Miles 1970).

RESPONSE OF THE GROUSE POPULATION

Grouse were counted in March–April and their breeding success (number of young reared per old bird) assessed in August, using dogs to find and flush the birds (Jenkins, Watson & Miller 1963). On the unburnt area, the density of breeding grouse was steady at 44–57 birds/km² from 1962 to 1968 (Table 3). On the burnt area it remained at 49–51 birds/km² from 1962 to 1964, increased suddenly to 78 in 1965, remained at this level in 1966–67, and then suddenly decreased to the original level in 1968. Although patches

TABLE 3
Number of grouse per km² in spring and number of young reared per old bird in August on the burnt and unburnt areas

Year	Burnt area Grouse/km² in spring	Young: 1 old in August	Unburnt area Grouse/km² in spring	Young: 1 old in August
1962	49	0·5	52	0·1
1963	49	1·1	44	0·8
1964	51	1·0	54	1·1
1965	78	0·3	57	1·1
1966	82	0·2	51	0·2
1967	73	0·1	53	0·4
1968	51	0·5	53	0·4

were burned for four seasons, the population was at a high density for only 3 years. The reason for this is not understood. Possibly burning in the 1963–64 season was ineffective because the patches had very short perimeters, and comprised only 4% of the study area.

There was no difference in breeding success on burnt and unburnt areas, and the increase in density on the burnt area was not preceded by a large increase in the production of young grouse. Previous studies (Jenkins *et al.* 1967) showed that changes in numbers from one spring to another were positively correlated with breeding success in the intervening summer on all moors except those overlying base-rich rocks, where spring stocks were generally very high anyway; on average, a production of 1·9–2·2 young per adult preceded a 50% increase in numbers. The production on the burnt area never approached this level and hence the large increase in breeding density was probably due to improved survival of the resident grouse and/or to immigration.

FERTILIZING

STUDY AREA AND PROCEDURE

Preliminary trials on small plots (Miller 1968) showed that nitrate fertilizer increased the growth and nitrogen content of heather shoots, and that these plots were grazed selectively by grouse. Although phosphate fertilizer raised the phosphorus content of the shoots, it did not increase heather growth and grouse did not graze these plots selectively. Therefore nitrate fertilizer was used for the large-scale experiment.

This was done at Kerloch on 32 ha of uniform heather moor at 170–230 m altitude. The area was almost isolated from other moorland by surrounding woodland and farmland (Fig. 2) and formed a homogeneous and well-defined experimental unit. Heather aged about 15 years occupied most of the area but *Trichophorum cespitosum* was an important subsidiary species in the north-east corner where the soil was a peaty podsol. *Ulex europaeus*, *Pteridium aquilinum* and various grass species were scattered over the more freely drained and less podsolized soils on the southern and western parts of the area. The vegetation was grazed by rabbits and hares but cattle and sheep had been excluded since 1963.

The study area was divided into two equal parts (Fig. 2), the southern half being an unfertilized control. Ammonium nitrate was applied to the northern half in May 1965 at a level of 105 kg of nitrogen per ha. No more fertilizer was applied during the experiment. Since grouse are territorial for most of the year (Watson & Jenkins 1964), only those birds with territories on the northern half of the area had unrestricted access to the fertilized heather.

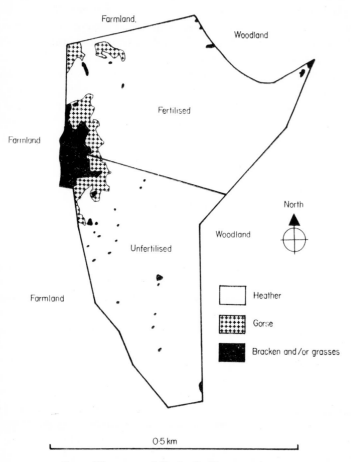

FIG. 2. The area used for the fertilizer experiment.

The densities of breeding grouse were similar on the two halves in 1962 and 1963. The study area was used for experiments on population removal in 1963–64 (Watson & Jenkins 1968), and in March 1965 each half was occupied by four cocks and four hens. These, and their progeny, were shot in summer 1965 so that by August there were no grouse on either area. This was done to eliminate any possibility of tradition influencing the selection of territories in autumn 1965. It also allowed a study of the colonization by grouse of two adjacent areas where heather growth differed quantitatively and qualitatively.

GROWTH AND CHEMICAL COMPOSITION OF HEATHER

Each area was divided into ten blocks for sampling. Heather growth was assessed in autumn by measuring to the nearest centimetre the length of the current year's long shoots (Miller *et al.* 1966) at twenty places in each block (Table 4). The heather on the fertilized area grew better than on the unfertilized area in 1965 ($P<0.01$) and again in 1966 ($P<0.10$). In 1967 and 1968 there was no difference in growth between the two areas.

TABLE 4

Mean length (cm) of leading long shoots of heather on the fertilized and unfertilized areas

Year	Fertilized area	Unfertilized area	S.E. of difference between means
1965	4·7	3·3	0·40†
1966	4·1	3·5	0·34
1967	5·4	5·2	0·42
1968	3·3	3·5	0·28

$†P<0.01$.

Samples of heather shoots were collected from the ten blocks in each area in April and November 1966, and in November 1967. Chemical analyses (Table 5) showed that in 1966 the heather shoots from the fertilized area contained more nitrogen than shoots from the unfertilized area in April ($P<0.001$) and in November ($P<0.05$). There was no difference between heather from the two areas in November 1967. Phosphorus in shoots from the unfertilized area was consistently higher than in those from the fertilized area ($P<0.10$ in April 1966, <0.001 in November 1966, <0.01 in November 1967). This was probably due to the differences between the soils of the two areas since the results of the pilot trials did not indicate that application of ammonium nitrate depressed the phosphorus content of heather shoots.

TABLE 5

Mean nitrogen and phosphorus content (% dry weight) of tips of heather shoots from the fertilized and unfertilized areas

Date		Fertilized area	Unfertilized area	S.E. of difference between means
April 1966	N	1·31	1·11	0·035§
	P	0·71	0·083	0·0058
November 1966	N	0·95	0·86	0·037†
	P	0·066	0·073	0·0038§
November 1967	N	1·15	1·13	0·026
	P	0·085	0·096	0·0037‡

†$P < 0.05$.
‡$P < 0.01$.
§$P < 0.001$.

RESPONSE OF THE GROUSE POPULATION

Immigrant grouse colonized each half of the vacant study area in autumn 1965. In the following spring the densities of birds on the two areas were the same (Table 6), despite the better growth and higher nitrogen content of the fertilized heather. Thus the territory size chosen by the incomers was not related to the length and quality of the heather shoots. This confirms the conclusion by Miller *et al.* (1966) that the size of territories chosen in autumn bears no relation to heather growth.

In summer 1966, the grouse on the fertilized area reared 1·8 young per adult, and by spring 1967 the breeding population had more than doubled (Table 6). By contrast, only 0·6 young per adult were reared on the control

TABLE 6

Number of grouse per km² in spring and number of young reared per old bird in August on the fertilized and unfertilized areas

Year	Fertilized area		Unfertilized area	
	Grouse/km² in spring	Young: 1 old in August	Grouse/km² in spring	Young: 1 old in August
1966	44	1·8	44	0·6
1967	112	0·9	44	0·0
1968	81	0·0	38	0·0

area where the breeding population in 1967 was the same as in 1966. Moreover, grouse on the control area hatched their eggs later and, when disturbed with their young, they showed less distraction behaviour (Watson & Jenkins 1964) in 1966 than birds on the fertilized area. In 1967–68 there were again differences between the two areas in breeding success and spring densities, although breeding was much poorer and the ensuing densities lower than in the previous year.

Thus the outcome of temporarily improving the growth and quality of heather was to increase the production of young grouse 12 months later, and to increase the density of breeding grouse 18 months later. This confirms the conclusions of Miller *et al.* (1966) and Jenkins *et al.* (1967) about the relation between the nutrition, breeding success and changes in population density of grouse.

MECHANISMS OF POPULATION CHANGE

Although burning and fertilizing both increased the density of breeding grouse, the mechanisms of increase differed. With fertilizing, the increase in density was preceded by improved breeding and conformed to the general pattern of changes in numbers described by Jenkins *et al.* (1967). By contrast, the increase in density on the burnt area occurred despite persistently poor breeding.

FERTILIZING

It is not clear whether the birds on the fertilized area bred better in response to the increased growth or to the improved quality of the heather. Moss (1969) concluded that the annual production of heather shoots in itself was not important in accounting for differences in breeding performance between base-rich and base-poor grouse moors. Instead, he emphasized the importance of the chemical composition of the birds' diet, particularly the phosphorus content, in the few weeks before eggs are laid. Nevertheless the possibility that heather growth, or some factor associated with it, affects breeding success from year to year within areas cannot be discounted entirely. Apart from the evidence from field surveys (Miller *et al.* 1966), correlations have been established between the mean temperature and the amounts of rainfall and sunshine (all of which affect plant growth) in one summer, and breeding success in the next (Miller, unpublished data).

Successful breeding is often preceded by early growth of heather in spring (Jenkins *et al.* 1967). The availability of new shoots at this time will increase the amounts of nitrogen, phosphorus and other nutrients in the diet of egg-laying hen grouse. There is indirect evidence (Moss 1967) that nitrogen and phosphorus are both deficient in the diet of grouse breeding in the wild. Heather shoots from the fertilized area contained less phosphorus in spring 1966 than shoots from the unfertilized area. The samples were collected on 19 April, 2–3 weeks before eggs were laid. Therefore, if the superior breeding success of the grouse on the fertilized area was due to the improved quality of their diet in spring, the significant nutrient was nitrogen, not phosphorus.

BURNING

Burning changes the productivity, chemical composition and structure of a heather sward. All may affect a grouse population, since the plant provides cover as well as food.

The yield of heather shoots from newly burnt areas is lower than from a mature stand. In fact productivity attains the maximum only when the ground cover of the plant is fully renewed and does not decrease thereafter (Miller, unpublished data); this may take several years if regeneration is slow, as it was on many of the patches on the burnt area. Therefore the increase in the density of grouse after burning was not due to an increase in the productivity of the heather.

Shoots from young heather are known to contain more nutrients than shoots from old heather (Thomas 1934; Thomas, Escritt & Trinder 1945). Hence the increase in breeding density after burning might have been due to the improvement in the chemical composition of the heather. If so, numbers must have increased as a direct consequence of the improved quality of the food because there was no intermediate stage of good breeding. This seems unlikely since there is evidence that grouse numbers are not adjusted to the food supply in this way (Miller *et al.* 1966; Moss 1969; and the results of the fertilizer experiment). Moreover, the density on the burnt area did not increase until 1964–65, despite the availability of young heather in the two previous years.

Possibly grouse reacted to the structural changes in habitat caused by burning. Open spaces would make walking easier and so perhaps make the habitat more attractive to grouse, which are essentially ground-dwelling birds. Burning may also have made it easier for a grouse to feed on the heather shoots. When feeding, a mature grouse stands about 25 cm above ground level, and at Kerloch most heather aged 10 years or more exceeded this height. Thus shoots on the young heather which sprang up on the burnt patches might have been more accessible than shoots on the original stands of old, rank heather.

ACKNOWLEDGEMENTS

We are grateful to Mr S. E. Allen and his colleagues in the Nature Conservancy Chemical Service for the chemical analyses, to Mr D. Welch for help in preparing Fig. 1, and to Miss A. M. Veitch, Mr G. M. Dier and Mr A. Walker for help in the field.

SUMMARY

The breeding density of red grouse (*Lagopus lagopus scoticus* (Lath.)) increased after their main food plant, heather (*Calluna vulgaris* (L.) Hull), was burned or fertilized. After fertilizing, there was an intermediate year of increased production of young grouse before the density rose, and so grouse numbers were not adjusted directly to the improved food. By contrast, burning did not result in better breeding, and grouse numbers did not increase until 3 years after burning began. The possible mechanisms of response by grouse populations to burning and fertilizing are discussed.

REFERENCES

JENKINS D., WATSON A. & MILLER G.R. (1963) Population studies on red grouse, *Lagopus lagopus scoticus* (Lath.), in north-east Scotland. *J. Anim. Ecol.* **32**, 317–76.

JENKINS D., WATSON A. & MILLER G.R. (1967) Population fluctuations in the red grouse *Lagopus lagopus scoticus*. *J. Anim. Ecol.* **36**, 97–122.

MILLER G.R. (1968) Evidence for selective feeding on fertilised plots by red grouse, hares and rabbits. *J. Wildl. Mgmt*, **32**, 849–53.

MILLER G.R. & MILES J. (1970) Regeneration of heather (*Calluna vulgaris* (L.) Hull) at different ages and seasons in north-east Scotland. *J. appl. Ecol.* **7**, 51–60.

MILLER G.R., JENKINS D. & WATSON A. (1966) Heather performance and red grouse populations. I. Visual estimates of heather performance. *J. appl. Ecol.* **3**, 313–26.

MOSS R. (1967) Probable limiting nutrients in the main food of red grouse (*Lagopus lagopus scoticus*). *Secondary Productivity of Terrestrial Ecosystems* (Ed. by K. Petrusewicz), pp. 369–79. Warsaw & Crakow.

MOSS R. (1969) A comparison of red grouse (*Lagopus l. scoticus*) stocks with the production and nutritive value of heather (*Calluna vulgaris*). *J. Anim. Ecol.* **38**, 103–22.

23

Picozzi N. (1968) Grouse bags in relation to the management and geology of heather moors. *J. appl. Ecol.* **5**, 483–8.

Thomas B. (1934) On the composition of common heather (*Calluna vulgaris*). *J. agric. Sci.* **24**, 151–5.

Thomas B., Escritt J.R. & Trinder N. (1945) The minor elements of common heather (*Calluna vulgaris*). *Emp. J. exp. Agric.* **13**, 93–9.

Watson A. & Jenkins D. (1964) Notes on the behaviour of the red grouse. *Br. Birds,* **57**, 137–70.

Watson A. & Jenkins D. (1968) Experiments on population control by territorial behaviour in red grouse. *J. Anim. Ecol.* **37**, 595–614.

DISCUSSION

D. C. Seel: What is lost in the way of nutrients from the ecosystem by burning? To put it another way, what do you think would happen to both the heather and the grouse if, instead of burning, you removed the heather by means of a mowing machine?

G. R. Miller: Over half of the nitrogen contained in the plant material is lost in smoke during a fire, and other minerals are readily dissolved from the ash. However, the soil organic matter retains much of the nutrient ions from the ash and the input of nutrients from rainfall is thought to be more than adequate (with the possible exception of phosphorus) to replace losses resulting from fire. There are disadvantages to cutting heather—it is slow, impracticable on rough ground and does not destroy litter and bryophytes at the soil surface; in fact, the debris from cutting adds to the litter and may inhibit regeneration of heather.

T. B. Reynoldson: Are there other grazers on heather, including insects, which may have influenced the amount of heather available to grouse in your experiments?

G. R. Miller: The only other consumers of heather present during these experiments were hares and rabbits; sheep and cattle were excluded, and there were no red deer. The impact of the hares and rabbits on the heather was negligible since there were only a few of them. In general, only sheep and cattle at high stocking density, and red deer and hares locally, can seriously reduce the amount of heather available to grouse.

O. W. Heal: Referring to Dr Reynoldson's question to Dr Miller, 'Are there any other major herbivores of *Calluna*?' there is some evidence from Moor House National Nature Reserve that small invertebrates, especially sap-sucking psyllids, consume more energy than do the sheep and grouse grazing *Calluna* on the blanket bog.

M. W. HOLDGATE: The burning experiment revealed a peak of grouse numbers in a narrow range of years, 3 to 6 after burning. If the structure of the growing heather determines the response, e.g. through the preferred feeding height of the birds, then on a moor where heather growth was slower (as at higher altitudes or further north) the response should be delayed (and vice versa where it was more rapid). Is there any evidence about this?

G. R. MILLER: It is not possible to detect this effect by survey since different proportions of moors are burned in different years and regeneration can vary from one season to the next. However, we are planning an experiment in which all burning will be completed within one year at high and low altitude. This should differentiate the effects of slow and fast regenerating heather on grouse numbers.

A. DUNCAN: Is it possible to measure consumption rates of grouse in nature and did you find any differences in rates between burnt and unburnt areas or in fertilized and unfertilized areas?

G. R. MILLER: Consumption of heather by red grouse has been measured by several different methods, giving figures of 80–120 g of dry matter per day. It is not known if there was any difference in consumption between treated and control areas.

H. V. THOMPSON: It seems that the burning experiment was unnecessarily complicated by burning the heather in such an intricate, patchwork fashion which created a large number of territorial boundaries. Was this necessary to the experiment?

G. R. MILLER: We burned a large number of small patches and strips because field observations had suggested that this might be the best management for grouse. Systematic burning to a predetermined plan is difficult because of the vagaries of wind and weather.

P. J. NEWBOULD: To what extent is heather grazed to its detriment by grouse?

G. R. MILLER: Heather is never grazed to its detriment by grouse. I estimate that grouse consume only 5–10% of the annual production of green shoots and flowers.

IRRUPTIONS OF CROSSBILLS
IN EUROPE

By I. Newton

The Nature Conservancy, 12 Hope Terrace, Edinburgh 9

INTRODUCTION

The aims of this paper are to re-examine current ideas on the movements of crossbills in the light of recent observations and ringing recoveries, and to discuss the movements in relation to food, breeding and moult. Unless otherwise stated, 'crossbill' in the paper will refer to the spruce-feeding common crossbill (*Loxia curvirostra* L.), though two other species are mentioned, the pine-feeding parrot crossbill (*L. pytyopsittacus* Borkh.) and the larch-feeding two-barred crossbill (*L. leucoptera* (Brehm)). All three make irregular mass migrations to areas outside their usual range. Such movements probably arise from all the main centres of distribution, as they have been recorded in North America (Griscom 1937), and the Himalayas, eastern Asia and Japan (Vaurie 1959), but have nowhere been better documented than in Europe. The literature on irruptive migration in birds has previously been reviewed by Lack (1954), Svärdson (1957), Wynne-Edwards (1962) and Ulfstrand (1963).

RESULTS

INVASIONS OF SOUTH-WEST EUROPE

Between 1900 and 1965 crossbills invaded south-west Europe at least sixteen times, at irregular intervals of between 1 and 17 years (Table 1). Often such invasions have occurred in two successive years, as in 1909–10, 1929–30, 1938–39, 1958–59 and 1962–63. Observations and ringing recoveries show that the migrants originate from that part of the boreal forest extending from Norway eastwards at least as far as the Urals. (The most easterly ringing recovery is from just this side of the Urals at 56° 40′ E—see Fig. 1.) In some years (1929, 1939, 1942), however, the birds came primarily from the eastern part of this area and in other years

(1927, 1930) primarily from the west, as shown by comparing the migration at different places (Table 2). The pattern of each invasion was much the same. The birds arrived in south-west Europe in summer and soon spread widely. Probably many perished, but some remained to breed in the following spring (e.g. Smith 1959; Davis 1964) and most then moved on again.

TABLE I

The dates of some recent crossbill invasions of south-west Europe

Year of invasion	Date of first arrivals in Britain	Intervals (years) between invasions
1903	9 July	
1909	23 June	6
1910	20 June	1
1927	May	17
1929	26 June	2
1930	2 July	1
1935	20 June	5
1938	?	3
1939	?	1
1942	May	3
1953	12 May	11
1956	3 June	3
1958	26 June	2
1959	12 July	1
1962	27 June	3
1963	10 July	1

Notes. Details from *Br. Birds* seriatim, Lack (1954, from Tischler 1941), Williamson (1954), and *Fair Isle Bird Obs. Bull.* (1960).

Smaller invasions might have occurred in 1911, when birds were recorded at Fair Isle (Williamson 1954), and in 1948, when birds appeared in Germany (Bub 1949).

THE ANNUAL CYCLE OF THE CROSSBILL

In northern Europe, the seeds of spruce (*Picea abies* L.) are formed in late June and remain in the cones until they are shed in April or May of the next year (Heikinheimo 1932). Crossbills obtain these seeds from cones at any stage of ripeness but, owing to their special bill structure, cannot pick up fallen seeds from the ground. In the period (up to 2 months) between the falling of one spruce crop and the formation of the next, common crossbills feed almost entirely from newly opening pine (*Pinus sylvestris*

TABLE 2

Years of heavy crossbill migration at three different places. Strong migration through Ottenby or Falsterbo (Sweden) indicates emigration mainly from Fenno-Scandia, and through the Col de Bretolet (Switzerland) and Rossitten (East Prussia) mainly from regions east of Fenno-Scandia

	Ottenby and Falsterbo	Col de Bretolet	Rossitten
1903			−
1909			+
1910			+
1927	+		−
1929	−		+
1930	+		−
1935	+		+
1938	−		+
1939	−		+
1942	−		
1953	+		
1956	+	−	
1958	+	−	
1959	−	+	
1962	+	−	
1963	+	+	
1964	−	−	
1965		−	

Notes. Records for Sweden cover the years 1927–1964, Bretolet 1955–1966, and Rossitten 1900–1940. Details from Lack (1954, from Tischler 1941), Roos (1965a, 1965b, 1967), Svärdson (1955), Ulfstrand (1963), and Orn. Beob. (1957–1967).

L.) cones (Juutinen 1953; Formosov 1960; Haapanen 1966). Usually pine and spruce occur in the same locality, so this change of feeding in early summer requires only short movements by the birds. By late June and July, however, when the pine cones are emptying, the birds rely increasingly on the fresh spruce crop (Formosov 1960), even though the cones are unripe at this stage. Spruce cone crops vary locally between areas in the same year and from year to year within the same locality, a good crop usually being followed by a poor one (Svärdson 1957). Some birds move locally every summer to areas where the new crop is good, so that in any one year the local density of breeding crossbills is closely correlated with the size of the spruce crop (Reinikainen 1937; Haapanen 1966). It is extremely rare for the pine crop to fail completely, so that wherever the birds find themselves, there are normally pine seeds available in summer. If spruce seeds are generally scarce, the birds eat pine seeds at other times of year, the buds and shoots of pine and spruce, and invertebrates from the leaves. Nevertheless, the distribution and movements of common crossbills are most closely associated with the spruce crop.

In some years, crossbills begin to breed soon after completing their moult

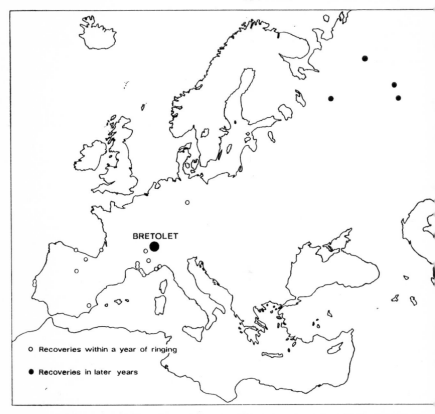

FIG. 1. Recoveries of crossbills ringed in Switzerland (mostly at Col de Bretolet) during the invasions of 1959 and 1963, showing that some birds returned to the boreal forest in a later year. The four concerned were classified at ringing as juvenile, first-year male, adult male, and male (unaged).

in September/October and they continue with successive broods until the following April/May (Formosov 1960, Newton, in prep.). In other years, they breed only in the latter half of this period, by which time the seeds have loosened in the cones. Either way, if the crop is good in the pine areas, the birds may also raise a brood in midsummer. Hence, under optimal conditions, a crossbill population may breed continuously for 9 months. In years when seeds are generally scarce, however, most birds do not breed at all (see later). The moult normally begins in late July or August, and the movements take place mostly between the end of breeding and the start of moult. The crossbill thus differs from other species of bird in making only one major movement each year.

The foregoing describes the annual cycle in a normal year. In an invasion year movement may begin earlier and end later than usual. Instead of dispersing in various directions, the birds move mainly in one direction; they also cover long distances, appearing up to 4000 km from their previous home range. Before they leave, they are unusually restless and excited (references in Lack 1954), and once underway, do not always halt when they meet with abundant food, but appear within a few days over their whole invasion area, including the furthest sea coasts and offshore islands.

Accepting that some birds move locally every year away from areas where food is locally scarce, the key question is why large numbers leave the regular range in some years but not in most.

THE CAUSE OF ERUPTIONS

Most authors agree in speculating that the adaptive value ('ultimate cause') of mass emigration is to avoid food shortage on the regular range, but there is dispute over the proximate factors which might release the flight. On one view, eruptions result directly from widespread crop-failure and are a continuation of the normal movements (Svärdson 1957), and on another view they result from 'overpopulation', and crowding is the proximate cause (Lack 1954).

There is no doubt that some eruptions have occurred in years of wide-spread crop-failure. In 1909, 1930 and 1935 spruce cones were scarce over almost the whole of Fenno-Scandia and northern Russia, and in all 3 years crossbills irrupted into south-west Europe. However, such wide-spread failure is rare, and when the crop is poor over part of this region it is usually good elsewhere. Within a smaller region, Formosov (1960) linked the four invasions which occurred between 1930 and 1940 with a poor spruce crop over part of northern Russia (though in 1930 the birds entering south-west Europe probably came mainly from Fenno-Scandia —see Table 2).

Three lines of evidence suggest, moreover, that eruptions of crossbills cannot be attributed to crop-failure alone, and that some other factor, probably the population level, is also involved. Thus, not all poor crops have resulted in eruptions; in some years emigration has begun before the new crop is ready; and it has often occurred in consecutive years from widely separated parts of the boreal forest.

In Sweden the spruce crop in different districts has been assessed annually for many years (see Svärdson 1957; Huss 1967), and over the country as a whole very small crops occurred this century in 14 different years (1901, 1903, 1929, 1930, 1932, 1935, 1938, 1939, 1943, 1947, 1955, 1957, 1961, 1963). Mass emigration of crossbills occurred in at most six of these years (1903, 1929, 1930, 1935, 1938, 1939), and in three of the six, the birds were probably mainly from further east. In the remaining 8 years of poor crops, there was no mass emigration. Further, big emigrations sometimes occurred in years of moderate crops (as in 1927, 1953, 1956, 1958, 1962); but never in years of exceptionally good crops (1913, 1915, 1921, 1928, 1931, 1934, 1942, 1945, 1954). It appears, therefore, that eruptions from Sweden have occurred in years of various crop-sizes, except the largest, and that by no means all poor crops have been followed by eruptions.

In those recent invasions that have been described in detail, most movement was in July and August, by which time the birds would be able to feed on the new spruce crop; and, to judge from the arrival dates in Britain, eleven of the invasions began in late June or July, only after the new crop became available (Table 1). However, another three started in May, so that in these the period of emigration spanned three successive crops (spruce–pine–spruce) and began before the new crop was ready. Elsewhere, migration in all months between early April and late December has been recorded in invasion years (Whitaker 1910; Roos 1965a; Svärdson 1955, 1957).

There is little quantitative information from the regular breeding range on crossbill numbers in different years; but, in addition to the local seasonal fluctuations already described, there are also larger changes occurring for longer and over a wide area. Often, several years when the birds are comparatively plentiful are followed by others when they remain scarce, even in good seed-years (though whatever the total population, the birds are always most numerous where the spruce crop is good). Further, before their main eruptions, crossbills are often said to be exceptionally numerous, and as mentioned, Lack (1954) suggested that the flights might be stimulated by high numbers as such.

Where invasions of south-west Europe have occurred in two successive years, they have often originated from mainly different regions (Table 2). In fact, emigration from Fenno-Scandia has often occurred in the same year as has emigration from further east (or in the preceding or the following year), thus showing that crossbills have repeatedly reached an eruptive state at about the same time in widely separated parts of the boreal forest. This cannot be fully explained in terms of food alone, in view of the regional variations in crop-size, and probably means that some other factor is involved; possibly the birds have repeatedly reached a population peak at about the same time over the whole area.

The hypothesis that best fits these various observations is that high numbers are a pre-requisite for eruptive movements, but that the size of the spruce crop modifies this. Once the population is high, emigration probably occurs in response to the first inadequate crop, and only an exceptionally good crop over a wide area will delay the flight another year. The more intense migratory state among erupting birds, characterized among other ways by a stronger directional tendency than usual, might be explained if an additional factor (crowding) contributes to its development in these years. In this way, the views of Lack, Svärdson and others are not incompatible. The problem remains, however, whether crowding alone will stimulate emigration, irrespective of food. The earliness of some irruptions suggests that it might, but none of these early movements has preceded good crops (which is needed to test this view), and it is possible that the birds can assess a developing crop before it is ready.

To conclude, the normal annual movements of crossbills are associated with food shortage and some eruptions have occurred in years of widespread crop-failure. However, not all poor crops have been followed by eruptions, and high numbers may also be a necessary pre-requisite, though confirmation is required. A small spruce crop may reduce breeding but does not necessarily mean that birds will starve, because alternative foods are normally available. On the other hand, when numbers are unusually high and all the birds remain, food-shortage will almost certainly accrue despite alternative foods, if not in one year then in the next. There will then be an advantage in emigration.

THE FATE OF BIRDS ON IRRUPTIONS

Since it reduces the population on the breeding grounds, mass emigration is presumably useful to those birds that stay, providing food is available there. The behaviour of the emigrants, in contrast, is frequently considered suicidal because they often reach areas devoid of conifers. Indeed, mass emigration is thought by some to be a means of population regulation when numbers exceed the food supply, the 'doomed surplus' moving out and wandering till death. On the other hand, Lack (1954) postulated that the movements might have survival value for the participants, as well as the residents, but at that time the only evidence that any birds attempted to return to their regular range consisted of birds flying in an appropriate direction at a date after the exodus.

In recent years ringing recoveries have provided the first concrete evidence that some crossbills return successfully to the boreal forest in a later year. These derive mostly from the 1740 birds ringed during the irruptions of 1959 and 1963 at the Col de Bretolet, in the Swiss Alps (Fig. 1). All but one of the seventeen caught again in the year after ringing were in latitudes to the south of Bretolet, whereas the four caught in later years were far to the north-east in Russia (the direction from which most of the birds had come). This suggests that crossbills in large irruptions do not return in the same year to their place of origin, but only in a later year. As is usual on the regular range, they apparently made only one movement each year. These results also show the long distances covered by migrating crossbills, as the most distant recoveries in different years are nearly 4000 km apart.

While on migration, crossbills eat a variety of unusual foods, but if they find areas of conifers they often remain to breed. Mostly they move on after breeding once, but some irruptions have resulted in the colonization of a new area. Usually these colonies persist for only a few years, but a stock of *curvirostra* has now maintained itself in the Scots pine plantations of East Anglia since the invasion of 1909. The stock in the New Forest is more recent and probably dates from the 1942 invasion. Both are in areas planted by man, but the presence of crossbills in the tropics (*L. curvirostra luzonensis* in northern Luzon, *L. c. meridionalis* in southern Annam, and *L. leucoptera megaplaga* on Hispaniola) shows that the birds have occasionally established themselves in natural forests following invasions. Nevertheless, the formation of colonies may be only an incidental result of irruptions.

THE DIRECTIONS AND DESTINATIONS
OF IRRUPTING CROSSBILLS

When Lack (1954) wrote, it was not known whether irrupting crossbills merely dispersed outwards in various directions from a centre or whether they moved mainly in one direction; and if the latter, whether the direction was the same in successive irruptions. The movements of 1942 and 1943 crossed Europe in opposite directions, and Svärdson (1957) suggested that such pendulum flights were regular, each westward movement being followed in the next year by a return to the east. There would be obvious advantages in such a system to a species which finds its food in the boreal forest, but the evidence for eastward movements from Scandinavia is 'virtually non-existent' (Ulfstrand 1963). The 1942 invasion westwards is one of the few in which a return flight was seen in the next year but, as discussed, ringing has shown that birds returned after the 1963 movement too. Probably the return flights are less obvious than the outward ones because the birds are fewer and more widely scattered by this time, and might not all go back in the same year.

Those common crossbills inhabiting the Russian forests have been seen irrupting in directions chiefly between west and south-west, and those from Fenno-Scandia between west and south. For Fenno-Scandia these directions are confirmed by the recoveries of birds ringed on several irruptions, all of which lie in the western half of Europe (see ringing reports for Norway, Sweden and Finland). Perhaps, therefore, far from moving randomly, crossbills irrupt in the same direction each time and thus have regular 'invasion areas', similar to the wintering grounds of normal migrants. Admittedly, the birds may move further in some years than in others, and their initial directional preference may disappear as the weeks pass and their migratory impulse wanes. (The appearance of odd vagrants in places like Iceland, Faeroe and Greenland is only to be expected, whatever the main directional tendency, as it occurs among other migrant species.)

The return of the birds after an outward movement is analogous to the 'homing' of other migrants, but unlike other species, crossbills often breed before returning. This means that, if they are to reach their ancestral home, young raised in invasion areas have to migrate in a direction opposite to that of their parents and that of all other individuals raised on the regular range. All experiments on other species of birds have shown that, if the young are displaced artificially, they set off in the direction normal for their population, even when this is wholly inappropriate under the experimental circumstances. Apparently, young birds are inherently 'programmed' for direction and distance, and can correct for displacement from a point they have previously experienced, but know nothing about the coordinates of areas which they have not yet visited (Matthews 1968 and *in litt.*). On these grounds, young crossbills reared in invasion areas are unlikely to reach the boreal forest, except by chance. Normally, however, they leave the immediate breeding area at the same time as the adults, but only ringing can show their fate.

FAT DEPOSITION AND MOULT

The fact that some crossbills move every summer raises the question of whether they lay down body fat before migrating. Figure 2 shows the weights of over a thousand birds caught on migration at the Col de Bretolet in the Swiss Alps in 1963. In both age-groups, there is considerable variation, some birds weighing up to 50% more than others. Wing-lengths

TABLE 3

Fat classes of migrating crossbills at Col de Bretolet in 1963. The figures show the percentage of birds in each class

		Amount of fat in clavicula		
	No. examined	Little or none	Filled	Bulging
Juveniles	349	44	49	7
Adults	638	45	49	6

Note. The totals differ from those in Fig. 2 and Table 5 because full data were not taken from every bird.

indicated that these birds came from a single population, and such a wide spread in the weights suggests that crossbills do deposit fat before migrating and that these birds were caught at different stages of fatness. This was confirmed by visual estimates (Table 3) of the fat present in the tracheal pit of each bird, which can be seen by blowing the feathers aside. Furthermore, ten more crossbills caught at other times of year in Switzerland weighed much less, on average, than birds caught on migration. The presence of extra fat on the migrants confirms that irruptions should not be regarded as the death wanderings of starving birds, but as an event for which the birds are physiologically adapted. It would, however, be useful to have monthly weights from the regular range in a year when the birds moved locally over shorter distances.

Generally, finches do not moult and migrate at the same time (Newton 1968). In normal years, most movement by crossbills would probably have ceased by the time the birds begin to moult in late July or August, but irruptive movements often continue into October, beyond the moult period, and sometimes into December (Roos 1965b). Some birds caught on migration at Fair Isle in August 1953 had started to moult, and others caught near Oxford in late October 1963 had finished or almost finished (Table 4). Unfortunately, the state of moult in the crossbills caught at the

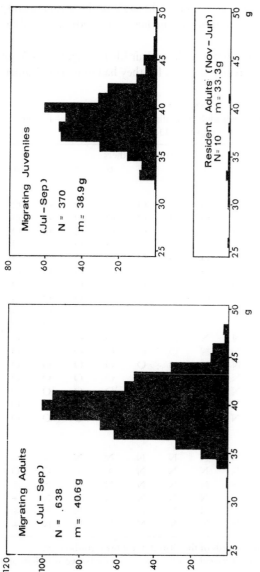

Fig. 2. Weights of crossbills. Those on migration were caught in 1963 at Col de Bretolet, Swiss Alps, and those resident were caught over several years elsewhere in Switzerland. Almost all the migrants were caught before mid-day, and most before 09.00.

Note. Wing length of migrants (mm): juvenile males, range 89–102, mean 97·1; juvenile females, range 89–100, mean 94·6; adult (yearling) males, range 90–102, mean 97·0; adult (yearling) females, range 89–100, mean 94·5; adult (older) males, range 91–104, mean 97·5; adult (older) females, range 89–100, mean 95·5; residents (7 birds), range 92–101, mean 96. To calculate the range of wing length, a few exceptionally long- or short-winged birds, comprising less than 1 % of the sample, were omitted because I considered these measurements were mis-takes or else from birds in moult.

Col de Bretolet was not documented, but odd notes taken at the time suggest that many of those moving in August and September had started to moult. Moulting juveniles were caught throughout the season (mainly 15 July to 6 September) and unmoulted (and hence unsexed) ones until the end of August.

In some of the migrating adults caught on Fair Isle in 1953, the moult of the flight feathers had apparently stopped, for they had some new feathers

TABLE 4

The state of moult in migrating adult crossbills; N—new feather, +—feather in growth, O—old feather

Sex	Date	Primary feathers								
		1	2	3	4	5	6	7	8	9
Fair Isle 1953										
♂	13 Aug.	N	N	O	O	O	O	O	O	O
♀	13 Aug.	N	N	O	O	O	O	O	O	O
♀	16 Aug.	N	N	+	O	O	O	O	O	O
♂	17 Aug.	N	N	N	N	O	O	O	O	O
♂	17 Aug.	N	N	O	O	O	O	O	O	O
♂	19 Aug.	+	+	O	O	O	O	O	O	O
♂	20 Aug.	N	O	O	O	O	O	O	O	O
♀	20 Aug.	+	O	O	O	O	O	O	O	O
♂	21 Aug.	O	O	N	N	O	O	O	O	O
♂	25 Aug.	N	N	N	N	O	O	O	O	O
♂	7 Sept.	N	N	N	O	O	O	O	O	O
♀	8 Sept.	N	N	N	O	O	O	O	O	O
♀	13 Sept.	N	N	N	O	O	O	O	O	O
♀	14 Sept.	N	N	N	O	O	O	O	O	O
♂	14 Sept.	N	N	N	N	O	O	O	O	O
♂	14 Sept.	N	N	N	N	O	O	O	O	O
♀	14 Sept.	N	N	N	N	N	O	O	O	O
Oxford 1963										
♀	23 Oct.	N	N	N	N	N	N	N	+	+
♂	24 Oct.	N	N	N	N	N	+	+	O	O
♂	24 Oct.	N	N	N	N	N	N	N	N	+
♂	24 Oct.	N	N	N	N	N	N	N	N	+
♀	24 Oct.	N	N	N	N	N	N	N	N	+
♀	26 Oct.	N	N	N	N	N	N	N	N	+

Notes. Details from the moult records of the British Trust for Ornithology.

On Fair Isle, birds that had not started moulting were caught between 19 June and 20 August.

Near Oxford, another eight adults caught from 23 to 26 October had finished the moult.

and some old, but none in growth (Table 4). In all, fifty-four adults were caught between 19 June and 14 September: three were moulting, thirty-seven had not started and fourteen showed arrested moult. Presumably, therefore, movement may in these circumstances suppress moult.

THE PROPORTION OF YOUNG IN INVASION FLOCKS

In many irruptive species, the young predominate in invading flocks, and among the adults females outnumber males (refs. in Lack 1954). This has given rise to the ideas (a) that irruptions follow good breeding seasons, (b) that the young emigrate in greater proportion than the adults, and (c)

TABLE 5

The number of adult and young crossbills caught or seen on migration in different years

Location	Year	Total adults and young	Adults (both sexes)	Adult male	Adult female	Young	Proportion of young (%)
North Sea	1927	81	9	2	7	72	88
Fair Isle	1935	99	54	17	37	45	45
Signilshär, Finland	1956	235	47	?	?	188	80
Fair Isle	1958, 59, 62	17	9	4	5	8	47
Britain (four localities)	1963	164	150	85	65	14	8
Col de Bretolet (Switzerland)	1963	1008	638	305	333	370	37
Stavanger (Norway)	1963	72	50	?	?	22	31

Note. Details from Murray 1928, Hildén 1960, Davis 1964, and unpublished Swiss records.

that more adult females than males leave. In the British literature, the preponderance of young in crossbill flocks is supported by many statements, but only one set of figures, from birds which settled on ships in the North Sea in 1927. Moreover, recent information is conflicting (Table 5).

Svärdson (1955) showed that the exodus from Sweden in 1953 did not follow successful breeding. In the winter of 1952/53, spruce seeds were scarce and the birds ate pine seeds instead: most birds did not nest, but remained in flocks throughout the spring. The exodus began in May and continued into October. Among ninety-nine birds caught on passage at Fair Isle (Table 5), there were more adults (54) than juveniles (45).

In 1963, the invading flocks contained even fewer young: among birds caught in Britain, adults outnumbered young by 11 to 1, and in Switzerland and Norway by 2 to 1. In contrast, young predominated in the ratio of 4:1 among birds migrating through Finland in 1956. Taking the data as a whole, the proportion of young in different invading flocks has varied between 8% and 88% (Table 5).

Among the adults, there were more females than males in the 1953 sample from Fair Isle ($P<0.01$), but in none of the other recent invasions did the sex ratio differ significantly from 1:1 (Table 5). However, many of the juveniles caught at Bretolet could be sexed because they had started to moult, and female juveniles (99) were more numerous ($P<0.05$) than males (59); the same was true of migrant young examined in Finland in 1956 (Hildén 1960). There is as yet no indication in the common crossbill that such figures might be influenced by different sex and age-groups moving at partly different times of year.

In the two-barred crossbill, the composition of the flocks invading Fenno-Scandia has also differed in different years. Thus, nearly all the birds seen in Norway in 1889 were adults (in Bannerman 1953, no details); yet all the birds in a flock of about 200 seen in Sweden in 1948 were juveniles (G. Notini, cited by Lack 1954); and among thirty-three migrants seen in Finland in 1956, thirty were juveniles (Hildén 1960). Further, there is one instance in this species of adult males migrating later than the females and young (Wynne-Edwards 1962).

In conclusion, observations on the common crossbill show that large-scale emigration does not always follow a good breeding season, and will occur in a population containing mainly adults. Except where the young are exceptionally numerous throughout an emigration, one cannot be certain that they are leaving in greater proportion than adults, unless there are figures from the breeding grounds as well. The view that more adult females than adult males leave is supported by a sample trapped during the 1953 invasion, but not in the 1963 one. In two invasions, however, juvenile females have outnumbered juvenile males. The proportion of young recorded in flocks of the two-barred crossbill may be influenced by the fact that different sex and age groups move at slightly different times.

DISCUSSION

The term 'irruption' has been used to cover not only the irregular emigrations of crossbills, but also the occasional unusually heavy migrations of regular migrants, such as the redpoll (*Acanthis flammea* (L.)), siskin (*Carduelis spinus* (L.)), pine grosbeak (*Pinicola enucleator* (L.)) and brambling (*Fringilla montifringilla* L.). All these finches feed from trees whose seed-crops fluctuate greatly from year to year, and the birds concentrate wherever their food is plentiful. However, the crossbill is the only irruptive migrant which feeds on the same type of food (conifer seeds) throughout the year, and migrates only once each year. The others depend on the critical seed-crop for only part of the year, and migrate twice each year, in spring and autumn. Further, whereas crossbills leave their home range only in exceptional years, the others migrate regularly every year, but in greatly varying numbers. The differences are linked with the ecological needs of the different species, the movements of the crossbill being adapted to annual fluctuations in food, and the others to seasonal, as well as to annual fluctuations (Svärdson 1957).

In conclusion, there is clearly a need for a detailed study of crossbills on the regular range. Ideally, a population containing marked birds should be measured in relation to food over several years. The endocrine and physical condition of a sample of birds should be examined throughout and, during an irruption, particular attention paid to the physiology, age and sex of the emigrants compared with the residents. There is also a need for more ringing, both on the regular range and in invasion areas.

ACKNOWLEDGEMENTS

My thanks are due primarily to Dr A. Schifferli of Schweizerische Vogelwarte Sempach, for putting the entire unpublished Swiss data on crossbills at my disposal and for helping with their interpretation. I am also grateful to Dr M. Simak for information on seed-crops, to the British Trust for Ornithology for their moult records, to Drs M. Radford and J. Pinowski for help with bibliography, to Mr D. McK. Scott for drawing the diagrams, and to Drs D. Jenkins and D. Lack for their comments on the manuscript.

SUMMARY

1.   Common crossbills differ from other migrant birds in making only one main movement each year. In most years the birds disperse over short distances, leaving areas where spruce seeds are scarce and concentrating where they are plentiful; but in eruption years they move mainly in one direction over long distances, appearing up to 4000 km from their home range. There were sixteen invasions into south-west Europe between 1900 and 1965, at intervals of between 1 and 17 years. Crossbills have often irrupted from widely separated parts of the boreal forest in the same or consecutive years.

2.   Mass emigration from Sweden has often occurred in years when the spruce crop was poor or moderate, but never when it was good, and not all poor crops have been followed by eruptions. Probably high numbers are a pre-requisite for eruptive movements, but the size of the spruce crops may modify this, a large crop delaying the flight another year. Crowding is considered a likely proximate cause contributing to the development of the migratory state. Only high numbers are likely to produce extreme food-shortage, the avoidance of which is assumed to be the 'ultimate cause' of eruptions.

3.   Ringing has shown that, after an irruption, some crossbills return successfully to their regular range in a later year, but the fate of the young raised in invasion areas is not known (apart from the few which initiate new populations).

4.   At least in eruption years, crossbills lay down body fat before migrating, like other migrant birds. In most years movements occur mainly in summer, between breeding and moult, but during irruptions they continue into the winter, during and beyond the moult period.

5.   Two recent eruptions have followed good breeding and contained more juveniles than adults, and two have followed poor breeding and contained more adults than juveniles. The earlier view, that more adult females than adult males migrate, is supported by a sample trapped during the 1953 invasion but not in the 1963 one. In two invasions, however, juvenile females outnumbered juvenile males.

REFERENCES

BANNERMAN D.A. (1953) *The Birds of the British Isles.* Vol. 1, 205. Edinburgh.
BUB H. (1949) Die Kreuzschnabel Invasion 1948 in Deutschland. *Orn. Mitt., Göttingen,* 1, 41–4.

DAVIS P. (1964) Crossbills in Britain and Ireland in 1963. *Br. Birds*, **57**, 477–501.
FORMOSOV A.N. (1960) La production de graines dans les forêts de conifères de la taiga de l'U.S.S.R. et l'envahissement de l'Europe occidentale par certaines espèces d'oiseaux. *Proc. Int. orn. Congr.* **12**, 216–29.
GRISCOM L. (1937) A monographic study of the red crossbill. *Proc. Boston Soc. nat. Hist.* **41**, 77–210.
HAAPANEN A. (1966) Bird fauna of the Finnish forests in relation to forest succession. *Suomal. eläin- ja kasvit. Seur. van. Julk.* **3**, 176–200.
HEIKINHEIMO O. (1932) Metsäpuiden siementämiskyvystä. *Metsätiet. Tutkimuslait. Julk.* **17**, 1–61.
HILDÉN O. (1960) Käpylintujen suurvaelluksesta 1956 ja erityisesti niiden ikasuhteista. *Ornis fenn.* **37**, 51–5.
HUSS E. (1967) Kottillgången 1966–1967. *Meddn från skogs Högsk.* 1p.
JUUTINEN P. (1953) Uber Nahrung und forstwertachaftliche Bedeutung des Fichlenkreuzschnabels (*Loxia curvirostra* L.). *Metsätiet. Tutkimuslait. Julk.* **41**, 1–41.
LACK D. (1954) *The Natural Regulation of Animal Numbers.* Oxford.
MATTHEWS G.V.T. (1968) *Bird Navigation.* 2nd edit. Cambridge.
MURRAY D.R.W. (1928) The 1927 irruption of the crossbill. *Br. Birds*, **21**, 227–8.
NEWTON I. (1968) The moulting seasons of some finches and buntings. *Bird Study*, **15**, 84–92.
REINIKAINEN A. (1937) The irregular migrations of the crossbills, *Loxia curvirostra*, and their relation to the cone crop of conifers. *Ornis fenn.* **14**, 55–63.
ROOS G. (1965a, 1965b, 1967) Notiser från Falsterbo fågelstation sommaren och hösten 1962, 1963, 1964. *Vär Fågelvärld* **24**, 257–71, 314–37; **26**, 256–65.
SMITH F.R. (1959) The crossbill invasion of 1956 and the subsequent breeding in 1957. *Br. Birds*, **52**, 1–9.
SVÄRDSON G. (1955) Crossbills in Sweden in 1953. *Br. Birds*, **48**, 425–8.
SVÄRDSON G. (1957) The 'invasion' type of bird migration. *Br. Birds*, **50**, 314–43.
ULFSTRAND S. (1963) Ecological aspects of irruptive bird migration in northwestern Europe. *Proc. Int. orn. Congr.* **13**, 780–94.
VAURIE C. (1959) *Birds of the Palearctic*, Vol. 1. London.
WILLIAMSON K. (1954) A synoptic study of the 1953 crossbill irruption. *Scott. Nat.* **66**, 155–69.
WHITAKER J.I.S. (1910) On the great invasion of crossbills in 1909. *Ibis*, 9th ser, **4**, 331–52.
WYNNE-EDWARDS V.C. (1962) *Animal Dispersion in relation to Social Behaviour.* Edinburgh & London.

DISCUSSION

V. C. WYNNE-EDWARDS: In support of the view that different sex and age groups of the two-barred crossbill may move at different times, in 1937 I took a film of about thirty migrants on board ship off the coast of Labrador, all of which were adult males.

I. J. PATTERSON: Would you care to speculate on which proximate factor might elicit these movements?

I. NEWTON: If emigration is in fact stimulated partly by crowding, this may presumably occur either when the birds are very numerous over wide areas, or when they have been unusually concentrated by an abundance of food in some areas and a shortage elsewhere. But whatever the proximate cause, there is the question of how long the birds must be exposed to it before leaving. The fact that movement begins earlier and ends later in invasion years suggests that their responsiveness covers a longer period than that immediately preceding the normal season of movement.

J. MORTON BOYD: (a) Should invasions not occur at times of plenty rather than at times of scarcity? (b) Have you any information on the body condition of the birds when invasions occur?

I. NEWTON: (a) Since the emigrants probably suffer heavy mortality, and many fail to breed, there can be no advantage in leaving the regular range unless conditions there are bad. If it is true that emigration occurs when numbers are high, it might often follow times of plenty (which permit successful breeding), but by responding in this way the birds leave before conditions can deteriorate. (b) Some birds caught on irruptions were heavy and contained much fat, others (especially after a long flight) were weak and starving, but clearly many birds are not starving when they leave their home range, otherwise they would not be so fat.

A. WATSON: The results a worker will get in a scientific problem vary according to the kind of methods he uses. I wonder if your findings are to some extent affected by the fact that you have been looking at migration, populations and food on a near-continental basis. It seems possible that this might tend to obscure regional or local variation. Perhaps some of the discrepancies in associating emigration with food shortage would not occur if populations were actually measured in relation to food on specific study areas, which is the research that I know you have suggested is necessary for a further advance with this problem.

Within a small region such as north-east Scotland, there is evidence of great local variation. For many years, Desmond Nethersole-Thompson has done a population study of Scottish parrot crossbills on study areas in the upper Spey Valley, and has compared this with a local forester's assessments of the pine-seed crop, which this species depends on for food. Population density varies greatly from year to year, with almost none present in poor years. The increases are far greater than can be accounted for even by 100% survival of all old birds and all young reared, and must be due to migration. He and I have compared notes on numbers from valleys in different parts of the east Highlands. In some years when the birds almost disappear from the Spey Valley, they appear and breed in large numbers in other parts of the region, sometimes only 20–50 km distant. Within the upper Spey Valley and elsewhere, the breeding density corresponds with the abundance of the seed crop in that year.

I. NEWTON: This situation is similar to that occurring in most years on the regular range of the common crossbill, the birds concentrating wherever their food is plentiful. It is the birds' tendency to vacate areas poor in seed which gave rise to the idea that irruptions resulted from widespread crop-failure, but, as explained, the situation is often more complex than this.

C. J. FEARE: What do crossbills eat on the regular range when the spruce crop fails?

I. NEWTON: There is always a small percentage of trees in seed, even though the crop is recorded as having failed. As mentioned in the paper, common crossbills will turn to pine, when spruce is scarce, at other times of year as well as in summer. They prefer the large mature pine cones, but if these are scarce they will feed from younger ones, even though their seeds are not fully formed. The birds also eat the buds and shoots of conifers, invertebrates from the foliage, and the fruits of other trees. Furthermore, their movements around the regular range help them to cope with a very patchy food situation; but they will breed only when spruce or freshly opening pine cones are available.

D. C. SEEL: If birds which are involved in an irruption migrate to central and southern Europe one year and return the next, they must be able to survive in these areas. Why should they then go back to northern Europe, and secondly, why is the normal range not more extensive than it is?

I. NEWTON: The boreal and montane conifer forests are the optimal habitats for the crossbill because food is normally available there. In my view, mass emigration occurs only when food is likely to become limiting there, under which circumstances some birds stand a better chance of surviving if they move out than if they stay. The emigrants survive in invasion areas mainly by eating alternative foods, for few of them find seeding conifers (though the situation has improved in Britain since reafforestation with these trees). Further, the majority fail to breed in invasion areas, but remain in flocks. In other words, conditions in invasion areas are less good for the birds, first because their chance of surviving is smaller there than in a normal year on the regular range, and second because very few can breed successfully. It should thus usually pay the survivors to return to the boreal forest. Viewed in this light, the normal range is probably as extensive as it could be. The sporadic breeding after irruptions occurs mainly in small areas of conifers (usually planted by man), which could not permanently support a population, but which happen to have a good seed-crop in the year in question.

D. C. SEEL: I would agree that there must be survival value in the crossbill moving away from northern Europe when food is likely to become short, if this means that the birds can survive better by so doing. However, there are also disadvantages; for example, most of the emigrants will lose a breeding season on their home range. Hence that part of the population which stayed behind and survived would be that much better off. Is it possible that the population consists of two components, one prone to stay in northern Europe, and the other prone to migrate?

I. NEWTON: Since some birds stay, while others leave, the population must consist of two components. The problem is what determines whether any particular individual will go or stay? Sex evidently has some influence, but this is not the whole story.

C. B. GOODHART: Is there any evidence of differences between residents and emigrants from the same population? In other words, are the emigrants a random sample of the whole population? One might expect that they would not be, and that, for example, the emigrants might, on average, be larger or smaller, or show greater variance in certain measurable characteristics. It should be easy enough to get such data and it might be worth it.

I. NEWTON: I think it unlikely that the emigrants form a random sample of the population. As mentioned, a start would be to compare the condition, age and sex of resident and emigrant birds. In theory, this is simple, but in practice it involves close liaison between ornithologists on the breeding range and in invasion areas. A study on the breeding range with marked individuals would at least show the condition of the birds before departure, though the research might have to continue for many years before the birds erupted.

D. M. STODDART: Periodic outbreaks of Norway lemmings are very similar to the outbreaks of crossbills—both are specialist feeders, and both irruptions allow some individuals to return the following spring. Some crossbills breed on migration; what happens to their young is not known. Lemmings usually breed on migration, and it seems likely that the young are among those returning to the mountain top the following spring. Gene flow between isolated populations is impossible under normal (non-irruption) conditions but very possible when irrupted animals meet in lowland areas. Such cannot be the reason for the irruption, but it may be an extremely important side effect, as important for crossbills as it is for lemmings. The proximate factor causing the eruption appears to differ in the two cases. It does not always appear to be absolute food shortage for the crossbills, but it may be an absolute shortage for lemmings. The emigrants in both species leave in good condition, with sufficient body fat. This generally good condition is shown by the breeding, en route, of the dispersing individuals, and further research may highlight the importance of this breeding in maintaining genetic variability.

K. R. ASHBY: Does any significant proportion of the populations of migratory locusts which erupt get back to the breeding grounds and contribute to the gene pool in later years?

G. BURNETT: In locusts it is not the same individuals which return after an irruption. Some populations leave behind a sedentary group when an irruption occurs and in the course of an outbreak later generations may return. The process is not quite the same in those species of locust which have very few outbreak centres; these are like the desert locust which may initiate swarming in many different parts of its range.

INTERACTIONS BETWEEN POPULATIONS OF SPOTTED HYAENAS *(CROCUTA CROCUTA ERXLEBEN)* AND THEIR PREY SPECIES

By Hans Kruuk

Serengeti Research Institute, Arusha, Tanzania

INTRODUCTION

The main question in this study in the Serengeti National Park and the Ngorongoro Crater in Tanzania, was whether spotted hyaenas, as the most abundant carnivores, affect populations of wildebeest (*Connochaetes taurinus albojubatus* Thomas), zebra (*Equus burchelli böhmi* Matschie) and Thomson's gazelle (*Gazella thomsonii* Günther), which are the three most common species there.

The effect of large carnivores on their prey populations has been well documented in very few cases; examples are studies by Murie (1944) and Mech (1966) with wolves, and Schaller (1967) with tigers. I have tried to consider to what extent hyaenas are responsible for the mortality of, and affect the population turn-over of, the most common ungulates in the study areas; conversely, to what extent these sources of food affect the hyaena population. An essential aspect is to compare data from the two areas, relating various observations of predators to differences in the prey populations between the areas, and vice versa.

This paper is a brief progress report, and for example there is not space here to go into the full details of counting methods or to describe qualifications in some of the data. The results will later be published more fully elsewhere (Kruuk, in preparation).

STUDY AREA AND FAUNA

Almost all observations were made from 1964 to 1967 on the plains of the Serengeti National Park and surrounding areas, and in the Ngorongoro

Crater. The areas have been described by Pearsall (1956), Grzimek & Grzimek (1959, 1960) and Watson (1967); Swynnerton (1958) published a list of species of large mammals. The Ngorongoro Crater floor covers approximately 250 km² at some 1800 m above sea-level, and consists of short grass plains with a lake, small patch of forest, and some brooks and marshes. It is inhabited by some 20 500 large wild animals, the majority of which are wildebeest, followed by zebra and then Thomson's gazelle (Turner & Watson 1964; Dirschl 1966). The area is surrounded by the steep crater wall, some 400 m high; there are some game trails leading up the wall and out of the crater, but generally the ungulates spend little time outside (Estes 1966) and the crater floor is a virtually self-contained 'ecological unit'. By contrast, in the neighbouring Serengeti, the same species of ungulates, which also make up the great majority of all large mammals there, migrate with a fairly regular yearly pattern over vast areas of flat grass-land and acacia-savannah. These migrations cover an area of 25 000 km², half of which is outside the National Park; approximately 940 000 large herbivores are involved (Watson 1967). The migratory animals spend the wet season out on the open Serengeti plains, and the dry half of the year in the wooded areas.

The spotted hyaena is a well-known scavenger on carcasses of animals which have died from disease or been killed by other carnivores, and it also lives off refuse around human settlements. But there have been sporadic reports on their hunting activities (e.g. Eloff 1964), and in the areas covered by this study they have been recognized as proper predators rather than scavengers (Kruuk 1966a & b). There are several other large carnivores in the areas under study, predominantly the lion (*Panthera leo massaica* Neumann), but also leopard (*Panthera pardus fusca* Meyer), wild dog (*Lycaon pictus lupinus* Thomas) and cheetah (*Acinonyx jubatus* Schreber), and three species of jackals (*Canis aureus bea* Heller, *C. mesomelas mcmillani* Heller, *C. adustus notatus* Heller). Several publications on the local food-habits of these other carnivores have appeared recently (Kühme 1965; Estes & Goddard 1967; Kruuk & Turner 1967; Schaller 1968).

METHODS

Although spotted hyaenas are abundant and occur in easily accessible habitats, their nocturnal habits make them frustrating objects of study and it is very time-consuming to collect even simple information. This is largely responsible for the scantiness of my data on several important issues. I followed hyaenas with a four-wheel drive vehicle, watched them at dawn, dusk and during moonlit nights, and gained further information from tracks or other evidence around kills, faecal analyses etc.

RESULTS

DENSITIES OF HYAENAS AND THEIR PREY SPECIES

Hyaenas usually spend the hours of daylight underground or in some cover, and are not easily noticed even when active. Hence aerial surveys, such as were made for counting ungulates in the area (Grzimek & Grzimek 1959; Turner & Watson 1964; Watson 1967) could not be used. Instead, a modified 'Lincoln-index' method was used (Southwood 1966), based on re-sighting of 51 individually marked hyaenas in Ngorongoro and 200 in Serengeti; but numbers were counted directly in some parts. Thus some 430 hyaenas were estimated in Ngorongoro and 3000 in the Serengeti.

TABLE I

Estimates of animal densities and number of prey available per hyaena

	Serengeti 25000 km²			Ngorongoro 250 km²		
	In the population	Per km²	Available per one hyaena	In the population	Per km²	Available per one hyaena
Spotted hyaena	3000	0·1	—	430	1·7	—
Wildebeest	360000	14·4	120	10240	41·0	24
Zebra	280000	11·2	93	4500	18·0	11
Gazelle	243000	9·7	81	5000	20·0	12

Ungulate estimates in Serengeti from Watson (1967); in Ngorongoro from Turner & Watson (1964), Dirschl (1966), Lemieux & Desmeules (pers. commun.).

There is a striking difference in the density of the hyaena populations in the two areas (Table 1). Similar differences occur in the density of lions, with some 0·2 per km² in Ngorongoro (my own observations) and some 0·1 per km² within the Serengeti National Park (Schaller, pers. commun.).

In Table 1, information has been compiled about the numbers of the hyaena's main prey species (see next section), collected by other workers. The Serengeti appears the less densely populated area, but nonetheless, the number of potential prey available per hyaena far exceeds that available to Ngorongoro hyaenas.

Like the ungulates in Serengeti and Ngorongoro, the two hyaena populations also differ in their pattern of distribution. The Ngorongoro hyaenas are residents, living in 'clans' of some thirty to eighty animals which defend a communal territory against other clans. Serengeti hyaenas may be resident, but many move over vast distances (marked hyaenas moved up to 100 km in 6 weeks), depending on the movements of the migratory herds.

TABLE 2

The yearly diet and food consumption of hyaenas

	Prey items per year			
	Proportion† (%)		Total no. taken by hyaena population‡	
Ngorongoro				
Wildebeest adult	48·9		1140	
Wildebeest calves	16·7		389§	
Zebra adult	12·4		289	
Zebra foals	5·7		133	
Gazelle	7·3		170	
Others	9·0		210	
	100		2331	
Serengeti‖				
Wildebeest adult	17·3	39·1	5592	9286
Wildebeest calves	2·1	4·7	679§	1116§
Zebra adult	17·5	18·0	5657	4275
Zebra foals	1·9	2·0	614	475
Gazelle adult	36·2	16·3	11701	3871
Gazelle fawns	12·4	5·6	4008	1330
Others	12·7	14·3	4105	3396
	100	100	32356	23749

† Calculated from Fig. 1.

‡ Calculated from the proportion of prey items, the yearly food consumption (see text) and weights of prey from Lamprey (1964), Sachs (1967), Watson (1967) or personal observations.

§ Underestimate, explained in the text.

‖ Evaluation of the Serengeti observations in the dry season, as explained in the text, gives two sets of figures as extremes.

DIET AND FOOD INTAKE OF HYAENAS

A total of 919 observations of 'kills' was made, involving 7695 participant hyaenas. Analysis of 810 faecal samples from the same areas largely confirmed results from direct observations; here I have used the results of direct observations where possible.

Because of seasonal changes in the accessibility of the terrain, it was not possible to collect the same number of observations in all months. There

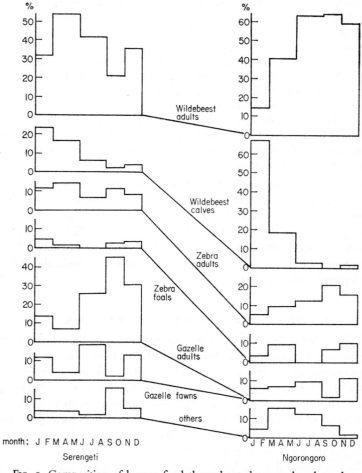

FIG. 1. Composition of hyaena food throughout the year, based on the percentage of each prey category in observations of kills and of scavenging.

are clear seasonal differences in the food of the hyaenas and I have therefore evaluated the yearly diet (Table 2) from the monthly percentages expressed in Fig. 1. Data for Thomson's gazelle and Grant's gazelle (*Gazella grantii grantii* Brooke) have been combined.

The seasonal trends in Fig. 1 are likely to be biased on two points. Firstly, when wildebeest calves become available as an extremely abundant and easily-caught source of food in January and February, hyaenas probably have a greater daily food intake than at other times of the year (the other prey species have their young in much less synchronized fashion). I have been unable to measure this increase in food intake; most likely it is at the cost of wildebeest calves only, and will therefore have little influence on further calculations of the yearly consumption of other kinds of prey.

Secondly, the wanderings of the Serengeti hyaenas pose problems in the dry season. Some hyaenas are then easily observed out on the open plains feeding on gazelle. Some of the plains hyaenas travel up and down to feed on herds of zebra and wildebeest in the bush sometimes more than 50 km away, but this is difficult to measure by direct observation. However, most hyaenas are in the less-observable areas of bush around the herds of migrating wildebeest and zebra. For estimates of food in the months of dry season in the Serengeti, I have, therefore, used the figures from faecal analysis, giving samples from the plains and from the bush areas as the extremes of the estimates.

After following one individual hyaena in Ngorongoro for 12 days, and a clan with seventy-six hyaenas for 7, 17, and 14 days, I estimated their killing rate at 2·51, 2·06, 1·78 and 1·75 kg per hyaena per day, or approximately 2·0 kg. In the Serengeti, this figure was more likely 3·0 kg per hyaena per day, although no accurate data could be obtained. These figures include wastage of parts of the carcass and are considerably below the hyaena's food-capacity.

NUMBERS AND PROPORTION OF PREY TAKEN

Table 2 shows the numbers of prey eaten per year by the whole hyaena population. Comparing this with Table 1, it appears that hyaenas in Ngorongoro ate each year 11·1% of the adult wildebeest population, 9·4% of all zebra, and 3·4% of the estimated population of gazelle. In the Serengeti, hyaenas ate yearly a much lower proportion (between 1·6% and 2·6%) of the adult wildebeest population than in Ngorongoro, and only 1·7% to 2·3% of all zebra, but ate a similar proportion (2·2% to 6·5%) of the total gazelle population.

For evaluating the number of wildebeest calves taken as hyaena food, direct observation or faecal analysis is not satisfactory because of the increased food intake of hyaenas during the very brief calving period. For instance I calculated from direct observation and faecal analysis over the whole year that 390 calves were killed per year in Ngorongoro. Yet in 1967, for example, some 3000 calves were born and did not survive there, and it was clear from other evidence that most were killed by hyaenas during the brief calving period. If each hyaena had killed an extra six calves during the calving season in the Crater, this would increase the estimated total kill of 390 to 2970 calves, which is similar to the total of 3000 missing in 1967. In the Serengeti, a similar increase in killing rate would raise our estimates to some 19 000 calves killed. Such short-term increases in killing rate are difficult to account for, but they have little or no effect on our estimates for other kinds of prey because they occur over such a short period.

EFFECTS OF PREDATION ON PREY POPULATIONS

The above differences in the proportions of the populations of ungulates taken by hyaenas in Ngorongoro and Serengeti are striking, and here I will consider especially the effect on the wildebeest. Direct observations show that nearly all wildebeest which die in Ngorongoro are killed by hyaenas, and that the wildebeest population is fairly stable. The number killed by hyaenas matches what little information I have on recruitment (in June 1967, the proportion of calves was 9·4% and of yearling wildebeest 13·6%, similar figures to the 11·1% of adult wildebeest killed by hyaenas). In the Serengeti, the mortality of adult wildebeest increased during the first half of the study period from 5·9% to 15·7% of the population per year (Watson 1967), but no data are available for 1967. It is clear that at least in 1965 and 1966, hyaenas caused only a small part (less than 17%) of the total mortality of adult wildebeest. Over the whole period of study, there are many observations from the Serengeti of wildebeest dying from causes other than hyaena predation.

I am unable to estimate precisely the effect of hyaena predation on the newborn wildebeest. Observations merely suggest that in Ngorongoro, hyaenas could account for the entire mortality of wildebeest calves and probably in fact do so. For the Serengeti, Sinclair (pers. commun.) calculated from Watson's data a mortality in 1964, 1965 and 1966 of 21 050, 42 340, and 59 840 wildebeest calves (25·5%, 46·0% and 45·0% of the number of calves present). The effect of hyaenas on the mortality of calves is therefore probably much less than in Ngorongoro; maybe they take between 15% and 50% of the annual mortality of calves. It is likely that

25

wildebeest recruitment is affected by hyaenas at least in some years. But the annual fluctuations in calf mortality are so great that it is improbable that they are caused by these predators.

We know even less about the populations of the other prey species than about the wildebeest. In general, the observations indicate that hyaenas appear to have similar effects on them as on wildebeest.

It is important to consider how much of the prey is killed by hyaenas rather than scavenged (Table 3). Many observations show that hyaenas prefer to scavenge if dead meat is available; the scavenging rate is determined largely by this availability. In the Serengeti, it is not unusual to

TABLE 3
Scavenging and killing by hyaenas

| | Total no. of observations | Proportion (%) | | |
		Hyaena-killed	Uncertain	Scavenged
Ngorongoro	297 (4404 hyaenas)	82·1	11·4	6·4
Serengeti	622 (3290 hyaenas)	53·0	21·7	25·2

Mann-Whitney test for difference in monthly kill/scavenging ratios between the two areas, $P < 0.002$.

find animals which have died from disease or unknown causes, or to see animals looking diseased or very thin; for every hyaena there are also more carnivores of other species. In Ngorongoro, it is extremely rare to find ungulates dying from causes other than predation, one rarely sees an animal ill, and there are relatively few other carnivores (the ratio of lions to hyaenas is approximately 1 : 1·5 in the Serengeti but 1 : 8 in Ngorongoro). Perhaps we should not wholly exclude the amount scavenged by hyaenas from our calculations of the effect of these predators on the ungulates. It is clear, however, that the greater importance of scavenging in the Serengeti makes the effect of hyaenas on ungulates there even smaller than postulated above.

The wildebeest being the prey species for which most data are available, I have compared the relative ages when they die in the two areas (Fig. 2). Hyaena victims in the Serengeti are on average older than in Ngorongoro, and the age distributions of wildebeest that died in the Serengeti (a) probably from disease or starvation, and (b) killed by hyaenas, are very similar. Serengeti wildebeest killed by hyaenas are the oldest animals; age classes VI and VII have very worn first molars and incisors, and some of the incisors

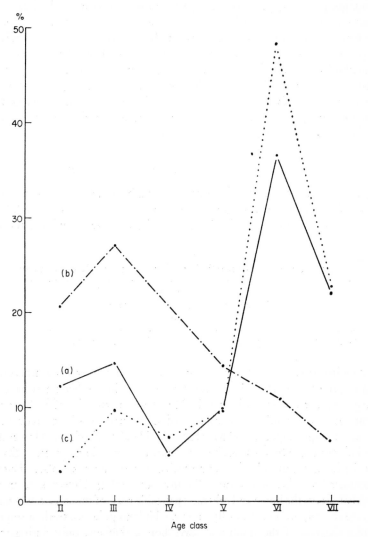

FIG. 2. Distribution of age classes of adult wildebeest, based on tooth wear (a) killed by hyaenas in the Serengeti, (b) killed by hyaenas in the Ngorongoro Crater, (c) died of disease, starvation or old age in the Serengeti. χ^2 (a), (b) = 17·0, $P<0·01$; τ (a), (c) = 0·69, $P<0·05$.

are often missing. (On the other hand, wildebeest killed by lions are younger (Schaller, pers. commun.).) One can predict from the hyaena's hunting methods that the 'weakest' wildebeest (i.e. the least able runners) will be most readily caught, and hyaenas show behaviour which probably has the function of testing a herd for stragglers. The difference in age distribution of wildebeest killed by hyaenas in the two areas probably reflects a difference in age distributions of the two living populations; Watson (1967) also suggested a more rapid turn-over in Ngorongoro than in Serengeti. In the Serengeti, there is simply a higher proportion of old animals available than in Ngorongoro.

PREY AFFECTING HYAENA POPULATIONS

With the smaller number of prey available per hyaena in the Ngorongoro, one might perhaps expect more intraspecific competition over food. To investigate this more closely, I recorded the number of hyaenas found

TABLE 4

Average number of hyaenas eating per adult carcass of wildebeest and zebra

	Ngorongoro	Serengeti
Wildebeest	25·7	14·8
Zebra	22·0	13·4

Mann-Whitney test for wildebeest, $P < 0.001$; for zebra, $P < 0.02$.

eating from a carcass of wildebeest or zebra, when no other large carnivores were present (Table 4). It is clear that hyaenas in the Serengeti have a larger meal from one kill than in Ngorongoro, and therefore have to compete less. This might be compensated for by an increase in the frequency of kills in Ngorongoro, but I have found it difficult to check this. However, as kills are made by those individuals which also obtain the greatest share of the food, it is most likely that no such compensation takes place. This is supported by other observations. For instance, Serengeti hyaenas are more likely to leave the bones and other less palatable parts of the carcass, whereas hyaenas in Ngorongoro usually eat every bit (except horns, teeth or sometimes the rest of the skull) and leave bones, skin etc. only during the wildebeest's calving season. Also, hyaenas in Ngorongoro compete aggressively over food more often than in Serengeti.

If this difference in intraspecific competition lasted a long time, and food supply were critical for the size and structure of the hyaena population, a difference in age structure between the Serengeti and Ngorongoro stocks of hyaenas might be expected. Indeed it appears from Fig. 3 that the teeth of Serengeti hyaenas are more worn than at Ngorongoro Crater. We may conclude that at present the adult Serengeti hyaenas are on average older than their Ngorongoro neighbours. Figure 3 compares

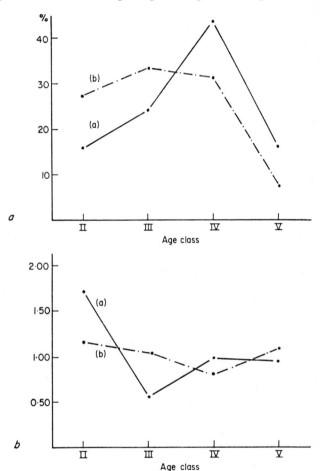

FIG. 3a. Distribution of age classes of adult hyaenas, based on tooth wear (a) Serengeti population, (b) Ngorongoro Crater population. χ^2 (a), (b) = 8·7, $P < 0.05$.

FIG. 3b.

$$\text{Ratio} \quad \frac{\% \text{ in age class of hyaenas found dead}}{\% \text{ in age class in live population}}$$

(a) Serengeti, (b) Ngorongoro Crater.

tooth wear in the living population with the tooth wear of skulls found in the field or of marked hyaenas assumed to have died. The difference in the age structure of the populations is probably due to differences in adult mortality—adult Ngorongoro hyaenas are almost equally likely to die irrespective of their age class, whereas mortality of Serengeti hyaenas is relatively low in 'middle age'. In both populations, mortality is high in the youngest age group—the age when hyaenas become independent of their mothers, and when they are lowest in dominance during fighting over food.

The causes of mortality are largely interspecific and intraspecific aggression; of twenty-two inspected freshly-dead hyaena, 36% were killed by lions, 14% by other hyaenas, 9% by man (herdsmen and poachers), and 41% died of starvation, disease or an unknown non-violent cause. All but one of the ten adult hyaenas found dead, or seen killed, had died through violence; whereas of the young and immatures, eight out of twelve, or 67%, died of non-violent causes. Most hyaenas killed by lions or by other hyaenas had died near a kill, but lions also mauled two hyaenas although there was no kill nearby. The samples were too small to indicate differences in the causes of mortality between the two areas; but it is clear that hyaena mortality is closely linked with competition over food.

During the dry season, Serengeti hyaenas may have to travel more than 50 km from their den to the nearest concentrations of game, and female hyaenas leave their suckling cubs for several days on end. At this time, cubs are not infrequently emaciated and have been found dead near the entrance of the den. It has been impossible to determine the extent of this mortality, as most cubs probably die down their holes whereas mere disappearance may mean that the female has moved them. Probably this mortality occurs especially in the Serengeti, not in Ngorongoro, and it may well be the important factor deciding the large difference in hyaena density. But there is also evidence in both areas that females are not reproducing as fast as would be possible, and there may be more to the difference in density than is obvious at present.

Some other possible factors influencing hyaena density have to be mentioned. Hyaenas show clear habitat preferences. These are most obvious in the Serengeti where the same prey densities in different habitats are accompanied by different numbers of hyaenas; the bush areas are less favoured. This is unlikely to explain large differences between the Ngorongoro Crater and the Serengeti, however; the Serengeti plains rarely, and then only locally and temporarily, carry the same hyaena densities as at Ngorongoro, yet the vegetation appears to differ only slightly. Good denning places also appear equally over-abundant in both areas.

The territorial system of hyaenas (Kruuk 1966a) in which large groups of animals occupy and defend large areas, is unlikely in itself to affect the numbers of hyaenas within these groups. It should, however, be considered in the relation between hyaenas and their food supply; most of the 'boundary clashes' are over kills, and really lethal fighting over food has been observed only between members of different clans.

DISCUSSION

A comprehensive study of relations between predators and prey with life-spans as long as the large mammals will have to include many years of observations. It is clear, therefore, that any conclusions reached in the present study can only be very tentative; nevertheless, it seems useful to consider some of the implications.

Hyaenas are shown to feed largely on wildebeest, gazelle and zebra, scavenging when dead meat is available, killing adults and young animals if need arises. It has also been shown that the Ngorongoro Crater carries a higher ungulate density, and a higher number of hyaenas (and lions) per unit prey, than the Serengeti. In the Serengeti, the low predator/prey ratio enables hyaenas to scavenge more than in Ngorongoro, and the animals killed are often very old. Mortality (at least amongst wildebeest) caused by hyaenas is low in the Serengeti compared with other causes of death; on the whole, hyaenas appear of little importance in the ungulates' ecology, except maybe as predators of the young. There is food in abundance for the adult Serengeti hyaenas, though probably some cubs starve because of the unfavourable seasonal distribution of the prey. In Ngorongoro, prey animals are killed in their prime, and hyaenas appear to be the major source of mortality amongst ungulates of all age classes; a high turn-over of the population has been indicated at least for the wildebeest, and the ungulate population almost certainly is 'under pressure' from the hyaenas. At the same time, however, the Ngorongoro hyaenas are kept at the particular level of numbers probably by food shortage; there is much more competition over food amongst them, mortality amongst hyaenas in general appears to be associated largely with interspecific and intraspecific competition, and adult hyaenas in Ngorongoro are on average younger than in the Serengeti. The difference in hyaena density between Ngorongoro and Serengeti may be due to a higher juvenile mortality in the latter area, caused by temporary food shortage as a consequence of the migratory habits of the prey.

There are indications, then, that the Serengeti hyaenas are being prevented from reaching a population level where they might exert a controlling influence on prey populations. In Ngorongoro, this is not the case and the present hypothesis is that the population turn-over of prey species there is very much influenced by hyaenas, and that the numbers of hyaenas in turn are influenced by their food.

If food supply asserts such pressure on the hyaena population, it is unlikely that hyaenas could solely determine the density of these same species of prey, even in Ngorongoro. Hyaenas may influence the population structure and turn-over of the prey species; they may also influence fluctuations in numbers of one kind of prey by eating more or less, or by switching to or from an alternative species. But the mean levels of population around which these fluctuations take place must be determined by something else. This might be, for instance, the herbivores' food supply; the grasslands of Ngorongoro Crater do, indeed, carry a higher biomass than most other grasslands in East Africa, and appear comparatively heavily used. It is possible that predation by hyaenas might finely adjust ungulate populations in the Ngorongoro Crater to their food supply, without causing over-grazing, by predator pressure reacting to small changes in the physical condition of the ungulates. In the Serengeti, as yet, there are no clues as to what determines the densities of the ungulate populations.

SUMMARY

1. In two areas in Tanzania, the Serengeti National Park and the Ngorongoro Crater, I am studying predation by spotted hyaenas (*Crocuta crocuta*) on wildebeest (*Connochaetes taurinus*), zebra (*Equus burchelli*) and gazelle (*Gazella thomsonii* and *Gazella grantii*).

2. Densities of both predator and prey are very different in the two areas; in the area with lowest densities (Serengeti), the number of hyaenas and other predators per unit potential prey is also lowest.

3. In the Serengeti, hyaenas account for only a small part of ungulate mortality; they also scavenge more and kill older prey than in Ngorongoro where they are by far the most important mortality agent.

4. There is more competition over food amongst Ngorongoro hyaenas and the adults die younger. In both areas, mortality appears closely linked to competition over food.

5. The hypothesis is put forward that hyaena populations in both areas are limited by food supply, but in different ways: the Serengeti population

by the food supply for the small hyaena cubs, which is limited because of the migrations of the ungulates; the Ngorongoro population by food for the adults.

6. Hyaenas may influence the age-structure and turn-over of prey populations but do not determine the mean levels of prey populations on different areas. The balance of predator and prey populations in Ngorongoro possibly provides a mechanism for a fine adjustment of the ungulate populations to their food supply. This must be achieved differently in the Serengeti.

ACKNOWLEDGEMENTS

Many observations were made together with my wife, who also corrected and typed the manuscript. I am grateful for permission to work in the areas from the Director of Tanzania National Parks, Mr J. S. Owen, and the Conservator of the Ngorongoro Conservation Area, Mr S. ole Saibull, and for hospitality in the Serengeti Research Institute from Dr H. F. Lamprey. Professor N. Tinbergen, F.R.S., and Mr J. S. Owen initiated this study and have since helped me in many ways. Many people assisted with information about kills etc., especially Dr G. B. Schaller. He, Dr P. J. Jarman and Mr A. Sinclair also criticized the manuscript. Dr J. M. Cullen assisted with calculation of population data. The study was financed by the Netherlands Foundation for the Advancement of Tropical Research (W.O.T.R.O.).

REFERENCES

DIRSCHL H.J. (1966) Management and development plan for the Ngorongoro Conservation Area. Rep. to Minist. of Agric. Forests and Wildlife, Tanzania.

ELOFF F.C. (1964) On the predatory habits of lions and hyaenas. *Koedoe*, **7**, 105–12.

ESTES R.D. (1966) Behaviour and life history of the wildebeest, (*Connochaetes taurinus* Burchell). *Nature, Lond.* **212**, 999–1000.

ESTES R.D. & GODDARD J. (1967) Prey selection and hunting behaviour of the African wild dog. *J. Wildl. Mgmt*, **31**, 52–70.

GRZIMEK B. & GRZIMEK M. (1959) *Serengeti darf nicht sterben*. Berlin.

GRZIMEK M. & GRZIMEK B. (1960) A study of the game of the Serengeti plains. *Z. Säugetierk.* **25**, (suppl.) 1–61.

KRUUK H. (1966a) Clan-system and feeding habits of spotted hyaenas (*Crocuta crocuta* Erxleben). *Nature, Lond.* **209**, 1257–8.

KRUUK H. (1966b) A new view of the hyaena. *New Scient.* **30**, 849–51.

KRUUK H. & TURNER M. (1967) Comparative notes on predation by lion, leopard, cheetah and wild dog in the Serengeti area, East Africa. *Mammalia*, **31**, 1–27.

KÜHME W. (1965) Freilandstudien zur Soziologie des Hyaenenhundes (*Lycaon pictus lupinus* Thomas). *Z. Tierpsychol.* **22**, 495–541.

LAMPREY H.F. (1964) Estimation of the large mammal densities, biomass and energy-exchange in the Tarangire Game Reserve and the Masai steppe in Tanganyika. *E. Afr. Wildlife J.* **2**, 1–46.

MECH L.D. (1966) The wolves of Isle Royale. *Fauna natn. Pks U.S.* **7**.

MURIE A. (1944) The wolves of Mount McKinley. *Fauna natn. Pks U.S.* **5**.

PEARSALL W.H. (1956) Report on the ecological survey of the Serengeti National Park, Tanganyika. Fauna Preserv. Soc., London.

SACHS R. (1967) Live-weights and body measurements of Serengeti game animals. *E. Afr. Wildlife J.* **5**, 24–36.

SCHALLER G.B. (1967) *The Deer and the Tiger*. Chicago & London.

SCHALLER G.B. (1968) Hunting behaviour of the cheetah in the Serengeti National Park, Tanzania. *E. Afr. Wildlife J.* **6**, 95–100.

SOUTHWOOD T.R.E. (1966) *Ecological Methods*. London.

SWYNNERTON G.H. (1958) Fauna of the Serengeti National Park. *Mammalia*, **22**, 435–50.

TURNER M. & WATSON M. (1964) A census of game in Ngorongoro Crater. *E. Afr. Wildlife J.* **2**, 165–8.

WATSON R.M. (1967) The population ecology of the wildebeeste (*Connochaetes taurinus albojubatus* Thomas) in the Serengeti. Unpublished Ph.D. thesis, Cambridge.

INTERRELATIONS OF A YOUNG
PLAICE POPULATION WITH ITS
INVERTEBRATE FOOD SUPPLY

By J. H. Steele, A. D. McIntyre,
R. R. C. Edwards and Ann Trevallion
Marine Laboratory, Aberdeen

INTRODUCTION

The Marine Laboratory is conducting a study of food chains in a sandy bay at Firemore in Loch Ewe, on the west coast of Scotland (Fig. 1). The site was chosen as presenting a relatively simple example of a marine food chain leading to a fish population. The study started in 1965 and is still continuing. Details of methods and results given briefly in this paper are

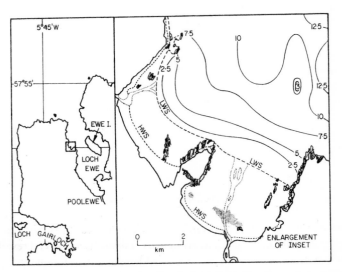

Fig. 1. Firemore Bay, and its location in Loch Ewe. Depths are in metres.

375

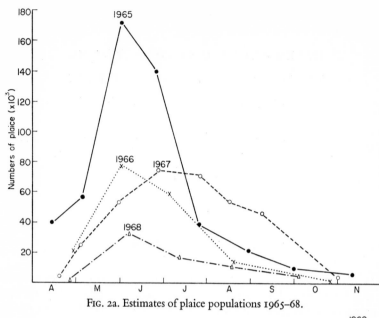

FIG. 2a. Estimates of plaice populations 1965–68.

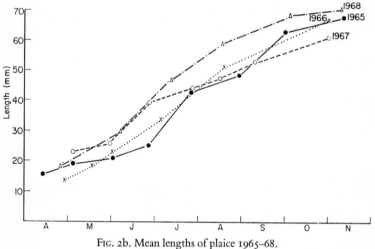

FIG. 2b. Mean lengths of plaice 1965–68.

discussed fully elsewhere (Edwards & Steele 1968; Edwards, Finlayson & Steele 1969; McIntyre & Eleftheriou 1968; Edwards, Steele & Trevallion, in preparation). The results are examined here as a basis for an attempt to describe some of the relations between two main components of the system, o-group plaice (i.e. plaice during their first year of life) and the macro-benthos on which they feed.

THE PLAICE POPULATION

NUMBERS AND BODY SIZE

After metamorphosis, the plaice (*Pleuronectes platessa* L.) settle on the sandy bottom near low-water mark at Firemore, but moving with the tide, cover an area from the tide-mark to a depth of 4–6 m below low-water. Settlement starts in this area in mid-April and continues till about the end of May. The numbers of fish, Fig. 2a, were estimated from nine tows of a

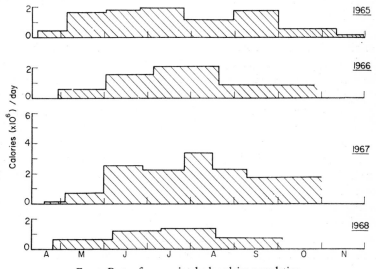

FIG. 3. Rate of energy intake by plaice population.

beam trawl at right angles to the shore-line, spaced equally across the bay (Edwards & Steele 1968). Assuming that the hauls represent random samples of the population, the 95% confidence limits, based on a logarithmic transformation of the data, are $0.6n$ and $1.7n$ of the mean n. Even with these limits there is evidence of significant differences in mortality particularly between 1965 and 1967 for which most data are available. The former year had a large population at the end of settlement compared with 1967; but a heavy mortality in 1965, unlike the negligible mortality in 1967 until September, resulted in a significantly lower September population in 1965 than 1967. From tows taken across the bay at different depths there is evidence that the fish remain within the bay until about the end of September after which emigration to deeper water may occur.

The lengths of the fish were measured (Fig. 2b) and the 95% confidence limits on the mean length are ± 2·5 mm. Variations in mean length up to the end of May depend mainly on differences in the timing and intensity of settlement. There are also differences in growth rate from June to September, particularly in 1967, when growth rate was appreciably lower than in the other years.

METABOLISM AND GROWTH RATE

Results are also available on the metabolism of plaice at different growth rates (Edwards, Finlayson & Steele 1968; Edwards, Steele & Trevallion, in preparation) which, together with the data on weight (derived from length) and numbers, provide estimates of the rate of energy requirement for the total population (Fig. 3). It can be seen that during June to August, the effect of an increase in size of individual fish, combined with a decrease in population, gives a relatively uniform rate of energy intake in any one year. This suggests that there are limits to the rate at which the fish population can take energy from the available food, so that the increasing growth of individual fish depends on the mortality rate of the population. This limit in growth rate is demonstrated by tank experiments (Edwards, Steele & Trevallion, in preparation) where fish with an abundant natural food (the siphons of *Tellina tenuis* da Costa) grow at 0·5 mm/day; on average the growth rate in the sea over approximately the same range of length (20–50 mm), is 0·3 mm/day. Further, in 1967, the growth rate in July and August was even lower than average, at 0·18 mm/day, implying a severe limitation on growth due to the combination of lack of food and very low mortality rate.

FOOD TAKEN BY O-GROUP PLAICE

The stomach contents of the plaice have been described for the years 1965–67 (Edwards & Steele 1968). The data, including those for 1968, are given in Fig. 4. The siphons of *Tellina* and the palps of polychaetes appear above the sand surface and are cropped by the smaller fish. These organs regenerate, a *Tellina* siphon regenerating completely in about a

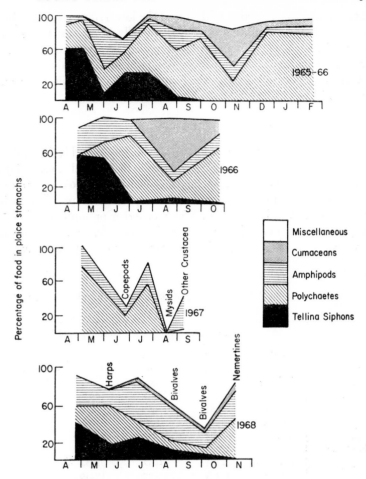

FIG. 4. Food of plaice (wet weight per 100 stomachs) 1965–68.

month, and so form a renewable food supply, to some extent within one year, certainly from year to year. The plaice also take whole polychaetes; smaller fish can take particularly one small species, *Spio filicornis* (O. F. Müller) and later in the year (e.g. December 1965) the larger fish take the larger polychaetes whole, although, as shown in Fig. 3, by this time the energy requirements of the population are very small. It appears that these two components, *Tellina* siphons and polychaetes, are preferred; the next component, small crustaceans (mainly amphipods and cumaceans) usually form a small fraction of the diet. (The low *Tellina* value on 1 June 1965 is considered abnormal since exceptionally heavy surf moved the fish more offshore where *Tellina* are not abundant.) Apart from those foods, the fish sometimes feed on pelagic or epi-benthic organisms such as calanoid copepods or mysids which are likely to occur in abundance sporadically.

BENTHOS

The macrobenthos was sampled at approximately monthly intervals in 1965 on four transects on the beach and two in the subtidal zone, at right-angles to the depth contours (McIntyre & Eleftheriou 1968). Thereafter, changes in the main components of the population were followed by sampling usually at least once each year in spring or autumn, either on complete transects or at selected 'indicator' stations. On this basis, estimates were made of the numbers and biomass of the main groups in the bay.

TABLE 1

Numbers ($\div 10^5$) and dry weights (g) of plaice food in the subtidal zones at Firemore

		Spring 1965	Spring 1966	Autumn 1967	Spring 1968
Magelona spp.	numbers	658	338	789	195
	weight	59378	28877	62482	15854
Chaetozone	numbers	420	363	852	158
setosa	weight	11119	6448	16975	1877
Spio	numbers	284	680	111	+
filicornis	weight	8776	6555	1882	+
Other	numbers	305	525	1032	570
Polychaeta	weight	82725	39679	103972	32512
Crustacea	numbers	2203	1952	1984	1789
	weight	26555	13957	17249	18305

Most of the benthic species which are preyed on by plaice occur in distinct zones in the bay and are mainly subtidal (Table 1). *Spio filicornis* is distributed over all depth zones, but the other polychaetes, *Magelona papillicornis* O. F. Müller, *M. filiformis* Wilson, and *Chaetozone setosa* Malmgren, are found in appreciable numbers only deeper than 2 m. Of the small crustaceans, cumaceans tend to be more numerous in shallow water, but the amphipods are common at all depths. *Tellina tenuis*, on the other hand, is found at maximum density on the beach just below the level of low-water springs: higher on the intertidal zone its numbers decrease rapidly and it disappears completely before the mid-tide level is reached. In a seaward direction, it is reduced to very low numbers at depths greater than 2 m, although a related species, *T. fabula* Gmelin, replaces it (at lower density) in deeper water. Estimates of total quantities of *Tellina tenuis* are given in Table 2, based on a complete survey in 1965 and indicator stations in the later years.

TABLE 2

Numbers ($\div 10^5$) and dry weights (g) of *Tellina tenuis* at Firemore

	Spring 1965	Spring 1966	Spring 1967	Spring 1968
Numbers	172	101	54	43
Weights	132001	59974	45473	35385

As already indicated, however, much of the plaice predation is on regenerative parts of animals, and Table 3 gives a more accurate appraisal of the plaice food, using where necessary the ratio of palps or siphons to the weight of the whole animal, to convert to what is actually eaten.

TABLE 3

Dry weights (g) of plaice food at Firemore

	Spring 1965	Spring 1966	Autumn 1967	Spring 1968
Magelona (palps)	4750	2310	4999	1268
Chaetozone (palps)	890	516	1358	150
Spio (whole)	8776	6555	1882	+
T. tenuis (siphons)	17200	10100	5400†	4300
Crustacea (whole)	26555	13957	17249	18305

† Spring value.

COMPARISON OF PLAICE FOOD
WITH BENTHOS POPULATIONS

Tellina AND POLYCHAETES

The first problem concerns the total absence of *Tellina* siphons from the stomachs in 1967 and their reappearance in 1968. In an earlier paper dealing with the years 1965–67 (Trevallion, Steele & Edwards 1969) it was proposed that the absence in 1967 was due to the decline in *Tellina* numbers. With the data for 1968 and also the data on polychaetes available, it is now possible, and appears more reasonable, to consider the ratio of *Tellina* to

26

TABLE 4

Ratio of numbers of *T. tenuis* to numbers of *Magelona + Chaetozone*

Year	1965	1966	1967	1968
Ratio	1:6	1:7	1:30	1:8

polychaetes as the basis for this change. Further, number rather than biomass is probably the most appropriate value since this determines the ratio of items of food presented. These ratios (Table 4) provide evidence for the hypothesis that below a certain ratio the *Tellina* siphons are omitted completely from the diet. As described elsewhere (Trevallion, Steele & Edwards 1969) this threshold can permit energy to be used for growth and reproduction rather than for the siphon regeneration which appears to have priority. Thus the elimination of *Tellina* siphons from the diet may enhance the probability of larval production to the *Tellina* population at low densities and so contribute to the long-term stability of this population.

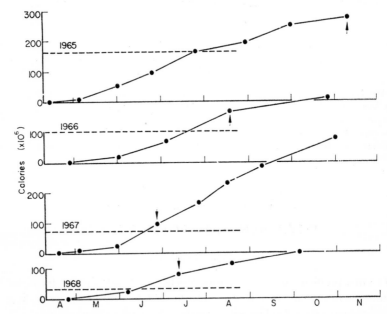

FIG. 5. Cumulative energy requirement of the plaice population. The dashed line represents energy available from *Tellina* and polychaetes, and arrows indicate the time at which feeding on these fell below 50% of the total food intake.

The other problem concerns the differences between years in the relative importance of *Tellina* plus polychaetes in the diet, as compared with epibenthic or pelagic species. It is apparent from the data in Tables 1 and 2 that there is a decline in the available siphons and palps, particularly when the absence of siphons in 1967 is included. In an attempt to quantify these effects, the rates of intake of energy in Fig. 3 have been integrated to provide a measure on any date of the energy consumed by the fish population until that date (Fig. 5). The dashed lines indicate the energy available initially from *Tellina* and polychaetes (using a factor of 5·2 cal/mg dry weight) and the arrows the time at which their percentage as weight first fell below 50%. The siphons and palps will regenerate after cropping has begun and thus continue to provide some food as indicated by the continuation of these in the diet after the points indicated by the arrows. Also other types of food are taken earlier in the season, apart from these items. However, this comparison does indicate how the available quantities of certain foods may influence changes with time in the species composition of the diet.

AMPHIPODS

The amphipods form a small but relatively constant part of the diet. The smallness of the proportion is surprising, since, in terms of biomass suitable to the fish as food, they form a large fraction of the total. This constancy is also surprising for another reason: it is maintained not only within years—when apparently the diet is forced to change as certain components decrease in abundance—but also from year to year when the initial proportions are different. At present, in the absence of appropriate field observations, it might be assumed that the fish are 'presented' with large numbers of amphipods so that the stimulus to feed on them always exists, but the fraction successfully taken is small and so their availability is effectively low but constant. This interpretation is supported by observations on plaice feeding in aquaria, where a large proportion of attacks on amphipods are unsuccessful.

DISCUSSION

PREY-PREDATOR RELATIONS WITHIN YEARS

This analysis of a prey-predator situation involves two quite different types of interpretation; one in terms of the energetics of the organisms and the other in terms of their behaviour. For the one invertebrate so far studied in detail, *Tellina tenuis*, laboratory experiments (Trevallion, Steele & Edwards 1969) have shown that, after the energy used in respiration, the energy demands of siphon regeneration, of growth and of reproduction are of the same order. From tank experiments with and without predation and with enhanced food supply, it appears that the normal food supply may not be sufficient for all three and that siphon regeneration has priority. In the absence of siphon regeneration the *Tellina* improve their condition and only then may successful spawning occur. However, for this to have a long-term stabilizing effect on the *Tellina* population it is necessary to include the behaviour pattern suggested for the fish, whereby *Tellina* escape fish predation when their density is low relative to the other main food component in the in-fauna.

The interest of this set of hypotheses is that one part of the stability is imposed from the higher levels of the food chain through behavioural mechanisms rather than by considerations of energy input at a lower level. Similar mechanisms of predation have been proposed by Tinbergen (1960) for birds and by Holling (1959) for small mammals. It might be interesting to consider whether, in other cases, the factors controlling a herbivore population may depend ultimately on predation behaviour rather than on food supply. The food is relevant in the present case since its concentration limits energy intake by *Tellina*. However, the details of the partition of energy by *Tellina* are essential, since we cannot assess the effects of the predators' behaviour on *Tellina* reproduction without knowing these energy requirements.

Similar considerations may apply to the cropping of polychaete palps, and the two preferred foods, siphons and palps, probably provide the most efficient method of taking energy from the limited food supply. For other items of food such as amphipods, it is apparent that quite different interactions occur and experimental work on these is needed.

For the plaice, the data on energy requirements are essential to an understanding of factors controlling their population. The results suggest that growth of individual fish appears to depend on the mortality rate of the population so that a roughly constant rate of energy intake into the population is maintained. The existence of known predators (Edwards & Steele 1968), the low mortality rates when predators are excluded in underwater cages (Edwards, Steele & Trevallion, in preparation), and the restriction of the population to the shallower water until late September, suggest that predation rather than migration or disease (McKenzie 1968) is the main cause of mortality.

Again, however, the mechanisms by which this mortality operates are probably behavioural. A possible hypothesis is suggested by the following observations. In summer 1965 when hatchery-reared plaice were released in mid-water and observed by divers, they remained in the water and were cropped rapidly by roundfish. By contrast, small plaice taken from the natural environment and then released in mid-water swim rapidly to the bottom and burrow into it. Also there is an apparent sequence in preferred food from the organs of stationary in-fauna through mobile epibenthos such as amphipods to the more pelagic items such as copepods; this suggests that searching time, i.e., time moving, is kept to a minimum. These two observations may result because the fish are exposed to predation only while moving just off the bottom. This is emphasized by the data for 1967. Predators may have been very scarce that year; at any rate, the plaice, by feeding more than usual on pelagic sources, were able to maintain an energy intake rather higher than average.

This very tentative explanation of how predation intensity in any year may depend on food supply in terms of searching time, is intended merely to indicate how the behaviour of individual plaice could control the energy intake of the population.

PREY-PREDATOR RELATIONS BETWEEN YEARS

A further feature of these hypotheses is the consideration of trends and differences between years. Predation on *Tellina* siphons effectively prevented reproduction of *Tellina*, and the *Tellina* component of the food available for plaice therefore decreased over the 4 years. During the same period, the extra concentration of feeding on polychaetes in 1967 coincided with, and so may have caused, the effective elimination of the one polychaete species (*Spio*) that was taken whole. Thus the generally decreasing trend in energy intake by the fish was associated with decreases in their two main food sources. On this must be superimposed fluctuations in predation; at present these are treated as random, but they can affect the trend in the energy intake as apparently in 1967. The possible consequence of this may be seen in the quasi-cyclical changes in abundance of *Tellina* (Stephen 1938; McIntyre 1969). Such changes could, in turn affect the abundance of plaice at the end of their first year of life.

In summary, these results suggest that the determination of energy flow is a necessary but not a sufficient condition to determine the fluctuations in the populations studied. They also indicate the need for more observation and experimental work on this apparently simple food chain.

SUMMARY

1.   Young plaice spend approximately the first 6 months after metamorphosis close inshore in sandy bays. The populations during 4 years have been studied in one small bay on the west coast of Scotland to show the relative effects of numbers settling in the bay, of mortality and of quantity of food available, in determining the numbers and growth rate during this period. The results suggest that mortality rate rather than settlement determines the numbers of plaice at the end of this period.

2.   Predation by plaice affects the density of populations of certain invertebrates. In particular, cropping of the siphons of a bivalve *Tellina tenuis*, by imposing a large demand on the energy intake of this organism, may inhibit its reproduction. This causes a decline in population density of *Tellina* below a threshold level where predation on them ceases and the fish switch to a mainly polychaete diet. This change should permit recovery of the *Tellina* population.

3.   It is proposed that these patterns of feeding behaviour illustrate how a carnivore population may produce quite large year-to-year fluctuations in its herbivore food while still permitting long-term stability in the herbivore species. At the same time they also suggest how fluctuations in the number of plaice recruiting to the offshore stocks may be dependent on the combined effects of food supply and mortality during this period.

REFERENCES

EDWARDS R. & STEELE J.H. (1968) The ecology of o-group plaice and common dabs at Loch Ewe. I. Population and food. *J. exp. mar. Biol. Ecol.* **2**, 215–38.

EDWARDS R.R.C., FINLAYSON D. & STEELE J.H. (1969) The ecology of o-group plaice and common dabs at Loch Ewe II. *J. exp. mar. Biol. Ecol.* **3**, 1–17.

HOLLING C.S. (1959) The components of predation as revealed by a study of small mammal predation of the European pine sawfly. *Can. Ent.* **91**, 293–320.

MCINTYRE A.D. (1969) The range of biomass in intertidal sand, with special reference to the bivalve *Tellina tenuis* da Costa. *Proc. 3rd Eur. Symp. mar. Biol.*, *Arcachon*, 2–6 *Sept.* 1968 (in press).

MCINTYRE A.D. & ELEFTHERIOU A. (1968) The bottom fauna of a flatfish nursery ground. *J. mar. biol. Ass. U.K.* **48**, 113–42.

MCKENZIE K. (1968) Some parasites of o-group plaice, *Pleuronectes platessa* L., under different environmental conditions. *Mar. Res.* **3**, 1–23.

STEPHEN A.C. (1938) Production of large broods in certain marine lamellibranchs with a possible relation to weather conditions. *J. Anim. Ecol.* **7**, 130–43.

TINBERGEN L. (1960) The natural control of insects in pine woods. I. Factors influencing the intensity of predation by song birds. *Arch. néerl. Zool.* **13**, 265–335.
TREVALLION A., STEELE J.H. & EDWARDS R.R.C. (1969) Dynamics of a benthic bivalve. *Marine Food Chains* (Ed. by J. H. Steele) pp. 285–95. Edinburgh.

DISCUSSION

J. PHILLIPSON: It is apparent from the rather limited information published on *Tellina* that in certain areas the bivalves are aggregated and in others overdispersed. Have you any evidence as to whether the distribution of young plaice accords with the distribution of their potential prey items, i.e. the *Tellina* siphons?

J. H. STEELE: In 1 year of sampling at Kames Bay, Millport, where *Tellina* is much denser than at Loch Ewe, Dr Edwards found a corresponding increase in density of o-group plaice.

M.W. HOLDGATE: In *Cardium* there is said to be considerable predation by crabs as well as flounders in the first year and by wading birds on older age classes. Is there any evidence about the role of such predators— especially birds—in the dynamics of *Tellina* populations?

J. H. STEELE: There is an annual mortality rate of approximately 35% in *Tellina* and this is due in part to predation by larger fish. Dr D. H. Mills examined the stomachs from specimens of all the bird species common in the area and found no evidence of predation on *Tellina*.

A. DUNCAN: Is there any evidence of difference in the calorific value of *Tellina* siphons and of whole pelagic crustacea? In 1967, the rate of increase of cumulated maintenance costs plus growth was greater than in other years and coincided with a greater component of pelagic crustacea in the food of the gut.

J. H. STEELE: The difference between the calorific value of siphons and crustacea is slight and the large energy intake in 1967 was due to the low mortality rate of the plaice.

K. L. BLAXTER: Is the grazing by plaice controlling the *Tellina* population?

J. H. STEELE: Heavy feeding on siphons may result in decreasing reproduction in *Tellina*.

J. H. LAWTON: How stable was the pattern of energy flow through the fish population compared with variations in population numbers?

J. H. STEELE: Energy flow was a great deal more stable than changes in population numbers.

A. MACFADYEN: Do *Tellina* differ from other bivalves in their dispersal powers, so that you are justified in arguing in terms of all recruitment coming from the population you are studying?

J. H. STEELE: Larvae probably spread throughout Loch Ewe but not beyond. There is evidence of synchronism in year-class strength among adjacent bays within the same loch.

THE CONCEPT OF ENERGY FLOW
APPLIED TO A WOODLAND
COMMUNITY

By G. C. Varley

Hope Department of Entomology
University Museum, Oxford

INTRODUCTION

Elton (1966) gives a masterly account of the complex pattern of the woodland community at Wytham, which includes hundreds of plant species and many thousands of animals. This provides a background for my attempt to collate the various census figures which are now available for some of

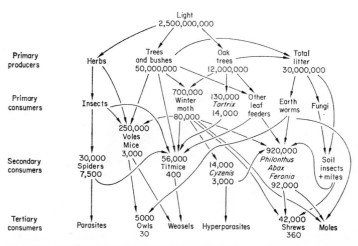

Fig. 1. A simplified energy chain for Wytham Wood. Estimates of consumption and production in kcal/ha/yr (see Table 2) are inserted above and below each link for which figures are available.

Wytham's commoner animals. I must stress that the various long and short term studies referred to here were made without any idea that they might be used in this way; the surveys were usually made independently on adjacent or sometimes overlapping areas of Wytham Wood. My calculations, preliminary as they are, suggest a number of interesting relationships which might be the basis for future work and I therefore present this study for your criticism.

Figure 1 is a simplified energy chain which shows the main species counted and omits or lumps many of the rest. The trophic level of polyphagous species like the small mammals and the birds is indeterminate. Titmice feed in winter partly on (a) beech mast and moss capsules, or on (b) females of winter moth (*Operophtera brumata* (L.)), or on (c) spiders and lady-bird beetles. These foods are at three different trophic levels (Betts 1955).

SEASONAL EFFECTS

In most years the general impression in Wytham Wood is that only a small proportion of the green plants is eaten by animals; but the woodland

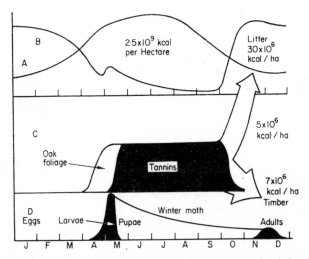

FIG. 2. Diagram of annual changes in (A) Solar energy received; (B) Quantity of litter present; (C) Quantity and quality of oak foliage; (D) Winter moth. Production is restricted to the larval period. Only larvae and adults are available to titmice; pupal numbers are reduced by predators.

trees were largely defoliated in May 1948, and badly damaged both in
1957 and 1965. Such enormous differences between years must be remem-
bered; but even more striking is the annual seasonal pattern. Figure 2
shows diagrammatically the seasonal change in the solar energy received
(2A). The oaks usually flush their leaves in April, but the precise date
varies from year to year by almost a month. There is a minor accretion of
litter in spring but the main fall of leaves and abscissed twigs is autumnal.
The litter (Fig. 2B) is virtually all consumed by midsummer and does not
accumulate. The earthworms at Wytham have not been counted, but such
litter feeders may be limited by food supply. The main standing crop of the
woodland is timber which is still commercially exploited. The foliage of
the tree is partly consumed, especially in May, by many kinds of caterpillar. The caterpillars (Fig. 2D) of most oak feeders finish feeding in May.
Most of them pupate in the ground where they are heavily attacked by
insect and mammal predators. The winter moth is available as food for
titmice for only the two short periods indicated in black. In the long summer period after winter moth larvae have left the oaks, the population of
leaf-feeding caterpillars is extremely low.

In most years the winter moth eggs begin to hatch 2 weeks or so before
most of the oaks have flushed their leaves. Only on the few trees which
regularly flush their leaves early do many winter moth larvae survive.
Few winter moth larvae establish themselves on late oaks and most perish
because they hatch before suitable food is available. The survival of only
those winter moth which hatch late should tend to select a race hatching
late enough to avoid this mortality. Two pieces of evidence show that any
advantage a larva may get from being late in hatching is offset by disadvantages it suffers as it becomes fully grown.

(1)   In 1965 M. P. Hassell and U. Ekanayake showed (Table 1) that although
oaks are some of the latest trees to flush, and many winter moth larvae feed
on other trees and bushes, development was completed soonest on oak!
Oak must be particularly suitable as food because moths from oak produce
more eggs. A shorter growth period also makes winter moth less susceptible
to the tachinid fly parasite Cyzenis (Hassell 1968).

(2)   Since winter moth hatches at a time apparently so unfavourable for
the exploitation of oak we suspected that if they hatched later they would
suffer even greater mortality. Feeny & Bostock (1968) and Feeny (1968)
found that young oak leaves contain up to 1% hydrolysable tannins. The
fully grown leaves harden and condensed tannins are laid down, mainly
in the palisade cells, until as much as 5% of the dry weight of the leaf is
tannin. Using an artificial diet to which 0·1%, 1% and 10% oak tannin

had been added, Feeny showed that growth of winter moth larvae was slowed down by 1% tannin and that the resulting pupae were underweight. On 10% tannin diet they did very poorly. The tannin apparently acts as an inhibitor of digestive enzymes and will inactivate bacterial amylase *in vitro*. MacGregor (1968) attributed a reduced growth rate in fifth instar winter moth larvae to the rise in the tannin content of the food.

TABLE I

Development, parasitization and egg production of winter moth (*Operophtera brumata* (L.)) in relation to the date of budburst of different food plants in 1965. Winter moth eggs began to hatch on 2 April

	Blackthorn *Prunus spinosa*	Hawthorn *Crataegus monogyna*	Hazel *Corylus avellana*	Oak *Quercus robur*
Date of bud burst	1 March	5 March	5 April	17 April
Date when winter moth larvae are fully fed				
median	25 May	26 May	27 May	24 May
latest	8 June	9 June	10 June	6 June
% parasitized by larvae of *Cyzenis*	22	6	14	6
Eggs per winter moth female	90	100	130	160

The mature oak leaves are largely protected from herbivores by this accumulation of tannins. The effect on feeding larvae is to slow or stop their development, so that they are longer at risk to parasites and predators. Larvae of the noctuid *Conistra vaccinii* (L.) feed on oak when they are small, but leave the trees in May to complete their feeding on the herb layer. Some caterpillar species feed in midsummer on the lammas shoots (i.e. secondary annual growth), which apparently have all the properties of young leaves and lack the high tannin content. Larvae of *Diurnea fagella* (Schiff.) feed in summer, but take until October to complete their growth. The tiny cryptic stick-like caterpillars of the light and common emerald moths, *Campaea margaritata* (L.) and *Hemithea aestivaria* (Hubn.) grow very slowly in summer, hibernate as small larvae on the bare twigs and rapidly complete their development on young leaves the following spring (Varley 1967a).

Census figures from various sources have been combined in Fig. 3. For the birds and some of the mammals two figures are available for each year.

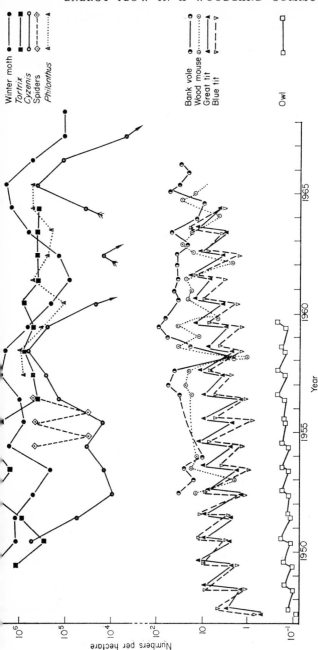

FIG. 3. Census figures for various species in Wytham Wood. The figures for the winter moth, *Tortrix*, and the parasitic fly *Cyzenis*, are the numbers of fully grown larvae/sq m under oaks, multiplied by 10^4. The spider figures were obtained within the oak census area and include all species and stages, counted twice a year on oak foliage. The population of the predatory beetle *Philonthus decorus* was estimated in the adult stage. The figures for titmice are from Marley Wood, about 2 km S.E. of Wytham Great Wood, where the oak study was made. The two points for each year represent the parental population nesting and the total population when the chicks have fledged. The census of mice and voles was in a big area of Great Wood. The owl census covered 400 ha, a large part of the Wytham Estate.

These figures represent the numbers of the parental population before breeding and the number immediately after. The difference represents a large part of the production since in these species adult size is reached rapidly. Long-lived species like the tawny owl remain rather constant in population density. The small mammals change by as much as fiftyfold and some of the insects by a hundredfold from the peak larval population in 1 year to the minimum larval population some years and generations later.

TABLE 2

Estimated numbers, energy consumption and production of oak and animals dependent on oak at Wytham Wood, Oxford

	A	B	C	A·B·C	
			Annual		
	Mean no/ha	Kcal/ oak or animal	energy consumption factor	Mean annual consumption kcal/ha	Mean annual production kcal/ha
Oaks only	30	10^7			12×10^6
Total litter	—	—			30×10^6
Winter moth	10^6	0·075	9	700000	80000
Tortrix	3×10^5	0·047	9	130000	14000
Cyzenis	15×10^4	0·019	5	14000	3000
Philonthus	60×10^4	0·05	10	300000	30000
Abax	6×10^4	0·45	10	270000	27000
Feronia spp.	13×10^4	0·27	10	350000	35000
Spiders—misc.	50×10^4	0·01	6	30000	7500
Great tit	5	30	270	40000	300
Blue tit	3·5	15	300	16000	100
Shrew	4	15	500	30000	300
Pigmy shrew	3	5	800	12000	60
Bank vole	36	30	150	160000	2000
Wood mouse	16	30	180	90000	1000
Tawny owl	0·16	600	54	5000	30

Note. The woodland area is diversified. Oaks are taken to cover one-sixth of the area and the mean population of *Tortrix* on oak is divided by six to get the population per hectare. In the few years in which winter moth has been studied on vegetation other than oak, its population per square metre equalled that on oak: so the oak figures have been used to represent the whole woodland area. *Cyzenis* was less numerous under the oaks than elsewhere and the mean population of *Cyzenis* has been determined from the numbers in the areas with and without oak. The figures for the predatory beetles *Philonthus decorus, Abax parallelopipedus* and *Feronia* spp. are estimates of mean adult numbers. The spider figures are the arithmetic mean of counts on the tree foliage. For the vertebrates the mean population for the year has been estimated. For sources of information and methods see text.

ENERGETICS

The picture is simplified if the differences between years are eliminated by taking the arithmetic means of the populations of each species (Table 2, column A). Column B gives an estimate of the calorific value of an average individual. This has been obtained either from the approximate relationship that the calorific value in kilocalories = wet weight $\times 1 \cdot 5$ or = dry weight $\times 5$. The population biomass (in kilocalories) can be obtained by multiplying column A $\times$ B.

To find the energy flow through the population we can use the relationship (in which all terms are expressed in calories) that the consumption

$$C = P + R + F + U \tag{1}$$

where P is the production, R the energy lost in respiration and F and U that contained in the faeces and urine.

For an insect like winter moth we have seen (Fig. 2) that production is restricted to the short period of larval growth. The annual consumption (C_a) is proportional to the maximum biomass (B_{max}) and we can write

$$C_a = B_{max} f \tag{2}$$

where f is the energy consumption factor, which is in effect C/P, the reciprocal of the production efficiency. To a first approximation the production $P = B_{max}$ which is the energy contained in the mature larval population.

For the birds and mammals the most important terms in equation (1) are R and F. Growth to adult weight in a bird like *Parus* is complete when the bird fledges about 2 weeks after hatching. For the remaining 50 weeks of the year the bird consumes food, but, except when the bird moults or reproduces, production is zero. The error will be small compared with the uncertainties of the census mean, if we neglect these short periods of growth and reproduction and calculate the annual consumption from data on adult food or respiration. It is convenient to use a new term, which can be regarded as a constant for a species, the annual energy consumption factor f_a. Then if the mean biomass $\bar{B}$ of the species is expressed in calories, the annual consumption

$$C_a = \bar{B} f_a \tag{3}$$

If a species consumes its own weight of food each day, and the calorific value of the food and the animal's body are the same, then $f_a = 365$.

The production of a population with a restricted breeding season can be estimated from the litter size or clutch size, the proportion which grow to adult weight, and the number of broods a year.

THE OAKS

These were studied by MacGregor (1968). The mean annual production of the three trees he studied, multiplied by the mean number of trees per hectare, gives a mean annual production of 12×10^6 kcal, much of which is accumulating as timber. The oak canopy covered perhaps one-sixth of the ground, and within the small census area for oaks the figure was about half.

WINTER MOTH AND GREEN TORTRIX

These have been counted by Varley & Gradwell (1968) and the energetics of winter moth were studied by MacGregor. The total energy consumption of winter moth and *Tortrix viridana* is much greater than that of the other species feeding on oak. The mean population per hectare for *Tortrix* has been adjusted because it feeds only on oak, whereas winter moth has comparable numbers on other trees and bushes.

The effect of partial defoliation on the production of the tree was estimated (Varley 1967b) to be at least ten times the consumption of the larvae; MacGregor has carefully re-investigated this important effect and finds that a figure of four times is more realistic. Some of this energy is lost directly as light to the herb layer, but some may be dissipated in the tree in the complex metabolic processes of re-foliation after damage.

CYZENIS ALBICANS (FALL.)

This is a tachinid fly, a specific parasite of winter moth. The figure in column A is the number of feeding larvae/ha. One larva of the fly eats one winter moth, which is killed as a pupa. The newly formed fly puparium weighs as much as a quarter the weight of a winter moth pupa, which indicates a production efficiency of 25%. The figure 5 has been entered in column C instead of 4 because the few flies which emerge must feed to mature their ovaries. *Cyzenis* is extremely variable in numbers. When winter moth populations rise to many hundreds to the square metre, over 20% may be parasitized by *Cyzenis*. The fly virtually disappears when winter moth is scarce. This is because *Cyzenis* eggs, containing larvae ready to hatch, are laid on leaves of oaks and other trees (Hassell

1968). Only if they are swallowed by winter moth can they develop. Otherwise the larvae die of starvation without hatching, and the chance of this happening is directly related to the winter moth population.

THE PREDATORY BEETLES

The commonest large predatory beetles which feed on the pre-pupae and pupae of winter moth in the litter layer are *Philonthus decorus* (Gr.) (Staphylinidae) and the carabids *Feronia madida* (F.), *F. melanaria* (Ill.) and *Abax parallelopipedus* (Pill. et Mitt.). Frank (1967a,b) found that *Philonthus* was a general feeder which devoured winter moth pupae only occasionally; the high population density of *Philonthus* (60/sq m) makes the number of winter moth taken considerable. He estimated for 1965 that 51 pupae/sq m (a little over 20%) were eaten by *Philonthus*, and that these four species of beetle took 95 pupae of the winter moth/sq m in that year. The figure of 10 in Table 2 column C is a guess since no detailed study has yet been made.

Dixon (1958) studied the feeding behaviour of a predatory lady-bird beetle, *Adalia decempunctata* (L.). The chance that a 'contact' between a newly hatched lady-bird larva and a first-stage aphid will be successful and give the predator a meal was only 0·05. The young aphid escaped nineteen times out of twenty, and bigger aphids almost always got away. The efficiency of the predator rose with each successive meal and with each succeeding moult. It is the availability of suitable prey for the predator's first stage which determines the initial survival, and from then on survival is high. Predators quickly become efficient enough to find the food they need and attack progressively larger and larger prey. Because they have so seldom been observed when small, this relationship requires confirmation for other predators, but it seems likely to be widespread.

The beetle populations at Wytham are remarkably stable from year to year, probably because the initial larval survival depends on small prey which feed on the litter. The survivors eat winter moth pupae if these are available, but are adaptable enough not to depend on this diet.

The disappearance of winter moth pupae, which we have termed 'pupal predation' has been proved to be density dependent both when the analysis is made between the different local concentrations within years and also when the analysis is made on the mean pupal predation and mean pupal population between years (Varley & Gradwell 1968). Predation might be expected to be a delayed density dependent factor if the increase and mortality of predator populations were dependent on the winter moth population, but there is only a slight sign of this effect. The predators must turn to other food when winter moth is scarce.

The adult beetles are long-lived. Pitfall trapping (Frank 1967a) suggests seasonal changes in activity but the beetles remain available as food for shrews during the whole year.

THE SPIDERS

These were studied by Turnbull (1960). Using fruit flies as food he showed that the conversion of food into growth was 27% for *Linyphia triangularis*, which would give a value for *f* of about $3\frac{1}{2}$ if the population represented peak biomass. However, his census figures and mean weights are for all species lumped together and include many juveniles. Production and consumption figures have been assessed by Kajak (1967) for spiders and her conversion factors are used in Table 2. There is no evidence for big population changes between years, and the argument used for the beetles is probably applicable here also; i.e. that the tiny newly hatched spiders must rely for their first meal on aphids, or on the minute diptera which abound as larvae in the litter.

THE TITMICE

Rapid changes in the population of birds like titmice makes the annual mean difficult to estimate. From Lack (1966, Table 2) the mean breeding population of *Parus major*, the great tit, can be calculated to be 2·5/ha. The numbers rise to a summer population of about 11/ha. Thus the production turn-over per breeding bird is more than $(11-2\cdot5)/2\cdot5 = 3\cdot4$. Allowing for nestling losses the production turn-over is likely to be at least 4. The mean annual production is $2\cdot5 \times 30 \times 4 = 300$ kcal/ha. To estimate the mean population for the year, on which food consumption must be based, one needs to interpolate between census figures. If the May population is 2·5, that of June is 12/ha falling quickly back to 2·5/ha next breeding season; the mean population must be about 5 birds/ha. The figure of 3·5/ha for the blue tit (*Parus caeruleus*) is derived in the same way.

The daily needs of a great tit would be satisfied by a diet of about 22 kcal (Gibb 1957), which could be provided by 300 5th instar winter moth larvae. When feeding young, each parent may take to the nest as many as 800 food items of this size (Royama 1966). Royama showed that the birds suddenly stopped taking winter moth larvae to the nest and turned to alternative larger prey items and our figures show that this happened whilst winter moth was still available.

Betts showed that in winter it is much more difficult for these birds to find food, and that stomachs of titmice from the Forest of Dean even contained moss capsules at this season. We examined stomachs from Wytham and found that in November and December some winter moths were eaten but predatory lady-bird beetles and spiders formed a large part of the food.

The nesting period coincides closely with the caterpillar maximum (Fig. 2) and second or other late broods may die of starvation in the nest, or suffer partial starvation and fledge under weight. Their survival is then reduced (Lack 1966).

THE SMALL MAMMALS

Hawkins & Jewell (1962) fed captive shrews on mealworms; they record the calorific value of the daily ration per gramme of body weight.

Sorex araneus (10g) takes more than its own calorific value of food per day, giving a value for the annual energy consumption factor $f_a = 500$. For the smaller *S. minutus* (3·5 g), f_a is as high as 800.

The two species of shrew between them appear to need on the average over 40 000 kcal per year. Buckner (1969) and Frank (1967a) studied them in the oak area and found that in summer *S. araneus* stored winter moth pupae in larders along their runs. *S. minutus* sometimes raided the nets on the tree trunks which accumulated winter moth females and other arthropods in November and December.

The production figures are based on the data for breeding in Southern (1964). The estimated mean consumption of the shrews is comparable to but less than the mean production for winter moth or for the predatory beetles; but these also provide food for a big unmeasured population of moles.

More detailed information about the diet of shrews and their changes in numbers are needed before we can decide whether shrew populations are more closely linked with the relatively constant litter fauna or with the highly variable tree defoliators.

The energetics of *Clethrionomys glareolus* have been studied by Ryszkowski & Petrusewicz (1967) and by Grodziński & Górecki (1967). From the former I derive a value for $f_a = 138$ and from the latter $f_a = 170$. I have compromised with a figure of 150 in Table 2. For *Apodemus sylvaticus* a figure of at least 180 seems to fit the information in Grodziński & Górecki (1967).

The census figures for these species at Wytham have been kindly provided by Mr Southern, who warned me that although the changes in Fig. 3 are largely real, the absolute values and therefore the mean in Table 2 should be regarded as provisional.

These species, in captivity, can be kept healthy on plant food, but if a live noctuid moth is introduced into their cage they pounce on it and treat it as a preferred food. In the field the animals are largely vegetarian (Chitty, Pimentel & Krebs 1968), but when insect food (such as winter moth prepupae) is available in May (Holišova 1967) they become insectivorous, although their vegetable food is still available.

The production of the small mammals is hard to measure. Changes in the census figures from 1956–66 suggest a mean production for *Clethrionomys* of at least 23 voles/ha. With a mean spring population of 18 pairs they might have produced at least two broods of 4 young per pair, which is a production of 140 voles/ha. If these are taken by owls and other predators when, on the average, they are half grown, this gives a production of 2000 kcal. Changes in the census figures for *Apodemus* give a minimum figure for mean production of 13 mice per year, but 8 pairs/ha could easily produce 80 mice/ha. Ryszkowski & Petrusewicz (1967) provide another approach to an estimate of the production by giving a figure for the 'turn-over' which is the ratio of production to standing crop. They give a figure of 4 for these small rodents which would provide figures for production double those in Table 2. If this value were constant, the logarithmic plot (Fig. 2) should show an annual rise of 0·6 (log 4) but in fact the rise between the two census figures is extremely variable.

THE TAWNY OWL

The population density of *Strix aluco* is very low and steady (Southern 1954). With a weight of 400 g, this is the heaviest species counted. The low annual energy consumption factor (54), although based on the needs of captive birds, probably applies to the wild, where the owl economizes in energy by flying little. It rests by day and hunts by night from perches. Nevertheless the mean annual energy consumption works out at 5000 kcal, which is more than the estimated mean production of the voles and mice. Figure 3 shows that the populations of these prey species were sometimes twice the mean level used in Table 2; but the minimum populations were a tenth of the mean. This fits with Southern's (1954) finding that the owls thrive when the small mammal population is above the average but when mice and voles are uncommon the owls turn to other food such as moles. Earthworms may also form a large item in the diet, but many owlets die of starvation in these conditions.

DISCUSSION

The estimates for the mean annual consumption and production for the major species which have been counted have been entered into Fig. 1. There are big gaps still unfilled. Census figures for many more caterpillar species which feed on oak in May and early June are available (Varley & Gradwell 1963) but although they would have filled the space in Fig. 3 the total consumption and production of these species is so much less than that of winter moth and *Tortrix* that their inclusion would have contributed little to the overall picture.

The figures for primary production in Fig. 1 are, of course, only approximate. Much of the production of the trees goes into timber which is not included in the diagram. The litter provides a very important and rather constant channel. In contrast the amount taken from the foliage by caterpillars and other insects is extremely variable, with a range of about thirty-fold.

Of the production of the primary consumers, at least 100 000 kcal/ha comes through the trees directly. If we assume 10% conversion, the energy flow through the litter is probably thirty times as great, and should be investigated.

Amongst the secondary consumers listed, even if we include the mice and voles which feed partly as primary and partly as secondary consumers, the dominant animals are the predatory beetles, which must get most of their food through the litter. At this trophic level many uncounted species could probably be quite important.

The shrews dominate the measurements at the tertiary level by a factor of ten, but probably moles would also have given high figures had we known their population.

Figure 1 does not in itself show clear evidence for food shortages. The estimated mean consumption for owls is greater than the figure for the production of the mice and voles, but a more generous estimate could have doubled this latter figure. Clearer evidence for food shortages comes from consideration of the effects of short term changes within the year in Fig. 2 and the superimposed changes between years in Fig. 3. When, in 1958, the combined mouse and vole populations were down to 3/ha, the total biomass of these species was equal to only about a week's food supply for the owls. As the small mammal population then rose again, the owls must have turned almost entirely to other foods, on which they survived, but did not breed.

In contrast, the population of the mature larvae of the parasitic fly *Cyzenis* which is specific to winter moth, is more variable in population than its host. Big reductions in *Cyzenis* numbers come whenever the winter moth population is below average. Then the vast majority of *Cyzenis* larvae starve without hatching from the egg, because hatching and feeding depend on being swallowed by a winter moth larva. *Cyzenis* mortality at this stage when plotted against *Cyzenis* population density, shows clearly a delayed density dependent relationship (Hassell, unpublished).

The big changes in winter moth population are also attributable to starvation which acts only temporarily. The majority of newly hatched larvae fail to get their first meal if, as is often the case, the oak buds have not yet opened by this time.

Starvation has a variety of different effects on the population dynamics of the species about which we have some information. When the available food supplies are very different in different seasons, the incidence of starvation is not density dependent, but produces big population changes which are not related to population density. Such catastrophic mortality may be a 'key factor' determining population change as has been demonstrated for the early larval mortality of winter moth (i.e. winter disappearance, k_1, Varley & Gradwell 1963, 1968). The mortality of nestling titmice and owls is, in years of shortage, related to clutch size. Within the nest, in any one year this mortality shows a density dependent relationship. However, if the overall mortality of nestlings is related in different years to the changing food supply, starvation will appear mainly as a variable, which will be responsible for big changes in the proportion of juveniles recruited to the autumn population. The breeding population of owls is extraordinarily constant, which implies a powerful regulatory influence after fledging and before the next breeding season. The surplus owls that disappear perhaps emigrate and die of starvation in less favourable habitats; if so, this mortality would be density dependent in relation to the Wytham population.

When mortality from starvation acts in a density dependent manner it is tending to stabilize the population. In other circumstances its effects are catastrophic and independent of density, leading to sharp reductions in the population. In the case of *Cyzenis* a comparison of the proportion of first-stage larvae starving and the population density of the species shows a delayed density dependent relationship which promotes cyclic changes in the population.

SUMMARY

1. The primary conversion of oak leaves by caterpillars is restricted to a very short season in May and June because the rise in tannins in the leaves makes them inedible after this.

2. The census figures from Wytham Wood for a few oak feeding insects, such as *Tortrix viridana*, winter moth (*Operophtera brumata*) and its parasite *Cyzenis albicans* (Diptera, Tachinidae) have been converted to numbers per hectare. A few figures are also available for spiders and for some predatory beetles.

3. Census figures are also available from nearby woodland areas for the great tit, blue tit, tawny owl, for woodmice and bank voles. Some figures are available for shrews.

4. Data from various sources have been used to derive estimates, based on the mean population figures, for the consumption and production of all these species.

5. The mean figures for consumption of a number of species are close to the mean figures for the production of the species on which they feed. The big variations in population numbers must lead to starvation and death unless the animal can find alternative food.

6. The parasitic fly *Cyzenis* is specific to winter moth, and its numbers changed by more than a hundredfold, being more variable than those of its host. The owl, being a general predator, changed its breeding population only by twofold during the period of study. Its breeding success is closely related to the numbers of mice and voles.

7. Depending on the circumstances, death caused by starvation may be shown to act as a density dependent factor, or as a delayed density dependent factor, or as a catastrophic factor (key factor) independent of population density.

REFERENCES

BETTS M. (1955) The food of titmice in oak woodland. *J. Anim. Ecol.* **24**, 282–323.

BUCKNER C.H. (1969) The common shrew (*Sorex araneus*) as a predator of the winter moth (*Operophtera brumata*) near Oxford, England. *Can. Ent.* **101**, 370–5.

CHITTY D., PIMENTEL D. & KREBS C.J. (1968) Food supply of overwintering voles. *J. Anim. Ecol.* **37**, 113–20.

DIXON A.F.G. (1958) The escape responses shown by certain Aphids to the presence of the Coccinellid *Adalia decempunctata* (L.). *Trans. R. ent. Soc. Lond.* **110**, 319–34.

ELTON C. (1966) *The Pattern of Animal Communities.* London & New York.

FEENY P.P. (1968) Effect of oak leaf tannins on larval growth of the winter moth, *Operophtera brumata. J. Insect Physiol.* **14**, 801–17.

FEENY P.P. & BOSTOCK H. (1968) Seasonal changes in the tannin content of oak leaves. *Phytochemistry,* **7**, 871–80.

FRANK J.H. (1967a) The insect predators of the pupal stage of the winter moth, *Operophtera brumata* (L.). (Lepidoptera: Hydriomenidae). *J. Anim. Ecol.* **36**, 375–89.

FRANK J.H. (1967b) The effect of pupal predators on a population of winter moth, *Operophtera brumata* (L.) (Hydriomenidae). *J. Anim. Ecol.* **36**, 611–21.

GIBB J. (1957) Food requirements and other observations on captive tits. *Bird Study,* **4**, 207–15.

GRODZIŃSKI W. & GÓRECKI A. (1967) Daily energy budgets of small rodents. *Secondary Productivity of Terrestrial Ecosystems* (Ed. by K. Petrusewicz), pp. 295–314. Warsaw & Crakow.

HASSELL M.P. (1968) The behavioural response of a tachinid fly (*Cyzenis albicans* (Fall.)) to its host, the winter moth (*Operophtera brumata* (L.)). *J. Anim. Ecol.* **37**, 627–39.

HAWKINS A.E. & JEWELL P.A. (1962) Food consumption and energy requirements of captive British shrews and the mole. *Proc. zool. Soc. Lond.* **138**, 325–8.

HOLIŠOVA V. (1967) Trophic relations in bioenergetic investigation. *Secondary Productivity of Terrestrial Ecosystems* (Ed. by K. Petrusewicz), pp. 331–47. Warsaw & Crakow.

KAJAK A. (1967) Productivity of some populations of web spiders. *Secondary Productivity of Terrestrial Ecosystems* (Ed. by K. Petrusewicz), pp. 807–20. Warsaw & Crakow.

LACK D. (1966) *Population Studies of Birds.* Oxford.

MACGREGOR K.A. (1968) The productivity relations between insects and oak trees. D.Phil. thesis, Oxford University.

ROYAMA T. (1966) The breeding biology of the great tit, *Parus major* with reference to food. D.Phil. thesis, Oxford University.

RYSZKOWSKI L. & PETRUSEWICZ K. (1967) Estimation of energy flow through small rodent populations. *Secondary Productivity of Terrestrial Ecosystems* (Ed. by K. Petrusewicz), pp. 125–46. Warsaw & Crakow.

SOUTHERN H.N. (1954) Tawny owls and their prey. *Ibis,* **96**, 384–410.

SOUTHERN H.N. (1964) *The Handbook of British Mammals.* Oxford.

TURNBULL A.L. (1960) The spider population of a stand of oak (*Quercus robur* L.) in Wytham Wood, Berks., England. *Can. Ent.* **92**, 110–24.

VARLEY G.C. (1967a) Estimation of secondary production in species with an annual life-cycle. *Secondary Productivity of Terrestrial Ecosystems* (Ed. by K. Petrusewicz), pp. 447–57. Warsaw & Crakow.

VARLEY G.C. (1967b) The effects of grazing by animals on plant productivity. *Secondary Productivity of Terrestrial Ecosystems* (Ed. by K. Petrusewicz), pp. 773–8. Warsaw & Crakow.

VARLEY G.C. & GRADWELL G.R. (1963) The interpretation of insect population changes. *Proc. Ceylon Ass. Advmt. Sci.* **18** (D), 142–56.

VARLEY G.C. & GRADWELL G.R. (1968). Population models for the winter moth. *Insect Abundance* (Ed. by T. R. E. Southwood), pp. 132–42. Oxford & Edinburgh.

DISCUSSION

D. Lack: Professor Varley showed that the oaks which produce their leaves early get badly eaten. What is the disadvantage of producing their leaves late? There is presumably a disadvantage, or all the oaks would come out later than they do.

G. C. Varley: They would miss the early fine weather.

N. Waloff: In spite of limitations of feeding time imposed by tannin in oak leaves, the insect fauna on oaks is richer than on most other species of trees. Would Professor Varley enlarge on this?

G. C. Varley: The great majority of insect species feed on the young leaves at the same time as the winter moths do. Later in the summer some leaf miners feed on mesophyll which has a relatively lower tannin content. Also there are many gall-formers. In comparing oaks with other trees I favour Southwood's hypothesis (1961, *J. Anim. Ecol.* **30**, 1–8) that the richness and diversity of the fauna of these trees is related both to their qualities as food and to the length of time they have lived here.

GENERAL DISCUSSION

C. M. PERRINS: The papers today have been concerned with effects of predators on their prey, and in some cases with the effects of the prey on their predators. I would first like to make some comments particularly about the herbivore-vegetation link. Firstly, I would like to ask how much information there is available about the long-term effects of heavy grazing by herbivores where the habitat is being altered to the detriment of the grazer? How do plants 'defend' themselves against grazing by structural and physiological adaptations which allow the plants to get back at the grazers—e.g. tannin production in oaks which prevents grazing? Perhaps also it is no accident that the seed crop is erratic in occurrence in some plant species, which is to the detriment of the seed-eaters such as the finches discussed by Dr Newton. Finally, we had several examples of animals having to go through periods of the year when food was short and having to rely on energy reserves built up in the body at other times of the year. When we study such animals we must be sure that we are covering a suitable length of time to include both the periods required to build up and store these reserves as well as the periods when they are being consumed by the animals.

J. D. CARTHY: There is a great deal of evidence on the stimuli by which insects recognize their food plants. It is plain that with many animals there are physical characters of the plant which render it unpalatable. Thus, *Cepaea* prefers dicotyledons to grasses, but prefers water extracts of the two equally; some physical factor causes the natural unpalatibility of the grasses. We know little of these factors in the choice exercised by herbivorous vertebrates. But we could find out whether the dispersion of the plants they prefer affects their ability to find them. Is there a threshold as there is for young plaice? Do they require the plants to be clumped before they can detect them? What is the minimum size of such a clump? These factors will have a strong influence on the effect which a grazing herbivore has on the plant community.

R. MOSS: Dr Carthy mentioned the effect of plant distribution on grazing, and specifically asked: 'Does a widely dispersed plant avoid grazing?'. Ptarmigan in Iceland feed selectively on widely dispersed *Polygonum* bulbils (Gardarsson & Moss, this symposium), and in this case the answer is negative. In response to Dr Perrins' point about plant/animal competition, red grouse appear to be well adapted to tannins—they grow well on heather containing 10% soluble tannins, whilst poultry chicks are adversely affected by 1% levels. Hence, if the production of tannin were a response by *Calluna* to grazing, grouse appear to have adapted in turn.

A. S. CHEKE: We must beware of drawing conclusions about the response of plant species to grazing if these are based on casual observations without direct study. On Dr Perrins' argument, one might argue that in Alaska, the overgrazing by deer caused a 'fight back' by the whole ecosystem—unpalatable plants become more frequent, and palatable ones disappear; the reactions of individual species of plant may not in any way be related to the grazing pressure, but to some other factors in the plants' metabolism.

R. H. V. BELL: There is reason to believe that the grass family as a whole is specifically adapted to regimes of heavy defoliation, i.e. by grazing or burning, for example, through the inaccessible growing points and storage organs. In addition, there is increasing evidence that the development and maintenance of grasslands is dependent upon these influences.

P. J. NEWBOULD: In terms of the effects of herbivores on plant production, there is a strong contrast between grass and trees. Grass, with an intercalary meristem, is well adapted to grazing from above. Individual leaves have a relatively short peak of photosynthetic activity. Moderate grazing usually increases the rate of primary production. Trees, with an apical meristem, have leaves with a relatively prolonged peak of photosynthetic activity, and consumption usually reduces the rate of primary production.

J. C. COULSON: How does an experiment such as that described today by Dr Miller determine the cause and effect relationship of, say, burning and fertilizer treatment, when the treatment appreciably affects the whole environment, and particularly the vast soil flora and fauna? Surely it is necessary to measure many more factors which can be influenced by the treatment before it can be concluded that the effect is produced in one particular way?

A. WATSON: Of course I admit that any experiment involving a change in or treatment of the environment will alter many things, which may or may not be important. However, there are so many variables in population ecology that we will never get out of the present confusion unless we experiment. Experimenting allows us to reject spurious relationships and associations of events, which we might think were cause and effect if the experiment were not done. At any rate, correlations and associations of events are often wrongly treated in the literature as if they were causal, even in cases where no experiments are considered or attempted. On a more positive level, an experiment often raises more questions than it solves, and many of these new questions and problems might well never have arisen otherwise. It is all very well to criticize experimentation, but I would put the ball back in your court by asking what alternative you propose to get us out of the present confusion in population ecology? I would maintain that this confusion has arisen and has continued precisely because too many people have been almost chronically unwilling to do experiments. Instead, the tendency is to maintain the same old field observations 'for another year to see what happens'.

PART IV · SUMMING UP AND GENERAL DISCUSSION

FEEDBACK FROM FOOD
RESOURCES TO POPULATION
REGULATION

By V. C. Wynne-Edwards

Department of Natural History, University of Aberdeen

In this paper I am more concerned with giving definition to a problem than solving it. The problem arises because all supplies of food are finite in terms of their yield per unit area and time, and because some of them in addition are vulnerable to over-exploitation, implying that their future productivity will be reduced. I want to show that the relationship between demand by consumers and production by suppliers of food is not usually either simple or direct, but tends on the contrary to be complex and involved.

I can approach the subject by means of an example, the usefulness of which is not diminished by the fact that I have used it before (Wynne-Edwards 1968). Beavers (*Castor canadensis*) where I have observed them in Canada depend for food very largely on the inner bark, twigs and leaves of poplars and willows, especially aspens (*Populus tremuloides*). In order to obtain these foods they cut down trees of widely varying size, though not often above 20 cm in diameter. The animal anchors its head firmly by digging the upper incisors into the bark or wood, and it chisels into the trunk with the procumbent lower ones, shifting its anchorage as often as necessary. The teeth are sharp tools and the wood is fairly soft, so that in an hour a beaver can bring down within reach enough food to last for 2 or 3 days. In summer the tree-bark diet is varied with water-lily roots and much other herbaceous vegetation. Tree-felling activity builds up to a peak in the autumn, when large numbers of logs and sticks are cut and floated piecemeal to the winter lodge. There they are stored under water and remain fresh and edible for 5 to 6 months. After the streams and lakes are frozen the beavers can enter the water from inside their snow-covered

413

lodge, swim out under the ice and retrieve food from the submerged stores.

Poplars and willows regenerate readily as suckers from their underground roots. It is difficult to estimate the mean age of the trees the beavers take because of their varied sizes. The great majority are probably in the range of 10 to 60 years, so that the mean is unlikely to be lower than 15 years, even allowing for the many saplings that are cut. Provision must consequently be made for a mean regeneration cycle of at least 15 years, which is approximately the same as saying that the annual food production is is only one-fifteenth of the standing crop. Though the beavers live within almost instant reach of enormous reserves, therefore, they must not remove more than one tree in fifteen in an average year, otherwise they will be consuming more than the total increment of the forest and before many generations the habitat will have been destroyed.

Beavers are not greatly subject to predation except by human trappers. Undisturbed populations are not given to rapid fluctuations in numbers though they can easily be decimated by over-trapping. Tree-felling rather tends to be concentrated at any one time in small patches, but the beavers' overall population density remains sufficiently low for them to occupy the same small watersheds indefinitely. Any sizeable area of beaver country could always support for the time being several times the population it actually carries; but if the food demand were allowed to build up and stay up above the long-term regeneration rate of the aspen groves, even though the consequences might not become apparent for several decades, it would certainly end in disaster.

There is no reason to doubt that beaver population density is primarily geared to the productivity of the habitat in terms of usable food, just as it has been shown to be on different areas in the red grouse (Watson & Moss 1970; Miller, Watson & Jenkins 1970). The relationship between the beavers and their food resource is only an extreme and therefore striking example of a common situation, in which the consumers are required to restrain their demands in order to safeguard future supplies.

Before I go on to analyse this situation further I must establish a second premise. It is that adaptations for self-regulation or population homeostasis in animals have now been sufficiently widely demonstrated for their existence to be no longer a matter of controversy. We do not know whether they are universal in the higher groups, but no laboratory population of arthropods or vertebrates has yet been found to lack them. Spontaneous regulation has been demonstrated in the laboratory, for example in *Daphnia*, *Drosophila*, *Tribolium*, various grain-infesting weevils and moths, in aquarium fishes such as guppies (*Lebistes reticulatus*), and in rats, mice and voles. In the wild there is a good example near at hand in the red grouse. In laboratory populations where the environment is under full control it can be shown that regulation generally depends on density-dependent changes in fertility and recruitment on the one hand, and in self-imposed mortality on the other. No intervention is necessary from external agents such as predators, parasites, disease organisms or extremes of climate.

To say that many animals have the power of population limitation is not to say that they always need to be using it. Homeostasis implies action to rectify an imbalance that has occurred, whether upward or downward. Populations can often be cut down below their ceiling level by environmental disasters, disease or predation. It has been shown that when mortality is experimentally imposed, for example on laboratory populations of the sheep blow-fly *Lucilia cuprina* (Nicholson 1955) and the guppy (Silliman & Gutsell 1958; Silliman 1968), they respond accordingly by a rise in the recruitment rate. When on the contrary the acceptable threshold of crowding is reached, stocks of guppies, house mice or flour beetles will reduce their recruitment to zero. Artificially super-crowded guppies actually eat one another until an acceptable density is restored (Breder & Coates 1932); the same is true of spider-crabs (Schäfer 1952).

Some wild populations are almost certainly kept down by predation for prolonged periods, like the present stocks of most of the commercially exploited fish in the North Sea; parasitism may do the same in insect populations. Others undergo a succession of episodes of rapid increase and catastrophic mortality because they inhabit harsh and erratic environments in which a state of population balance can rarely if ever be achieved. Still others are prevented from exploiting the available food supply to the full because the habitat is deficient in some independent respect, for example in allowing access or providing cover or water to drink. The number of breeding tits and pied flycatchers can be increased in some places in this country by providing nest boxes; new lobster colonies have been established in eastern Canada by dumping rocks on the sandy sea-floor to create artificial reefs.

When developing the theme of population homeostasis I have more than once been asked how self-regulation can be reconciled with the famous incident of the mule deer in the Kaibab National Forest in Arizona, first described by D. I. Rasmussen (1941; see also D. R. Klein 1970). In 1907 a campaign was begun there to conserve the game by exterminating the remaining wolves and pumas and at least reducing the much more numerous coyotes. As a result the deer population rose over the next 17 years from very roughly 4000 to 100 000. By 1924 the herds were inflicting such massive devastation on their food resources that in the two following winters 60 000 deer starved to death. Numbers continued to decline for a long time after that, though by 1939 they were levelling off again at around 10 000 head. A somewhat similar train of events affected the moose on Isle Royale in Lake Superior after the deliberate extermination of wolves, but in that case the wolves later recolonized the island across the ice in winter, and a population balance between wolves, moose and vegetation has now been restored (Mech 1966).

In interpreting these cases one must be on the watch for a possible trap for the unwary. One of the commonest mechanisms of population homeo-stasis is the exclusion of surplus individuals from the social group. These outcasts either emigrate, or suffer and generally die where they are, depend-ing on the circumstances. Many of them fall victims to predators, disease and starvation as a secondary result of their social rejection. In the red grouse Jenkins, Watson & Miller (1964) were able to show that winter predation on surplus non-territorial birds was seven times as heavy as it was on those that were established territory owners; and mortality from disease and starvation is mainly confined to the surplus group.

In the red deer in Scotland we can see another socially-determined differential in winter mortality, which falls heaviest on young animals and lightest on the hinds; the stags come in between. This arises at least in part from the undisputed right of the hinds to possess the best feeding. The young animals are proportionally most numerous at the bottom end of the feeding hierarchy. If there were still wolves to prey on the deer they would probably kill the animals that become weakened as a result of social discrimination; but nowadays many of them manage to struggle through the winter and survive to contribute to population growth and over-crowding.

When we look at predators in this light we see them playing a subservient part in the population control of deer (and grouse), acting as useful and in fact indispensable executioners to the surplus animals that are presented to them by the social machinery. It is important to recognize the difference between this secondary role, in which the deer depend on the predators, and the contrasting situation in which the predators take the initiative and are in primary control of the prey. Some sort of compromise could exist between predators and their prey, whereby the predators took what they were offered but also a little more as well, as a spur to better production by the prey. To my knowledge the first part of this relationship, namely the predators taking what they are offered, has been clearly demonstrated so far only in the grouse predators in Glen Esk, by Jenkins, Watson & Miller (1964). But we know also that extra mortality will in many cases stimulate better recruitment and growth, both from experiments of the kind I have already mentioned and from exploiting wild populations of fishes, fur-seals and whales.

THE RENEWAL OF
FOOD RESOURCES

Recruitment and renewal are important alike to consumers and producers of food. If one has to consider the food resources of animals in general it is obvious that they embrace an immense variety of types. They can be more or less nutritious, more or less easy to find, eat and digest, more or less indispensable to the consumer, and so on. There are however some simple generalizations to be made, both as to the nature of the food source and the way in which it is renewed and becomes available for use.

True foods consist either of living organisms, their attached parts and products, or else of the dead and detached organic materials that can be grouped broadly as detritus and carrion. Both these categories could of course be subdivided into animal and vegetable components. In addition there remain as minor components of diet any inorganic matter that may be directly ingested and assimilated; being energy-free and non-renewable they need not concern us in the present context.

When living organisms supply the food it is useful to distinguish (1) whether whole individuals are killed by the consumer when it feeds or (2) whether it takes expendable or regenerative parts that entail little or no damage to continued production. In the category of whole organisms one must include for example the prey taken by carnivores, the hosts of many parasitic insects, and seeds of plants which the consumer digests and destroys. In the alternative category there are a variety of vegetative parts of plants, the fluids sucked from plants and animals after their integument has been harmlessly pierced, the dermal materials taken from their hosts by ecto-parasites, and so on, always provided that the consumer does not kill the supplier by its actions nor terminate its productivity.

Foods differ to an important extent in how much of the year they remain available, and whether they are steadily or only intermittently renewed. For example, the grasses eaten by herbivores may grow continuously for long periods, and honey is similarly renewed day by day by flowers while they are in bloom. Though the consumers have to move about as they feed they can return to the same site as soon as enough new material has accumulated to make it worth while to crop it again.

At the other end of the scale are the food resources that produce discrete standing crops, for the most part at annual intervals. Examples are the seeds and fruits of many plants, the leaves of deciduous trees and the life stages of many insects. In practice there is of course an overlap between continuous and discontinuous renewal: herbs and grasses may grow only for part of the year and then die back; prey animals may be continually present but vary seasonally in abundance. The same applies with equal force to detritus. The consequence is that for one reason or another a very large number and variety of animals have to rely on a succession of standing crops, whether their food is animal, vegetable or detrital.

CONDITIONS IMPOSED ON THE CONSUMERS

When we consider the conditions that must govern food consumption, we should notice that there are some kinds of food that can be totally consumed with impunity, others where a capital stock has to be left intact if future supplies are to be assured, and still others that are super-abundant. Superabundance can be defined as meaning that while the supply lasts the demand by the consumers does not catch up with it, so that an excess is lost and channelled elsewhere. We should not use super-abundance to describe another familiar but different situation in which the consumers begin to exploit a newly available standing crop, very plentiful at the outset but not ephemeral, so that there is time ahead in which it will in fact gradually be used up. Yet another situation of abundance is to be seen in the aspen trees consumed by the beavers; though the standing crop here is vast and never depleted its renewal rate is slow, and it cannot be regarded as superabundant unless a substantial part of what is freely available eventually goes to waste.

Examples of foods that can be completely consumed are the numerous forms of organic detritus and carrion, and the nectar and fruit which plants have evolved to entice mobile animals and make them perform useful services in return. Even such wholly available foods as these entail a need for limiting consumer demand unless they are superabundant. Possibly there are micro-organisms with rapid reproductive powers and the faculty of becoming dormant and indestructible at virtually a moment's notice, which can afford to multiply as fast as their food assimilation rate will allow whenever the habitat provides a nutritive medium. Metazoan animals on the other hand require a longer period of sustained feeding in order to complete their life-cycle and it follows that they have at the outset a forward demand for food. This makes it more or less essential to place an upper limit on the total consumer demand so that it does not exceed the rate of food production. Thus when there are too many vultures for the supply of carrion a dominance hierarchy intervenes and ensures that some of the population are adequately fed and can sustain the stock; the rest must go elsewhere or starve. When there are too many bees for the supply of nectar and pollen they proceed to draw on the food reserves they have already stored, at the same time cutting down on the production of brood so as to reduce their onward demand.

Active animals require at least a minimum quota of food, and in times of protracted food shortage social mechanisms which ensure that at least some individuals receive it, are essential to the stock's survival. In these circumstances there is no safe alternative to eliminating from the feeding group the surplus individuals that cannot adequately be fed.

I have been dealing first with categories of food all of which is available for consumption without prejudice to future supplies. There is another almost exactly opposite situation in which the consumers cannot possibly deplete their food resources because only a tiny fraction of it ever comes within their reach. An example can be seen in sedentary aquatic animals like barnacles, which filter minute particles of food out of a virtually limitless ocean of water. What restricts the consumer is its need for attachment to a substrate in a position where food in sufficient quantity is brought to it by the water currents, and where it has elbow room to spread its appendages for fishing. Vantage points tend in consequence to be colonized at the maximum density at which the individuals animal can operate without interference. Even here adaptations are known to exist which control the original density of larval settlement when a new colony is founded (Knight-Jones & Stevenson 1951).

The situations to be considered next are those in which all the food must not be taken because to do so would imperil its ability to regenerate in future. In order to husband these resources a limit has again to be imposed on the consumption rate, for two reasons instead of only one, because the future viability of both the consumers and the suppliers is at risk. We can take as an example a situation where the food becomes available as a standing crop, as happens with the seeds and hibernating insects and spiders on which many of our wintering birds depend. It has to be eked out for the next 5 or 6 months until alternative foods become available, and some of the stock must be left unused in order to renew the supply. It would be a mistake to assume that a suitably low mean daily consumption rate could invariably arise, and the resources be efficiently exploited in a competitive world, by accident, which must be taken to include the intervention of disease and predation. It is typical of the situations in which having a homeostatic regulation of demand reveals its survival value.

A similar situation exists where the food consists of the leaves of ever-green trees, whether they are the broad-leaved trees of the tropical rain-forest or the narrower leaves of conifers. Each leaf remains in service for several years, so that the rate of replacement is much slower than it is for deciduous trees. Klomp found (1958, p. 33) in his studies of the pine looper moth (*Bupalus piniarius* L.), that there were powerful density-dependent restrictions on the insects' growth and fecundity although the pine needles on which the larvae feed were always present in excess. There was evidence of self-regulation of consumer demand although he could never at any time demonstrate that there was actual competition for food.

There is another good illustration in the Canada porcupine (*Erethizon dorsatum*) which like the beaver consumes the inner bark of trees, in this case mostly conifers. It does not have to cut the trees down like the beaver, but instead climbs up and, sitting on one of a whorl of branches, gnaws what Spencer (1964) describes as a cat-face on the trunk beside it. This prevents the conduction of nutrients down the tree, so that they tend to accumulate immediately above the wound, a fact which undoubtedly attracts the porcupine to return later and enlarge the scar. Some trees are girdled and killed in the process, but the majority are not and the tree in the course of time heals by a process of inrolling bark and wood from the sides of the wound. Spencer's remarkable study was made in Colorado, in an extensive forest dominated by the piñon pine (*Pinus edulis*), the oldest of which were between 500 and 900 years old. It is an arid region where tree-ring chronology is facilitated by great differences in annual growth that depend on variations in soil moisture. Using sawn sections of trunks and branches and samples taken from living trees with a forester's augur, he sampled more than 2000 trees in order to obtain an index of scar frequency, through a time period extending back more than 100 years. It showed since 1870 three clear peaks of porcupine abundance with lows before and after, and a less marked peak about 1840. The peaks varied in height and in the intervals between them; each took about as long to build up as it did to decline. Their underlying cause is quite unknown, although it is not likely to be due to predation since porcupines are seldom tackled by predators, particularly in the southern part of their range. The earliest scar sampled had been made in the year 1693 so that what is strongly reflected is the fact that, in spite of population highs, the porcupines had been living in the forest for 250 years without destroying it.

This consumer/supplier relationship is different only in degree from those for animals which browse and graze. Pasture fertility can notoriously be damaged through over-grazing by domestic livestock, with grievous and lasting consequences if it leads to soil erosion. The Kaibab deer reveal that over-grazing is not confined to farm animals, and wild herbivore populations require to be fully protected from it if they are to survive.

Carnivorous predators are similarly placed. Their food is whole organisms. Short-lived and seasonally reproducing prey may be superabundant for part of the year. On the other hand predators that depend on slow-growing fish or large mammals, as for example sharks and lions do, are faced with essentially the same situation as the beavers. An average lion in the game reserves of Kenya has been shown to kill game at a rate of 20 kg per day; its prey consists chiefly of wildebeest, zebras and Thomson's gazelles (Wright 1960). The average demand is estimated at 36 kills per lion per year. We do not know the exact turnover rates of the prey populations, but they can be roughly estimated, and it would appear that a minimum standing stock of over one hundred head of game is required to support a lion, without allowing for what is killed additionally by leopards, cheetahs and hyaenas. A ratio of about 200:1 is suggested independently by Wright's census figures for prey species and lions in the small National Park at Nairobi. In practice the lions' own food requirements are much more than filled by the prey it kills, and this suggests that, from the lions' point of view, prey is not scarce nor difficult to obtain; the left-overs go to feed an abundant fauna of carrion-eaters.

Under more or less primeval conditions, where these can still be found, the lions appear to maintain a population density comfortably below the carrying capacity of the habitat. Fairly generous margins have been suggested in some of the other situations I have described and they may be quite common in nature, especially perhaps where the consumers are long-lived and not very mobile, and cannot quickly adjust their demands to annual fluctuations in supply, or where a great safety margin is required by the consumers because of inherent difficulties in allowing for the demands made by competing species. Homeostatic adaptations may therefore tend to be prudential and to budget for a moderate super-abundance of food.

THE NEED FOR ADAPTIVE
FEEDBACK

I could easily extend the evidence to show that animal populations do not always or even generally obey a 'Parkinson's Law' and expand to eat the food available. It is not necessary to labour the point that adult population densities of vertebrates rather seldom exceed a level that allows ample feeding under average conditions. Red grouse stocks thrive where heather is plentiful, and most of the starvation mortality that has been shown to occur among them is of their own contriving, affecting individual birds that have been squeezed out in the social process of population control.

Catastrophic starvation of course occurs in animals as well, sometimes on a large scale. On land it often results from severe drought or frost. It is by nature an uncontrollable accident, and as such it is dangerous to the stock's survival. If it leaves any survivors they normally proceed to make good the losses and attempt to re-establish a self-determined population density in place of the one that has been forcibly imposed from outside.

Two things I hope I have made clear so far. One is that starvation will not automatically intervene as a convenient safety-valve to eliminate a population surplus whenever this occurs, and leave unharmed the appropriate number of survivors; if the elimination is to be controlled it is essential that the population itself should be taking the initiative. The second is that the growth of consumer demand generally needs to be stopped before any food shortage occurs; I shall refer again to certain exceptions to this in a minute. By the time general starvation sets in the resource will already have been overtaxed, in some cases so severely that it will fail to yield a succeeding crop of the size required. The population-limiting mechanisms that can be studied in action, such as the feeding hierarchy, are contrived to operate while food is plentiful and while the consumers are in normal health and condition. It is seldom in fact that gross hunger can be relied upon to provide the initial stimulus for homeostatic responses in social physiology and behaviour.

Before I take this proposition further I must refer again briefly to the category of animals whose food is completely consumable without harming its renewal. These were the carrion and detritus feeders and the eaters of fruit and honey. Some of them enjoy a day to day accession of supplies, for example the scavenging birds and mammals, the consumers of some forms of detritus or of secretions like honey and honeydew. Any excess of demand over supply will in these cases show up as an immediate shortfall, and hunger could therefore serve as a trigger for the homeostatic machinery; but the results would still be better, and the mechanism more efficient, if it could be set going in advance and avert the shortage altogether.

THE NATURE OF THE FEEDBACK

Our knowledge of the nature of the feedback systems that control the operation of population homeostasis is as yet only fragmentary. The homeostatic processes I have in mind include changes in mutual tolerance or aggression, in territory size, the amount of emigration, the age of sexual maturation, changes in fertility and reproductive success, in cannibalism and other socially-promoted forms of mortality both of young stages and adults.

We heard this morning in the paper by Miller, Watson & Jenkins that in the red grouse there is a strong correlation on different areas between mean territory size and the abundance and nutrient quality of the heather, the plant that provides the grouse with its staple food. Provided heather is the dominant plant on the moor, grouse population density is higher where the underlying soil is richer than where it is poorer; the heather the soils support varies correspondingly in its content of nitrogen, phosphorus and other minerals (cf. Miller, Jenkins & Watson 1966). Grouse numbers within a moor increase where the moor has been improved by burning in small patches, which makes it possible for individual territories to contain long heather for shelter and young heather of higher nutrient status for feeding (Picozzi 1968). The application of nitro-chalk fertilizer to an experimental area of 16 ha resulted 2 years later in roughly trebling the population density of grouse, compared with its original level and with the level on the control area alongside.

Taken together, these results offer a convincing demonstration that the quality of the food supply has an effect on territory size. Watson (1964) has shown additionally that territory size is correlated with the aggressiveness of the owner, and (1967) that aggressiveness itself can be increased by experimentally raising the testosterone level in the blood or lowered by an implant of oestrogen. It is a fair presumption that these findings are all physiologically related, and that in nature the plane of nutrition affects the endocrine responses of the cock grouse, as it is known to do in various other animals.

Moss (1967) investigated the extent to which grouse are selective in what they pick and eat from the heather plant. They have usually no shortage of energy-giving food, but in order to get the nitrogen and phosphorus they require they have to concentrate on the tips of the green shoots. These tips are a tiny fraction of the heather plant. One begins to see that, although grouse live at a low density on relatively vast expanses of heather, they have a more limited supply of certain nutrients than one might at first suspect. Body-weight in vertebrates is normally associated with nutrition, and it has been known for some years now that the mean body-weight of grouse in autumn varies from year to year and place to place, and that it is significantly correlated both with clutch-size and reproductive success earlier in the same year (Jenkins, Watson & Miller 1963, p. 370).

Although the grouse may seldom suffer from hunger in the crudest sense, there are no obvious objections to the hypothesis that the intake of critical nutrients could have far-reaching physiological consequences, and set up a train of events that resulted in determining the level of aggressiveness, the size of territories, the success of breeding and other components of the homeostatic machinery. One might expect to find a similar trigger mechanism in other animals, although it does not follow that homeostasis is always or necessarily mediated in this way.

Feedback is different for example in flour beetles (*Tribolium* spp.). In laboratory cultures they live in a medium consisting initially of fresh wheat flour, which provides a sufficient source of food. Part of the density-dependent control of population growth depends on physical interaction, including cannibalism and interference between individuals; but another important part depends on the production by adult flour beetles of a secretion of a gas containing quinones, which impairs fertility and larval development. As the population grows and the flour medium becomes conditioned by an increase in the rate of secretion, further growth in numbers is brought to a halt (Ladisch, Ladisch & Howe 1967). No corresponding pheromone is known in the granary weevil (*Calandra granaria*), but nevertheless the females are inhibited from laying their eggs in all the wheat grains available; they always leave a proportion of them intact and undamaged, perhaps originally as a reserve to provide the next year's crop. The granary weevil was one of the first organisms in which self-regulation of populations was conclusively demonstrated, nearly 40 years ago (MacLagan 1932).

Pheromones inhibiting growth are produced by a variety of aquatic organisms, including species of mollusca, fish and amphibia. Pheromone production is primarily conditioned by crowding, but it is not known whether it is additionally influenced by the prevailing nutritional plane (for references see Wynne-Edwards 1962, p. 557).

Silliman's (1968) recent experiments with guppies have shown that the ceiling population density attained by these fish is not dictated by the food supply alone, but results from a compromise between two factors, the daily ration of food provided and the direct effect of crowding itself. In a series of tank experiments he found that in populations with twice and three times the daily food rations that were provided for the controls, all in identical tanks, the densities became stabilized at 1·8 and 2·5 times the control level. In other words, three times the food could produce only two and a half times the population of fish. A closely similar compromise effect was demonstrated long ago in *Drosophila* (Robertson & Sang 1944). It means that in these organisms the amount of space available has a share in providing the feedback which leads to the regulation of recruitment and mortality.

CONCLUSION

We are still a long way from solving the kind of problem with which I began, namely what kind of mechanism it is that allows beavers, or for that matter lions, hyaenas, sharks and a host of other animals, to restrict their numbers in such a way as to use only the annual income from their food resources and not dig into the capital which is available and free for the taking. It is not difficult to see that such animals can fairly readily acquire information about the degree of crowding that exists in their own populations and how it is changing. I have drawn attention elsewhere (Wynne-Edwards 1962) to the widespread existence of what I have called epideictic displays, which could serve as indicators of the prevailing population density and might therefore condition the participating individuals accordingly. But the degree of crowding that animals will tolerate is also often influenced by the state of the food supply, as we have noticed in the red grouse, the guppies and *Drosophila*. One expects animals to live at higher densities in richer and more productive habitats than they do in marginal and unproductive ones; and there is no reason to doubt that food is for most species the ultimate determinant of population density. The point I particularly want to bring out is that the availability of food, or the weekly food intake, often cannot itself serve as a proximate indicator or feedback, because there are so many kinds of consumers like beavers that must always have plenty in reserve and consequently can never experience density-dependent nutritional effects. Their plane of nutrition may vary from time to time or place to place, but it cannot possibly serve as a warning that the invisible threshold has been reached, and the rate of food consumption has overtaken the rate of its renewal.

For this reason it seems necessary to postulate that some animals at least are capable of going another stage beyond habitat selection, itself an immensely complex operation, and of undertaking quantitative habitat appraisal in terms of food resources. For the grouse, habitat appraisal can perhaps be registered automatically by the extent to which its everyday nutritional requirements are physiologically satisfied. But for the beaver it must I think require some kind of completely abstract survey of the amount of standing timber and its suitability and productivity as a food supply. The beaver must perform the equivalent of stocktaking, and though the reaction is no doubt completely automatic and involves no rational process, it seems likely to depend on visual cues, like so much of habitat selection. The physiological consequences must be the same as in the grouse, determining the population size, its recruitment rate, and the level of aggressive behaviour. Beyond this I am not prepared to speculate.

REFERENCES

BREDER C.M. & COATES C.W. (1932) A preliminary study of population stability and sex ratio of *Lebistes*. *Copeia*, 1932, 147–55.

JENKINS D., WATSON A. & MILLER G.R. (1963) Population studies on red grouse, *Lagopus lagopus scoticus* (Lath.) in north-east Scotland. *J. Anim. Ecol.* **32**, 317–76.

JENKINS D., WATSON A. & MILLER G.R. (1964) Predation and red grouse populations. *J. appl. Ecol.* **1**, 183–95.

KLEIN D.R. (1970) Food selection by North American deer and their response to over-utilization of preferred plant species. *Animal Populations in relation to their Food Resources.* (Ed. by A. Watson), pp. 25–46. Oxford.

KLOMP H. (1958) Larval density and adult fecundity in a natural population of the pine looper (*Bupalus piniarius* L.). *Arch. néerl. Zool.* 13 suppl., 319–34.

KNIGHT-JONES E.W. & STEVENSON J.P. (1951) Gregariousness during settlement in the barnacle *Elminius modestus. J. mar. biol. Ass. U.K.* **29**, 281–97.

LADISCH R.K., LADISCH S.K. & HOWE P.M. (1967) Quinoid secretions in grain and flour beetles. *Nature, Lond.* **215**, 939–40.

MACLAGAN D.S. (1932) The effect of population density upon the rate of reproduction with special reference to insects. *Proc. R. Soc.* B, **III**, 437–54.

MECH L.D. (1966) The wolves of Isle Royale. *Fauna natn. Pks U.S.* **7**, 1–210.

MILLER G.R., JENKINS D. & WATSON A. (1966) Heather performance and red grouse populations. I. Visual estimates of heather performance. *J. appl. Ecol.* **3**, 313–26.

MILLER G.R., WATSON A. & JENKINS D. (1970) Responses of red grouse populations to experimental improvement of their food. *Animal Populations in relation to their Food Resources* (Ed. by A. Watson), pp. 323–35. Oxford.

MOSS R. (1967) Probable limiting nutrients in the main food of the red grouse (*Lagopus lagopus scoticus*). *Secondary Productivity of Terrestrial Ecosystems* (Ed. by K. Petrusewicz), pp. 369–79. Warsaw & Crakow.

NICHOLSON A.J. (1955) Compensatory reactions of populations to stresses, and their evolutionary significance. *Aust. J. Zool.* **2**, 1–8.

PICOZZI N. (1968) Grouse bags in relation to the management and geology of heather moors. *J. appl. Ecol.* **5**, 483–7.

RASMUSSEN D.I. (1941) Biotic communities of the Kaibab plateau. *Ecol. Monogr.* **3**, 229–75.

ROBERTSON F.W. & SANG J.H. (1944) The ecological determinants of population growth in a *Drosophila* culture. I. Fecundity of adult flies. *Proc. R. Soc.* B, **132**, 258–77.

SCHÄFER W. (1952) Der 'Kritische Raum', Masseinheit und Mass für die mögliche Bevölkerungsdichte innerhalb einer Art. *Zool. Anz.* 16 Supplementband, 391–5.

SILLIMAN R.P. (1968) Interaction of food-level and exploitation in experimental fish populations. *Fishery Bull. Fish Wildl. Serv. U.S.* **66**, 425–39.

SILLIMAN R.P. & GUTSELL J.S. (1958) Experimental exploitation of fish populations. *Fishery Bull. Fish Wildl. Serv. U.S.* **58**, (No. 133), 214–52.

SPENCER D.A. (1964) Porcupine population fluctuations in past centuries revealed by dendrochronology. *J. appl. Ecol.* **1**, 127–49.

WATSON A. (1964) Aggression and population regulation in red grouse. *Nature, Lond.* **202**, 506–7.

WATSON A. (1967) Social status and population regulation in the red grouse (*Lagopus lagopus scoticus*). *Royal Society Population Study Group*, No. **2**, 1966, 22–30. The Royal Society.

WATSON A. & MOSS R. (1970) Spacing as affected by territorial behavior, habitat and nutrition in red grouse. A.A.A.S. Symp. 1969. *The Use of Space by Animals and Men* (Ed. by A. H. Esser), in press.

WRIGHT B.S. (1960) Predation on big game in East Africa. *J. Wildl. Mgmt*, **24**, 1–14.

WYNNE-EDWARDS V.C. (1962) *Animal Dispersion in relation to Social Behaviour.* Edinburgh & London.

WYNNE-EDWARDS V.C. (1968) Population control and social selection in animals. *Biology & Behaviour Series: Genetics* (Ed. by D. E. Glass), pp. 143–63. New York.

GENERAL DISCUSSION

H. KLOMP: During this symposium I have seriously been asked by one of you: 'Are you a vertebrate or an invertebrate?' My prompt answer in consternation was: 'I am sorry, but I come from the continent and there we all are more or less intermediate'. But after 2 days, which seems to be not too long a period for slowly-thinking animals like tunicates, I began to realize that there really may be something wrong with people having grown up in an overcrowded country like Holland, where the air and the water are heavily polluted with all kinds of disgorged wastes of industry and motor cars. Therefore, it may be justifiable to call you up to protest against the advancing deterioration of the environment before, so to speak, the British people start to evolve back as well.

I would now like to make a contribution on the function of territories in determining the numbers of animals. Dr Watson asks: (a) is spacing behaviour the immediate factor limiting numbers, and is this behaviour independent of, or is it governed by, nutrition; or (b) is this behaviour simply the way in which animals, whose numbers have apparently been limited by some other factor, space themselves subsequently?

A theoretical approach to these questions may be expressed as:

$$D_1 = D_0 \times R_0, \text{ where } D_0 \text{ is density at time } t_0$$
$$D_2 = D_1 \times R_1, \quad \text{where } D_1 \text{ is density at time } t_1$$

etc.   etc.

where R is the net rate of reproduction in successive generations.

$$D_n = D_0 \times R_0 \times R_1 \ldots \times R_{n-1}$$
$$= D_0 \times \prod_{i=0}^{n-1} R_i$$

If there is no trend in density, the geometric mean of R $(\overline{R})$ must be at, or very near to, the value of unity. If $\overline{R} > 1$, population density has an increasing trend, and if $R < 1$, there is a decreasing trend.

Any advantageous recurrent mutation arising in the population will give its carrier a net rate of reproduction larger than that of the non-mutant genotypes, and therefore $\overline{R}$ of the population will become progressively more than unity. Consequently, if no density dependent mechanism is operating, the numbers of animals will steadily increase. This will ultimately result in collective suicide with the consequent disappearance of the advantageous gene. Without regulation of numbers, therefore, evolution would not be possible, or in other words, regulation of numbers is a prerequisite to the occurrence of evolution.

If we consider territorial behaviour as a characteristic of a species with survival value, then its evolution can have taken place only because the numbers in the population were limited, e.g. by competition for food. This might mean that the latter process is still operational in territorial species, and that the territorial behaviour simply spaces out the animals.

J. GREENWOOD: There is one logical objection to Dr Klomp's argument being accepted as absolutely definite. It is a small objection, depending on an unlikely assumption, but it does mean that absolute reliance cannot be placed on Dr Klomp's argument. A mutation could arise which gave the bearer an increase in $\bar{R}$ and at the same time caused it to decrease the $\bar{R}$ of the non-mutant genotype to an exactly corresponding extent, so that the overall $\bar{R}$ remained constant. This would lead to evolution in the absence of population control.

H. KLOMP: In rare cases this might happen, but it seems unlikely that evolution has progressed along these pathways only.

A. WATSON: Population regulation in the past, with intraspecific competition for food as the most obvious mechanism, does not necessarily entitle us to assume that once territorial behaviour has been evolved the original intraspecific competition will continue in exactly the same way and that territorial behaviour is now merely a mechanism of spacing. Territory might well have several functions in any one species.

I did not claim in my paper that the evolutionary reason for territorial behaviour is that it sets an upper limit to populations, or that population regulation cannot be achieved without territorial behaviour being involved. Furthermore, it is time that these problems should be tackled experimentally—the trouble about evolutionary reasons is that we seldom have enough causal facts about the situation as it is right now. These facts are necessary if we are to extrapolate from the present situation and have a meaningful discussion about evolutionary reasons in the past. The appropriate null hypotheses are available and should be put up and tested in a variety of species.

V.C. WYNNE-EDWARDS: Would Dr Watson care to suggest how territory size might have been evolved and adjusted to serve two different functions at the same time?

A. WATSON: The possible functions of territorial behaviour have been fully discussed elsewhere, particularly in a review of the literature by Prof R. A. Hinde (1956, *Ibis*, **98**, 340–69). An example of two such functions could be (1) population limitation and (2) a means of allowing males and females to meet and breed without undue interference.

J. H. LAWTON: A comment on the removal experiments carried out by Dr Watson on red grouse: would it not also be worth while doing addition experiments? If this resulted in the population going back to the original number, it would support the idea that territory was helping to limit the number of individuals able to settle and remain on the area rather than that territory size was a result of the area simply being divided up by the number of animals present.

A. WATSON: There are practical difficulties in trying addition experiments with birds, because they can fly so easily to other areas; small rodents are perhaps more suitable for this kind of experiment. But in any case there is a surplus of red grouse around in all areas every year, and these birds do not get territories. This refutes the possibility that territory size is merely a result of the ground being divided up by the total number of birds present.

R. C. BIGALKE: Can we reasonably expect a universal answer to the question of whether territoriality must have evolved before a homeostatic mechanism? Must the answer not vary according to whether the animal species (or group) considered is one inhabiting (a) a stable environment where catastrophic peaks and troughs of climatic conditions (rainfall or temperatures, etc.) and hence of the availability of plant food, are extremely rare, or (b) an unstable environment where such environmental viscissitudes are common and where, accordingly, the animal cannot easily and sensibly 'plan ahead', because great fluctuations are the norm? It may be worth while examining this question by comparing equivalent species inhabiting areas which differ in this way.

IRENE WERTH: If territorial behaviour in vertebrates could have evolved in connection with parental care, and was reproductively successful in this way, could it not now be acting in the reverse way by limiting reproduction to those individuals which hold territories?

V. C. WYNNE-EDWARDS: Territory-holding is fairly common among invertebrates where there may be no parental care at all, e.g. among the Crustacea. Territoriality is just one of the many ways in which population regulation may be brought about. In some groups, territoriality appears whilst in others there are, for example, hierarchies. Competition for food does not automatically eliminate the population surplus—

you have got to have a hierarchy first, and there are a number of species where the hierarchy is built in as an adaptation, i.e. in red deer stags whose antlers automatically create differences in status in any social group.

H. KLOMP: I should like to ask Professor Wynne-Edwards what kind of situation requires a hierarchy to develop first?

V. C. WYNNE-EDWARDS: All that is required is a situation to develop in which it is necessary to get rid of a surplus of animals. A hierarchy is *one* mechanism which will help to do this.

H. KLOMP: I can hardly imagine that ichneumonid parasites, for example, could have a hierarchical mechanism.

V. C. WYNNE-EDWARDS: I do not say that such animals will have a hierarchical mechanism but what is required is some mechanism which, by eliminating the surplus, prevents populations from rising to starvation level. The elimination need not occur immediately but may take place the following year, for example.

G. C. VARLEY: Klomp cannot conceive of insect parasites with a hierarchical system, but the parasite *Nemeritis* seems to have a homeostatic mechanism built in to behaviour, which reduces its area of discovery in relation to parasite population density in a steady manner over three orders of magnitude. This can stabilize parasite-host oscillations which are inherent in the Nicholson & Bailey (1935, *Proc. zool. Soc. Lond.* 1935, 551–98) system where the area of discovery is a constant. Possibly the function of this reduction of efficiency with increasing population is that it reduces the chance of local extinction.

D. LACK: I have a general point about regulation. The interaction between a predator population or a prey population tends to be unstable so that we tend to be troubled theoretically because in natural habitats (though not necessarily in those modified recently by man) we normally see stability. The answer, I suggest, is illustrated by Dr Klein's account of how wolves were introduced by man to an island with deer and, after initial increase, the wolf population died out. I suggest that this sort of thing has happened millions of times in the past. We do not normally see the unstable cases, because the animals concerned are extinct. All we see are the biased sample of populations which, for one of probably many reasons, have not become extinct in this kind of way. So, of course, we see those which are, in the long term, stable—they are the only ones left.

J. C. COULSON: It is probable that territoriality and hierarchies exclude poor-quality individuals, either from the population or from the breeding group. This need not result in a proportional loss of breeding potential to the population, as the excluded animals, at least in some species, produce few young per individual if conditions are altered to allow breeding. Studies in progress on the kittiwake (*Rissa tridactyla*) suggest that mortality and reproductive success differ several-fold between breeding adults of good and poor quality.

V. C. WYNNE-EDWARDS: Has Dr Coulson tried removing his high-quality birds (as Dr Watson has done with the red grouse), and found that others took their places?

J. C. COULSON: I have done the reverse by giving them more room to breed and these other birds do come in to breed.

K. R. ASHBY: I should like to point out that the harm to a species of overeating its food resources will vary greatly with circumstances. An insectivorous animal may suffer a reduction in numbers for only a small number of years as the prey is capable of quick recovery. A herbivorous mammal in an extreme climate may, by overgrazing, exterminate itself over a large area and cause long-term damage to the habitat. The case of the reindeer in western Alaska described by A. S. Leopold and F. Fraser Darling (1953, *Wildlife in Alaska*, New York) is a case in point.

M. J. WAY: In insects, there may be a striking change in the type of individuals produced in response to population density (e.g. aphids may produce long-distance migrants as opposed to short-distance migrants). This changing quality in response to conditions seems to be important in insects, but what happens in birds? It does not seem to happen with grouse but do, for instance, crossbills or great tits change in behaviour in a way which is likely to buffer the effects of changing density or food supply?

A. WATSON: There appears to be no evidence of a direct behavioural response in the red grouse, involving a different quality of bird in relation to a change in the current conditions of the food supply. However, we do have evidence that grouse of different quality, involving their subsequent survival, aggressive behaviour and territory size, are produced in indirect response to a change in the food supply of their parents. In this sense, the grouse may not be all that different from some insects and some principles may therefore be fundamentally similar even in widely different taxonomic groups.

M. J. COE: Professor Wynne-Edwards has referred to the work of Wright on lions in East Africa, and Dr Watson has told us that territory may have many functions. The game populations of the Nairobi National Park have been assessed since 1960. In this area, we have a population of territorial lions preying on three main species of game—zebra, wildebeeste and hartebeeste. Although these predators are removing up to 15% of the biomass per annum, there are a number of environmental factors that can operate drastically on prey numbers. In 1960, the numbers were between 9000 and 13 000 under conditions of fairly good rain and food, but a drought in 1961 led to the death of half of these animals. Since this time, the numbers have remained relatively constant.

The numbers of individual species have changed since the mortality in 1961, and in this case, it appears that the change in relative abundance of the prey species is almost entirely due to lion predation. When the animals were abundant, the lions' main prey was wildebeeste, and after the drought, their numbers remained constant at the new lower level for several years due to the lions. During this time, zebra and hartebeeste numbers recovered well and increased steadily. We now find that in the last 2 years that the lion is switching its main predatory activities to zebra and hartebeeste. One suspects here that it takes some years and perhaps even a generation for a prey image to change and have an appreciable effect on a prey population. In this way, we see that territoriality spaces lions and that they can control the numbers of certain species. Also, that the prey species will fluctuate under the influence of environmental factors and to a lesser extent on changing prey images of lions.

N. WALOFF: In insects, we recognize territoriality in time as well as in space in the form of polymorphic strains which may, for example, be polymorphic in rates of development.

J. GREENWOOD: Following on Dr Waloff's point, one sometimes imagines that polymorphism is an exceptional phenomenon. In fact, all populations may turn out to be polymorphic—and possibly highly polymorphic, remembering that recent work on *Drosophila* has shown that its populations are polymorphic at perhaps 30% of their gene-loci. This is relevant to Dr Coulson's point: any population is immensely variable in any particular character. The question of how this variability is maintained in the population is perhaps the biggest unsolved question facing population geneticists today. When we ask about the maintenance of variability in the factors affecting population regulation, we are, therefore, only considering a small part of a very much bigger question, one which even the population geneticists cannot answer.

H. N. SOUTHERN: In our discussions, we have been concerned with the quality of food—especially the food of herbivores. The feeling among ecologists at present is that we ought to find out from the specialists whether we are going about measuring the quality of food in the right way and whether we are measuring the right things?

K. L. BLAXTER: Food quality is a biological measurement, which varies with the species of animal and the function which it is undertaking. Lactation, for example, demands entirely different nutrient needs to pregnancy in maintenance of weight, locomotion, and so on. Chemical analysis of food is a guide in many instances but is not an absolute measure. Furthermore, the quantity ingested is important; the percentage nutrient content is seldom a sufficient measure of the nutritive worth of a diet, though obviously it is of value.

P. R. EVANS: We have heard much about lower limits of certain nutrients in foodstuffs. Yet I understand that high percentages of, e.g. certain amino-acids in food intake can be damaging. Would Dr Blaxter care to comment on this?

K. L. BLAXTER: The response curves to increasing levels of a nutrient with the exception of energy usually show ceilings and often a reduction in response and toxicity at high levels. An example is vitamin A where the response increases, remains stable, and then declines. Polar bear liver as a sole diet is indeed toxic for man for this very reason. Furthermore, one has to consider the balance between nutrients, that is the interrelations between the proportional supply of the thirty to forty specific nutrient essentials a mammal or bird needs in its diet.

A. DUNCAN: Could I ask Dr Blaxter to comment on the relevance of metabolic studies to the subjects we have been discussing in the last 3 days?

K. L. BLAXTER: Oh, I think they are very relevant!

EXCURSIONS

MARINE LABORATORY

During the morning of Friday 28 March, the Society visited the Marine Laboratory of the Department of Agriculture and Fisheries for Scotland. A short talk was given by Mr B. Parrish who introduced members to the fisheries research work of the laboratory and to the dozen or so demonstrations spaced out in various parts of the laboratory.

The demonstrations included investigations of herring races, endocrine studies, diseases of salmonid fish, plankton as indicators of ecological conditions, mortalities of fish eggs and larvae, reproduction in diatoms, tagging techniques, fish parasites, fish counting by sonar, interstitial fauna of sand, fluorescence microscopy of sand grains, and instrumentation for research on fishing gear. The laboratory library and permanent demonstration halls were also open to the Society. About forty attended.

J. H. Fraser

DEESIDE AND THE BRAEMAR HILLS

About forty people came on this all-day excursion. The first stop was at Glen Tanar, where we saw very fine natural regeneration of pine on an estate where the owner has been practising extensive forestry without clear felling.

At lunch time, we stopped below Braemar and climbed a sparsely tree-dotted hillside to look at the natural pine forest of Ballochbuie. Meanwhile a very bold cock grouse continually kept calling only 40 m distant, as he threatened us invading his territory. We were lucky to see a capercaillie flying into the wood.

In Glen Clunie, grouse were showing territorial behaviour at the roadside and the party got good views of this. The bus stopped at the Cairnwell road summit at 600 m. Later we gathered on the steep slope of Cairnwell hill, where we had good views of ptarmigan in winter plumage, grouse in pairs, and several mountain hares in white winter coat; finally we came within 10 m of a pair of ptarmigan. Most of the ground was deep under snow but enough vegetation was visible to see something of this arctic-alpine community. Damage to this habitat, caused locally by construction operations and by skiers and walkers near the ski lifts, could not be seen because of deep snow.

We then moved back down Glen Clunie to Braemar, but again left the bus en route to see hundreds of red deer stags and hinds, in separate groups on different bare hillsides. From Braemar we went up the Dee Valley to Linn o Dee, a spectacular sight with the gorge ensheathed in thick ice, and then stopped in the open parkland pine forest of lower Glen Lui, near Mar Lodge in the foothills of the Cairngorms. Here we walked for 1 km and had good views of many red deer stags at fairly close range, including a few which had already shed their antlers. The lack of tree regeneration and severe grazing of ground vegetation was well seen in this area, which supports high concentrations of deer during winter snow. John Forster was lucky to spot a great grey shrike on a dead tree, and every-one else was able to observe it. The bus driver then summoned us for the drive down Deeside. We had one more stop at Muir o Dinnet to see spectacular regeneration of young birch and pine which are taking over from what was originally a heather moor.

We were fortunate to have Peter McIntyre, Forestry Commission District Officer for Deeside, with us for the whole excursion as he is so knowledgeable about the history and geomorphology of Deeside as well as about its forestry. Thanks are due to him for giving up a whole working day to be with us.

Adam Watson

CULTERTY FIELD STATION

The bus party which had earlier visited the Marine Laboratory later lunched at Dunbar Hall and then came by bus to Newburgh. On arrival, the members were shown the Field Station and its grounds with one saltwater and two freshwater ponds containing a collection of waterfowl. Exhibits of the various ecological studies being carried out at the Station were on show and some time was spent examining these and discussing the studies with the research workers concerned.

The party was then ferried across the River Ythan by catamaran and we walked over the sand dunes of the Forvie Nature Reserve and up the edge of the Ythan Estuary. This gave a good opportunity to see a well-developed dune succession and a variety of estuarine habitats with their associated birds, the most noticeable being the large populations of eider duck and oystercatcher on the mussel beds.

I. J. Patterson

LIST OF REGISTERED
PARTICIPANTS

ANDERSON Mr A.   Culterty Field Station, Newburgh, Aberdeenshire
ANDERSON Mr J.M.   20 Gladsmuir Road, Archway, London N19
ASHBY Dr K.R.   Department of Zoology, University of Durham
ATKINS Miss E.F.   Department of Biology, Paisley College of Technology
BADCOCK Miss R.M. Department of Biology, University of Keele
BAILEY Mr G.N.A.   Department of Zoology, University of Exeter
BAILEY Dr R.S.   The Marine Laboratory, Victoria Road, Aberdeen
BALL Mr M.E.   Nature Conservancy, Edinburgh
BEAVER Dr & Mrs. R.A.   Department of Zoology, University College of North
   Wales, Bangor
BELL Mr. M.J.V.   Ministry of Defence, London
BELL Mr R.H.V.   Department of Zoology, University of Manchester
BELLAMY Mr L.S.   Department of Zoology, University College of North Wales,
   Bangor
BERRIE Dr A.D.   Department of Zoology, University of Reading
BIGALKE Dr R.C.   P.O. Box 662, Pietermaritzburg, S. Africa
BINNEY Miss C.   School of Biological & Environmental Studies, New University
   of Ulster, Coleraine, N. Ireland
BJARNOV Mr N.   Hillerødsholms Alle 45, 3400 Hillerød, Denmark
BLACKMAN Mr R.A.A.   Dove Marine Laboratory, Cullercoats, Tynemouth
BLAXTER Mr J.H.S.   Department of Natural History, University of Aberdeen
BLAXTER Dr K.L.   Rowett Research Institute, Bucksburn, Aberdeen
BLAZKA Dr P.   Hydrobiological Laboratory, Vltavska 17, Prague 5
BLOCK Dr W.   Department of Zoology, University of Leicester
BOYD Dr J.M.   The Nature Conservancy, Hope Terrace, Edinburgh 9
BRAMLEY Mr P.S.   Riverside Cottage, South Perrott, Beaminster, Dorset
BRAY Dr R.P.   The Game Research Association, Fordingbridge, Hants
BREMNER Mr D.   Malham Tarn Field Centre, nr. Settle, Yorks
BRITTAIN Mr J.E.   Department of Zoology, University College of North Wales,
   Bangor
BROADHEAD Dr E.   Department of Zoology, University of Leeds
BROMLEY Miss H.J.   Schouw 18, Lemmer, Friesland, The Netherlands
BROWN Mr N.M.D.   Department of Zoology, University of Durham
BRUTON Miss M.   Parsons Field House, Green Lane, Durham
BURNETT Dr G.F.   Department of Agriculture, University of Aberdeen
BUSTARD Dr H.R.   R.S.B.S., Australian National University, Canberra
BUTTERFIELD Mrs J.   39 Old Elvet, Durham

CAMERON Dr R.A.D.   Department of Biological Sciences, Portsmouth College of Technology
CAMMELL Mr M.E.   23 Upper Nursery, Sunningdale, Ascot
CAMPBELL Mr R.   25 Henry Road, Oxford
CARTHY Dr J.D.   Field Studies Council, 9 Devereux Court, WC2
CHADWICK Dr M.J.   Department of Biology, University of York
CHAMBERS Mr M.   Schouw 18, Lemmer, Friesland, Holland
CHEKE Mr A.S.   Edward Grey Institute, Botanic Garden, Oxford
CHERRETT Dr J.M.   University College of North Wales, Bangor
CHINNERY Mr L.E.   New University of Ulster, Coleraine, N. Ireland
COAKER Mr T.H.   National Vegetable Research Station, Wellesbourne
CODY Mr C.B.J.   52 Farnham Road, Newtonhall Estate, Durham
COE Dr M.J.   Animal Ecology Research Group, Botanic Garden, Oxford
COLDREY Miss J.M.   Edward Grey Institute, Botanic Garden, Oxford
CORBET Dr Sarah A.   Westfield College, London NW3
CORNWALLIS Mr L.   Edward Grey Institute, Botanic Garden, Oxford
COTTON Dr M.J.   Department of Biological Sciences, University of Dundee
COULSON Dr J.C.   Department of Zoology, University of Durham
COWLEY Dr J.J.   Queen's University, Belfast, N. Ireland
CROTHERS Mr J.H.   Nettlecombe Court Field Centre, Williton, Taunton, Somerset
CURRIE Mr A.   Hill Farming Research Organisation, Yetholm, Roxburgh
DAWSON Mr and Mrs D.G.   Edward Grey Institute, Botanic Garden, Oxford
DIXON Dr A.F.G.   Department of Zoology, University of Glasgow
DREW Mr J.A.   7 Merton Road, Southsea, Hampshire
DUCKETT Mr M.J.   39 Elvet Crescent, Durham
DUFFEY Dr E.   Monks Wood Experimental Station, Abbots Ripton, Huntingdon
DUNCAN Dr A.   Royal Holloway College, Englefield Green, Surrey
DUNN Mr E.K.   Department of Zoology, University of Durham
EADIE Mr J.   Hill Farming Research Organisation, Yetholm
EDGAR Dr W.D.   Department of Zoology, University of Glasgow
ELLIOTT Dr R.J.   The Nature Conservancy, Attingham Park, Shrewsbury
EVANS Mrs D.M.   3 Windmill Hill, Durham
EVANS Dr P.R.   3 Windmill Hill, Durham
FERNS Mr P.N.   Department of Zoology, Hatherly Biological Laboratories, Prince of Wales Road, Exeter
FINCH Mr S.   National Vegetable Research Station, Wellesbourne
FLEGG Dr J.J.M.   British Trust for Ornithology, Beech Grove, Tring, Herts
FLOWERDEW Mr J.R.   Animal Ecology Research Group, Botanic Garden, Oxford
FORSTER Mr J.A.   The Nature Conservancy, Blackhall, Banchory, Kincard
FORSYTH Mr J.   Department of Extra-Mural Studies, Queen's University, Belfast
FRANKLIN Mr A.   M.A.F.F. Fisheries Laboratory, Burnham-on-Crouch, Essex
FRAZER Dr J.F.D.   The Nature Conservancy, 19 Belgrave Square, SW1
FRASER Miss S.M.   Department of Zoology, University of Durham
FREWIN Miss A.   The Graduate Society, 38 Old Elvet, Durham
FRY Dr C.H.   Department of Natural History, University of Aberdeen
FRYER Dr G.   Freshwater Biological Association, Ambleside, Westmoreland
GARDARSSON Mr A.   Museum of Natural History, Reykjavik, Iceland
GARLAND Miss A.   Menaifron, Craig-Y-Don Road, Bangor, N. Wales

GEORGE Mr D.R. Department of Zoology, University of Durham
GIMINGHAM Dr C.H. Department of Botany, University of Aberdeen
GIRLING Mr D. 38 Old Elvet, Durham
GLUE Mr D.E. British Trust for Ornithology, Tring, Herts
GOLDSPINK Mr C. Limnologisch Instituut, 6ij Oosterzee(Brug), Friesland, Netherlands
GOODHART Dr C.B. Department of Zoology, Downing Street, Cambridge
GORDON Mr I.J. Hatherly Laboratory, University of Exeter
GORMAN Mr M. Culterty Field Station, Newburgh, Aberdeenshire
GOSS-CUSTARD Dr J.D. Department of Psychology, University of Bristol
GOULDER Dr R. Freshwater Biological Association, Ambleside, Westmorland
GRANT Mr D.R. Hawkslee, St. Boswells, Roxburghshire
GREENE Mr M.F. Department of Zoology, University of Glasgow
GREENWOOD Mr J.J.D. Department of Biological Sciences, University of Dundee
GRIFFITHS Mr M.E. Biophysics Research Unit, University of Technology, Loughborough, Leics
GRIME Dr J.P. Department of Botany, University of Sheffield
GRIMSDELL Mr J.J.R. Marshall Laboratory, Cambridge
GWYNNE Mr D.C. 22 Braehead Crescent, Hardgate, Dunbartonshire
HAMILTON Mr R.M. Orielton Field Centre, Pembroke, South Wales
HARRIS Dr M.P. Edward Grey Institute, Botanic Garden, Oxford
HART-JONES Miss B. Parson's Field House, Green Lane, Durham
HATTO Miss J. School of Plant Biology, University College of N. Wales
HEAL Dr O.W. Merlewood Research Station, Grange-Over-Sands, Lancs
HEALEY Dr I.N. Department of Zoology, University of London, King's College, WC2
HEPPLESTON Dr P.B. University of Aberdeen, School of Agriculture
HOLDGATE Dr M.W. The Nature Conservancy, 19 Belgrave Square, SW1
HOLMES Dr R.T. Department of Biological Sciences, Dartmouth College, Hanover, New Hampshire, U.S.A.
HOLTER Mr P. Peter Bangsvej 125, 2000F, Copenhagen, Denmark
HOROBIN Mr J.C. Department of Zoology, University of Durham
HUGHES Mr G. 5 Leighton Place, Leighton Street, Pietermaritzburg, Natal, South Africa
HUGHES Dr R.D. C.S.I.R.O., Division of Entomology, P.O. Box 109, Canberra City, Australia
HUTCHINSON Mr C. 15 Cil Coed, Coed Mawr, Bangor
HUXLEY Mr C.R. Department of Zoology, University of Southampton
HUXLEY Mr T. Countryside Commission, Perth
JACKSON Mr J. Department of Zoology, University of Glasgow
JENKINS Dr D. The Nature Conservancy, Hope Terrace, Edinburgh, 9
JEWELL Dr P. Department of Zoology, University College, Gower Street, WC1
JOHNSTONE Mr G.W. 1 Webster Coates Avenue, Edinburgh 12
JONES Dr N.V. Department of Zoology, Hull University
KAY Dr R. Rowett Research Institute, Bucksburn, Aberdeen
KESSLER Mr and Mrs A. Zoological Laboratory, Free University, De Boelaan 1087, Amsterdam, Holland
KING Dr CAROLYN M. Animal Ecology Research Group, Botanic Garden, Oxford

KITCHING Mr R.L. Animal Ecology Research Group, Botanic Garden, Oxford

KLEIN Dr D.R. Wildlife Research Unit, University of Alaska, College, Alaska 99701, U.S.A.

KLOMP Dr H. Department of Zoology, Binnenhaven 7, Wageningen, Netherlands

KRUUK Dr H. Serengeti Research Institute, P.O. Box 3134, Arusha, Tanzania

LACK Dr D. Edward Grey Institute, Botanic Garden, Oxford

LANGHAM Dr N. Marine Laboratory, Victoria Road, Aberdeen

LAWTON Mr J.H. Animal Ecology Research Group, Botanic Garden, Oxford

LAUGHLIN Mr K.F. Department of Biology, University of Stirling

LEACH Mr J.H. Culterty Field Station, Newburgh, Aberdeenshire

LINN Mr I.J. Hatherly Laboratories, Exeter

LITTLEWOOD Mrs C.F. Department of Forestry and Natural Resources, University of Edinburgh

LLEWELLYN Mr M. Department of Biology, Queen Elizabeth College, Campden Hill, W8

LLOYD Mr. D.E.B. Culterty Field Station, Newburgh, Aberdeenshire

MACAN Dr T.T. Freshwater Biological Association, Ambleside, Westmorland

MACAULAY Mr E.D.M. Entomology Department, Rothamsted Experimental Station, Harpenden, Herts

MCDONALD Mr I. Department of Zoology, Reading University

MACFADYEN Professor A. New University of Ulster, Coleraine

MCINTYRE Mr A.D. Marine Laboratory, Victoria Road, Aberdeen

MACLAGAN Dr D.S. Department of Zoology, West of Scotland Agricultural College, Glasgow C2

MCNAUGHTON Mr F.C. Oceanographic Laboratory, Edinburgh

MCNEILL Dr S. Imperial College Field Station, Sunninghill, Ascot

MAHER Dr W. Department of Biology, University of Saskatchewan, Saskatoon, Canada

MARTIN Mr G.H.G. Portsmouth College of Technology

MASON Mr C.F. Animal Ecology Research Group, Botanic Garden, Oxford

MATHEWS Mr C.P. Department of Zoology, University of Reading

MEAD Mr C.J. British Trust for Ornithology, Tring, Herts

MEESE Mr G.B. 29 Kepier Court, Gilesgate, Durham

MILLER Dr G.R. The Nature Conservancy, Blackhall, Banchory, Kincard

MILLER Mrs L.M. Gilbank, Schoolhill, Banchory, Kincard

MILNER Dr C. University of Saskatchewan, Saskatoon, Canada

MORGAN Mr N.C. The Nature Conservancy, Hope Terrace, Edinburgh

MOSS Dr R. The Nature Conservancy, Blackhall, Banchory, Kincard

MYCOCK Mr E.R. New University of Ulster, Coleraine

NEWBOLD Mr C. New University of Ulster, Coleranie

NEWBOULD Professor P.J. New University of Ulster, Coleraine

NEWTON Dr I. The Nature Conservancy, Hope Terrace, Edinburgh 9

NICHOLSON Mr I.A. The Nature Conservancy, Hope Terrace, Edinburgh 9

O'DONNELL Mr T.G. Animal Ecology Research Group, Botanic Garden, Oxford

OWEN Dr M. The Wildfowl Trust, Slimbridge, Glos

OXLEY Mr E.R.B. School of Plant Biology, University College of North Wales, Bangor, Caernarvonshire

PAGE Mr A.   13 Ogwen Terrace, Bethesda, Caernarvonshire

PARMA Mr S.   Limnological Institute, 'Vijverhof', Nieuwersluis(U), Netherlands

PARRISH Mr B.B.   Marine Laboratory, Aberdeen

PARSONS Mr J.   Department of Zoology, University of Durham

PATTERSON Dr I.J.   Culterty Field Station, Newburgh, Aberdeenshire

PEAKIN Dr G.J.   Department of Biology, Goldsmiths College, Lewisham Way, SE14

PERRIN Mr M.R.   Department of Zoology, Hatherly Laboratories, Exeter

PERRINS Dr C.M.   Edward Grey Institute, Botanic Garden, Oxford

PHILLIPS Miss B.J.   Parsons Field House, Green Lane, Durham

PHILLIPSON Dr J.   Animal Ecology Research Group, Botanic Garden, Oxford

PIEARCE Mr T.G.   Department of Zoology, University College of North Wales, Bangor

PORTER Mr W.B.   1 Cragside, Witton Gilbert, Durham

PRATER Mr A.J.   Merlewood Research Station, Grange-Over-Sands, Lancs

PRESTT Mr I.   Monks Wood Experimental Station, Abbots Ripton, Hunts

PRICE-JONES Dr D.   Jealott's Hill Research Station, Bracknell, Berks

PROCTER Mr J.   Merlewood Research Station, Grange-Over-Sands, Lancs

RATCLIFFE Dr D.A.   Monks Wood Experimental Station, Abbots Ripton, Hunts

REAY Mr P.J.   College of Technology, Portsmouth

REDFERN Miss M.   Department of Biological Sciences, Portsmouth College of Technology

REES Mr D.I.   c/o The Nature Conservancy, Penrhos Road, Bangor

REYNOLDSON Dr T.B.   Department of Zoology, University College of North Wales, Bangor

RICHARDS Mr J.   Animal Ecology Research Group, Botanic Garden, Oxford

RINGELBERG Dr J.   Laboratory of Animal Physiology, Kruislaan 320, Amsterdam

ROCHARD Mr J.B.A.   Pest Investigation Unit, D.A.F.S., Edinburgh 12

RODGERS Mr A.   Department of Zoology, Box 30197, Nairobi

RUSSEL Dr R.J.   Skovbakkevej 7A, 8220 Brabrand, Denmark

SCOTT Mr D.A.   Edward Grey Institute, Botanic Garden, Oxford

SEEL Dr D.C.   Department of Zoology, University of Southampton

SEFTON Mr A.D.   Department of Zoology, University College of North Wales

SIMS Mr R.E.   11 Wearside Drive, Durham

SMITH Miss L.F.   22 Elvet Crescent, Durham

SMYLY Mr W.J.P.   Freshwater Biological Association, Ambleside, Westmorland

SOLOMON Mr M.G.   Imperial College Field Station, Ascot, Berks

SOUTHERN Mr and Mrs H.N.   Animal Ecology Research Group, Botanic Garden, Oxford

SPENCER Mr S.R.   Imperial College Field Station, Ascot, Berks

STAINES Mr B.W.   Nature Conservancy, Blackhall, Banchory, Kincard

STEELE Dr J.H.   Marine Laboratory, Victoria Road, Aberdeen

STILEMAN Mr R.   12 Bloomfield Terrace, SW1

STODDART Dr D.M.   Animal Ecology Research Group, Botanic Garden, Oxford

STRANDGAARD Dr H.   Game Biology Station, Kalø, Rønde, Denmark

SUMMERS Mr C.F.   c/o Department of Botany, University of Aberdeen

SUTTON Dr S.L.   Department of Zoology, University of Leeds

TAYLOR Mr J.C.   Infestation Control Laboratory, Worplesdon, Surrey

TAYLOR Mr T.J.   Department of Biological Sciences, Sir John Cass College, EC3

TETT Mr P.   Marine Station, Millport, Isle of Cumbrae

THOMAS Mr G.   R.S.P.B., Sandy, Beds.

THOMPSON Mr H.V.   Infestation Control Laboratory, Worplesdon, Surrey

THORPE Mr J.E.   Freshwater Fisheries Laboratory, Pitlochry

TITTENSOR Mr and Mrs A.M.   Department Forestry and Natural Resources, University of Edinburgh

USHER Dr M.B.   Department of Biology, University of York

VANPRAET Mr G.   P.O. Box 36362, Nairobi, Kenya

VARLEY Professor G.C.   Hope Department of Entomology, Oxford

WAKE Mr J.R.   Department of Biological Sciences, University of Dundee

WALOFF Dr N.   Department of Zoology, Imperial College, SW7

WALLWORK Dr J.A.   Department of Zoology, Westfield College, NW3

WATSON Dr A.   Nature Conservancy, Blackhall, Banchory, Kincard

WAY Mr M.J.   Imperial College Field Station, Ascot, Berks

WEBB Mr M.J.   16A Scotforth Road, Scotforth, Lancs

WEBB Dr N.R.   Nature Conservancy, Furzebrook

WEIR Dr J.S.   Department of Zoology, University of Leicester

WERTH Miss I.   Department of Zoology, University of Leeds

WILLIAMS Mr G.   Department of Zoology, University of Reading

WILSON Miss M.   Parsons Field House, Green Lane, Durham

MING-HUNG-WONG Mr   c/o Graduate Society, Durham City

WOOD Mr T.A.   2 Westfield Road, Aberdeen

WYNNE-EDWARDS Professor V.C.   Department of Natural History, University of Aberdeen

YOUNG Dr J.O.   Department of Zoology, University of Liverpool

AUTHOR INDEX

447

GEOGRAPHICAL INDEX

SPECIES INDEX

SUBJECT INDEX